Formal Models of Operating System Kernels

Iain D. Craig

Formal Models of Operating System Kernels

Springer

Iain D. Craig, MA, PhD, FBCF, CITP

British Library Cataloguing in Publication Data
A catalogue record for this book is available from the British Library

ISBN-13: 978-1-84996-592-7 e-ISBN-13: 978-1-84628-718-3

9 8 7 6 5 4 3 2 1

Springer Science+Business Media
springer.com

To a very special friend—
Eheu fugaces labuntur anni

Preface

The work that this book represents is something I have wanted to do since 1979. While in Ireland, probably in 2001, I sketched some parts of a small operating system specification in Z but left it because of other duties. In 2002, I worked on the sketches again but was interrupted. Finally, in April, 2005, I decided to devote some time to it and produced what amounted to a first version of the kernel to be found in Chapter 3 of this book. I even produced a few proofs, just to show that I was not on a completely insane tack.

I decided to suggest the material as the subject of a book to Beverley Ford. The material was sent on a Thursday (I think). The following Monday, I received an email from her saying that it had gone out for review. The review process took less than 2 weeks; the response was as surprising as it was encouraging: a definite acceptance. So I got on with it.

This book is intended as a new way to approach operating systems design in general, and kernel design in particular. It was partly driven by the old ambition mentioned above, by the need for greater clarity where it comes to kernels and by the need, as I see it, for a better foundation for operating systems design. Security aspects, too, played a part—as noted in the introductory chapter, if a system's kernel is insecure or unreliable, it will undermine attempts to construct secure software on top of it. Security does not otherwise play a part in this book.

As Pike notes in [24], operating systems has become a rather boring area. The fact that two systems dominate the world is a stultifying problem. There *are* good ideas around and there is always new hardware that needs controlling. The advent of ubiquitous computing is also a challenge. I would be very pleased if formal models helped people define new models for operating systems (the lack of implementation problems *is* a real help—I have used formal models as a way of trying out new software ideas since the late 1980s).

Of course, I hope that people from formal methods and operating systems, as well as computer science more generally, will read this book. I would like to think that it is a demonstration that system software can be modelled and specified formally, endowing it with all the benefits of formal methods.

What makes this book different are the facts that it contains proofs of properties and that it is broader in scope. The majority of the studies in the literature omit proofs ([14] discusses proof but includes none). It seems to me that proof is necessary for, otherwise, one is just describing systems in just another fancy notation.

This book was written in a relatively short period of time (May–December, 2005). Every effort has been made to ensure that it is error-free. The way I approached the process of writing it was intended to reduce errors. Steve Schuman has also read the entire text and the proofs. However, I cannot say that the text does not contain any errors. For the mistakes that occur, I apologise in advance.

Acknowledgements

First of all, I would like to thank Beverley Ford. Next, I would like to thank Helen Desmond for running the project so smoothly. Steve Schuman promoted the project, gave extremely useful advice on how to pitch it and read the various intermediate versions of the manuscript (some a little chaotic) and checked the proofs. My brother, Adam, once again produced the artwork with remarkable speed and accuracy. For those who are not mentioned above and who helped, my apologies for omitting to mention you. Your help *was* appreciated.

Iain Craig
North Warwickshire,
January, 2006

Contents

List of Figures

1

Introduction

Dimidium facti qui coepit habet; sapere aude.
– Horace, Epistles, I, ii, 40

1.1 Introduction

Operating systems are, arguably, the most critical part of any computer system. The kernel manages the computational resources used by applications. Recent episodes have shown that the operating system is a significant thorn in the side of those desiring secure systems. The reliability of the entire operating system, as well as its performance, depends upon having a reliable kernel. The kernel is therefore not only a significant piece of software in its own right, but also a critical module.

Formal methods have been used in connection with operating systems for a long time. The most obvious place for the application of mathematics is in modelling operating system queues. There has been previous work in this area, for example:

- the UCLA Security Kernel [32];
- the work by Bevier [2] on formal models of kernels;
- Horning's papers on OS specification [3]
- the NICTA Workshop in 2004 on operating systems verification [23];
- Zhou and Black's work [37].

Much of the formal work on operating systems has been verificational in nature. That is, given some working software, an attempt is made to justify that software by constructing a formal model. This is clearly in evidence in the NICTA Workshop [23] papers about the L4 kernel [13, 31]. Formal methods in this case are used in a *descriptive* fashion. Certainly, if the model is good enough, it can be used to reason about the reliability of the implemented software; it can also be used to clarify the relationships between the modules in that software. However, the descriptive approach requires an adequate model, and that can be hard to obtain.

What is proposed in this book is a *prescriptive* approach. The formal model should be constructed *before* code is written. The formal model is then used in reasoning about the system as an abstract, mathematical entity. Furthermore, a formal model can be used for other purposes (e.g., teaching kernel design, training in the use and configuration of the kernel). C. A. R. Hoare has complained that there are too many books on operating systems that just go through the concepts and present a few case studies—there are a great many examples from which to choose; what is required, he has repeatedly argued, is detailed descriptions of new systems[1].

The formal specification and derivation of operating system kernels is also of clear benefit to the real-time/embedded systems community. Here, the kernels tend to be quite simple and their storage management requirements less complex than in general-purpose systems like Linux, Solaris and Windows NT. Embedded systems must be as reliable as possible, fault tolerant and small. However, a kernel designed for an embedded application often contains most of the major abstractions employed by a large multiprogramming system; from this, it is clear that the lessons learned in specifying a small kernel can be generalised and transferred to the process of specifying a kernel for a larger system. Given the networking of most systems today, some of the distinctions between real-time and general-purpose systems are, in any case, disappearing (network events must be handled in real time, after all).

For the reasons given in the last paragraph, the first specification in this book is of a kernel that could be used in an embedded or real-time system. It exports a process abstraction and a rich set of inter-process communication methods (semaphores, shared buffers and mailboxes or message queues). This kernel is of about the same complexity as μC/OS [18], a small kernel for embedded and real-time applications.

1.2 Feasibility

It is often argued that there are limits to what can be formally specified. There are two parts to this argument:

- limits to what can profitably be specified formally.
- *a priori* limits on what *can* be formally specified.

The first is either a philosophical, pragmatic or economic issue. As a philosophical argument, there is Gödel's (second) theorem. As an economic argument, the fact that formal specification and derivation take longer than

[1] This is not to denigrate any of them. Most contain lucid explanations of the concepts. The point is that they tend only to repeat the principles and sketch well-known systems.

traditional design and construction methods is usually taken as an argument that they are only of "academic interest". This argument ignores the fact that the testing phase can be reduced or almost entirely omitted because code is correct with respect to the specification. (Actually, a good testing schedule can be used to increase confidence in the software.) In the author's experience, formally specified code (and by this is meant specification supported by proofs) works first time and works according to specification.

The existence of a formal model also has implications for maintenance and modification. The consequences of a "small patch" are often impossible to predict. With a formal model, the implications can be drawn out and consequences derived. With informal methods, this cannot be done and users are disappointed and inconvenienced (or worse).

As to what can profitably be specified, it would appear that just about any formal specification can be profitable, even the swap program. There was a notion a few years ago that only safety-critical components should be formally specified; the rest could be left to informal methods. This might be possible if the dependencies between safety-critical and noncritical components can be identified with 100% accuracy. The problem is that this is not often undertaken. Again, formal methods reveal the dependencies.

So what about the argument that programmers cannot do formal specification, that only mathematicians can do it? One argument is that we should be teaching our people rather more than how to read syntax and hack code; they should be taught abstractions right from the start. This is not something one readily learns from lectures on syntax and coding methods; it is not even something that can be learned from lectures on design using informal tools or methods (waterfalls, 'extreme' programming, etc.). Much of the mathematics used in formal specifications is quite simple and its use requires and induces clearer thinking about what one is doing and why. There is a clear problem with the way in which computer scientists are trained and with the perceptions, abilities and knowledge of many of those who train them.

The second argument is that it is just impossible to specify everything in a formal way. Programs, and processes for that matter, are structured entities that can be described in formal ways. This is admitted by the other side, but there are things like compilers, operating systems, command and control systems and a whole list of other kinds of systems that simply cannot be formally specified. The reason usually given is that they are too complex or complicated. Operating systems have the additional problem that they deal with hardware and "you can't specify that". There are many possible answers; for example, to point to hardware specification languages, to point to specifications of hardware or to point out that a piece of hardware can be modelled in an abstract fashion. Critics object that the specification will be too abstract to be of use—it cannot capture every aspect of the hardware device. This is true: abstractions do not capture *every* detail, only the relevant ones.

For example, a model of a disk drive might include read and write operations and might contain a mapping from disk locations to data. Such a model would be of considerable use. The objection from the doubter is that such a model does not include disk-head seek time. Of course, seek times are relevant at low levels (and temporal logic can help—the specification says "eventually the disk returns a buffer of data or a failure report").

The next objection is that it is impossible to model those aspects of the processor required to specify a kernel. And so it goes on.

The only way to silence such objections is to go ahead and engage in the exercise. That is one reason for writing this book: it is an existence proof.

1.3 Why Build Models?

It has always been clear to the author that a formal specification could serve as more than a basis for refinement to code. A formal specification constitutes a formal model; important properties can be proved *before* any code is written. This was one of the reasons for writing [10]. In addition to that book, formal models and proofs were used by the author as a way of exploring a number of new systems during the 1990s without having to implement them (they were later implemented using the formal models). The approach has the benefit that a system's design or, indeed, an entire approach to a system, can be explored thoroughly without the need for implementation. The cost (and risk) of implementation can thereby be avoided.

In the case of operating systems, implementation can be lengthy (and therefore costly) and require the construction of drivers and other "messy" parts[2]. The conventional approach to OS (and other software) design requires an implementation so that properties can be determined empirically. Determining properties of all software at present is a wholly empirical exercise; not all consequences of a given collection of design decisions are made apparent without prolonged experience with the software. The formal approach will never (and should never) obviate empirical methods; instead, it allows the designer to determine properties of the system *a priori* and to justify them in unambiguous terms.

The production of a formal model of a system poses the same problems as does a conventional design and implementation. Interfaces have to be defined, as must behaviours. However, a formal model affords the opportunity to state the design in an unambiguous form in which properties can be stated

[2] *OS Kit* from the University of Utah—see the Computer Science Department's Web site—is a considerable aid in constructing new systems by providing Interrupt Service Routines (ISRs), drivers and other basic components that can be slotted together to form a substrate upon which to build the upper layers of an operating system. OS Kit is a software kit, not a formal specification or modelling tool.

as propositions to be proved. The proof of such properties makes an essential contribution to the exercise by justifying the claims. Proofs provide more insight into the design, even if they seem to be proofs of obvious properties (there are lots of examples above). The point is that the statement of a property as a proposition to be proved makes that property explicit; otherwise, it will remain implicit or just another line in the formal statement of the model.

The properties proved as part of formal modelling reveal characteristics of the software in a way that cannot be obtained by implementation—it can be construed as an exploration without the expense (and frustration) of implementation. This is, of course, not to deny implementation: the goal of all software projects is the production of working code. The point is that formal models provide a level of exploration that is not obtained by a purely empirical approach. Furthermore, formal models *document* the system and its properties: they can serve as information, inspiration or warnings to others.

A further advantage of the formal approach is that it always leaves implementation as an option. With the conventional approach, implementation is a necessity.

1.4 Classical Kernels and Refinement

The focus in this book is on what might be called the "classical" operating system kernel. This is the kind of kernel that is amply documented in the literature (the books and papers cited in this paragraph are all good examples). It is the approach to kernel design that has evolved since the early days of computers through such systems as the TITAN Supervisor [34], the THE operating system [19] and Brinch Hansen's RC4000 supervisor [5]; it is the approach to kernels described in standard texts on operating systems (for example, [29, 11, 26] to cite but three from the past twenty years).

The classical operating system kernel is to be found in most of the systems today: Unix, POSIX and Linux, Microsoft's NT, IBM's mainframe operating systems and many real-time kernels. In days of greater diversity, it was the approach adopted in the design of Digital Equipment's operating systems: RSTS, RSX11/M, TOPS10, TOPS20, VMS and others. Other, now defunct manufacturers also employed it for their product ranges, each with a different choice of primitives and interfaces depending upon system purpose, scope and hardware characteristics. Such richness was then perceived as a nuisance, not a reservoir of ideas.

The classical approach regards operating system kernels as layered entities: a layer of primitives must be defined to execute above the hardware, providing a collection of abstractions to be employed by the remainder of the system. Above this layer are arranged layers of increasing abstraction, including storage management, various clocks and alarms. Finally, there comes the layer in which file management, database interfaces and interfaces to network services appear. At the very top of the hierarchy, there is usually a mechanism that permits user code to invoke system services; this mechanism has been variously called SVCs, Supervisor Calls, System Calls, or, sometimes, Extracodes.

This approach to the design of operating systems can be traced back at least to the THE operating system of Dijkstra *et al.* [19]. (It could be argued that the THE system took many current ideas and welded them into a coherent and elegant whole.) The layered approach makes for easier analysis and design, as well as for a more orderly construction process. (It also assists in the organisation of the work of teams constructing such software, once interfaces have been defined.) It is sometimes claimed that layered designs are inherently slower than other approaches, but with the kernel some amount of layering is required; raw hardware provides only electrical, not software, interfaces.

The classical approach has been well-explored as a space within which to design operating system kernels, as the list of examples above indicates. This implies that the approach is relatively stable and comparatively well-understood; this does not mean, of course, that every design is identical or that all properties are completely determined by the approach.

The classical model assumes that interacting processes, each with their own store, are executed. Execution is the operation of selecting the next process that is ready to run. The selection might be on the basis of which process has the highest priority, which process ran last (e.g., round-robin) or on some other criterion. Interaction between processes can take the form of shared storage areas (such as critical sections or monitors), messages or events. Each process is associated with its own private storage area or areas. Processes can be interrupted when devices are ready to perform input/output (I/O) operations. This roughly defines the layering shown in Figure 1.1.

At the very bottom are located the ISRs (*Interrupt Service Routines*). Much of the work of an ISR is based on the interface presented by the device. Consequently, there is little room in an ISR for very much abstraction (although we have done our best below): ideally, an ISR does as little as possible so that it terminates as soon as possible.

One layer above ISRs come the primitive structures required by the rest of the kernel. The structures defined at this level are exported in various ways to the layers above. In particular, primitives representing processes are implemented. The process representation includes storage for state information (for storage of registers and each process' instruction pointer) and a representation of the process' priority (which must also be stored when not required by the scheduling subsystem). Other information, such as message queues and storage descriptors, are also associated with each process and stored by operations defined in this layer.

Immediately above this there is the scheduler. The scheduler determines which process is next to run. It also holds objects representing processes that are ready to execute; they are held in some form of queue structure, which will be referred to as the *ready queue*. There are other operations exported by

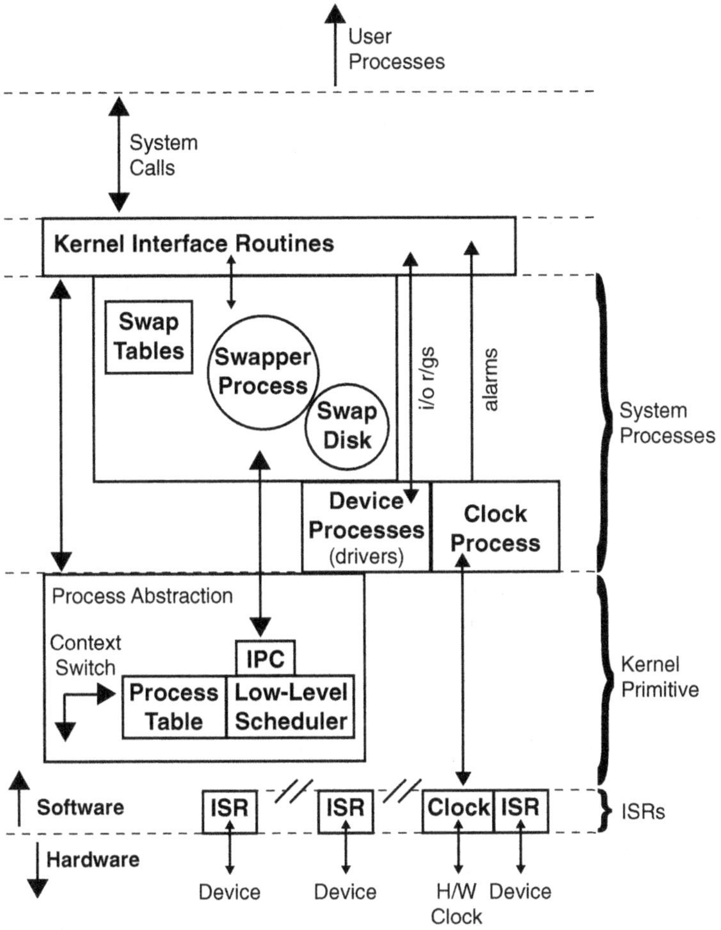

Fig. 1.1. *The layers of the classical kernel model.*

the scheduler, for example, removal of a ready or running process from the ready queue or an operation for the self-termination of the current process. Context switches are called from this layer (as well as others).

Above the process representation and the scheduler comes the IPC layer. It usually requires access not only to the process representation (and the process-describing tables) but also to the scheduler so that the currently executing process can be altered and processes entered into or removed from the ready queue. There are many different types of IPC, including:

- semaphores and shared memory;
- asynchronous message exchange;
- synchronous message exchange (e.g., rendezvous);

- monitors;
- events and Signals.

Synchronisation as well as communication must be implemented within this layer. As is well-documented in the literature, all of the methods listed above can perform both functions.

Some classical kernels provide only one kind of IPC mechanism (e.g., THE [19], SOLO [6]). Others (e.g., Linux, Microsoft's NT, Unix System V) provide more than one. System V provides, *inter alia*, semaphores, shared memory and shared queues, as well as signals and pipes, which are, admittedly, intended for user processes. The essential point is that there is provision for inter-process synchronisation and communication.

With these primitive structures in place, the kernel can then be extended to a collection of system operations implemented as processes. In particular, processes to handle storage management and the current time are required. The reasons for storage management provision clear; those for a clock are, perhaps, less so.

Among other things, the clock process has the following uses:

- It can record the current time of day in a way that can be used by processes either to display it to the user or employ it in processing of some kind or another.
- It can record the time elapsed since some event.
- It can provide a *sleep* mechanism for processes. That is, processes can block on a request to be unblocked after a specified period of time has elapsed.
- It can determine when the current process should be pre-empted (if it is a pre-emptable process—some processes are not pre-emptable, for example, some or all system processes).

In addition to a storage manager and a clock, device drivers are often described as occurring in this layer. The primary reason for this is that processes require the mechanisms defined in the layers below this one—it is the first layer at which processes are possible.

The processes defined in this layer are often treated differently from those above. They can be assigned fixed priorities and permitted either to run to completion or until they suspend themselves. For example, device drivers are often activated by the ISR performing a V (*Signal*) operation on a semaphore. The driver then executes for a while, processing one or more requests until it performs a P operation on the semaphore (an equivalent with messages is also used, as is one based on signals).

The characteristics of the processes in this layer are that:

- They are trusted.
- Their behaviour is entirely predictable (they complete or block).
- They run for relatively short periods of time when executed.

The only exception is the storage manager, which might have to perform a search for unallocated blocks of store. (The storage manager specified in Chapter 4 does exactly this.) However, free store is represented in a form that facilitates the search.

Above this layer, there comes the interface to the kernel. This consists of a library of system calls and a mechanism for executing them inside the kernel. Some kernels are protected by a binary semaphore, while others (Mach is a good, clear example) implement this interface using messages. Above this layer come user processes.

Some readers will now be asking: what about the file system and other kinds of persistent, structured storage? This point will be addressed below when defining the scope of the kernels modelled in this book (Section 1.7).

The classical model can therefore be considered as a relatively high-level specification of the operating system kernel. It is possible to take the position that all designs, whether actual or imagined, are refinements of this specification.

As a high-level specification, the approach has its own invariants that must be respected by these refinements. The invariants are general in nature, for example:

- Each process has a state that can be uniquely represented as a set of registers and storage descriptors.
- Each process is in *exactly one* state at any time. One possible set of states of a process is: *ready* (i.e., ready to execute), *running* (executing), *waiting* or (*blocked*) and *terminated.*
- Each process resides in at most one queue at any time[3].
- Each process can request *at most* one device at any one time. This is a corollary to the queues invariant.
- Each process owns one or more regions of storage that are disjoint from each other and from all others. (This has to be relaxed slightly for virtual store: each process owns a set of pages that is disjoint from all others.)
- There is exactly one process executing at any one time. (This clearly needs generalising for multi-processor machines; however, this book deals only with uni-processors.)
- When a process is not executing, it does nothing. This implies that processes cannot make requests to devices when they are not running, nor can they engage in inter-process communications or any other operations that might change their state.
- An *idle process* is often employed to soak up processor cycles when there are no other processes ready to execute. The idle process is pre-empted as soon as a "real" process enters the scheduler.

[3] It might be thought that each process must be on exactly one queue. There are designs, such as the message-passing kernel of Chapter 5, in which processes do not reside in queues—in this case, when waiting to receive a message.

- The kernel has a *single* mechanism that shares the processor *fairly* between *all* processes according to need (by dint of being the unique running process) or current importance (priority).
- Processes can synchronise and communicate with each other;
- Storage is flat (i.e., it is a contiguous sequence of bytes or words); it is randomly addressed (like an array).
- Only one user process can be in the kernel at any one time.

These invariants and the structres to which they relate can be refined in various ways. For example:

- Each process can share a region of its private storage with another process in order to share information with that other process.
- User processes may not occupy the processor for more than n μseconds before blocking. (n is a parameter that can be set to 1 or can vary with load.)
- A process executes until either it has exceeded its allocated time or a process of higher priority becomes ready to execute.

Multi-processor systems also require that some invariants be altered or relaxed. The focus in this book is on single-processor systems.

It is sometimes claimed that modern operating systems are interrupt-driven (that is, nothing happens until an interrupt occurs). This is explained by the fact that many systems perform a reschedule and a context switch at the end of their ISRs. A context switch is always guaranteed to occur because the hardware clock periodically interrupts the system. While this is true for many systems, it is false for many others. For example, if a system uses semaphores as the basis for its IPC, a context switch occurs at the end of the P (*Wait*) operation if there is already a process inside the critical section. A similar argument applies to signal-based systems such as the original Unix.

Because this book concentrates on classical kernel designs and attempts to model them in abstract terms, each model can be seen as a refinement of the more abstract classical kernel model. Such a refinement might be partial in the sense that not all aspects of the classical model are included (this is exemplified by the tiny kernel modelled as the first example) or of a greater or total coverage (as exemplified by the second and third models, which contain all aspects of the classical design in slightly different ways).

Virtual store causes a slight problem for the classical model. The layers of the classical organisation remain the same, as do their invariants. The principles underlying storage and the invariants stated above also remain invariant. However, the *exact* location of the storage management structures is slightly different.

The storage management of a classical kernel is a relatively simple matter: the tables are of fixed size, as are queue lengths (the maximum possible queue length is just the number of processes that can be supported by the kernel). The kernel stack (if used) will tend to be small (it must be allocated in a fixed-size region of store, in any case). The basics of virtual storage allocation and deallocation are simple: allocation and deallocation are in multiples of fixed-sized pages.

The problem is the following: the kernel must contain a page-fault handler and support for virtual storage. Page tables tend to be relatively large, so it makes sense to use *virtual store* to allocate them. This implies that virtual storage must be in place in order to implement virtual storage. The problem is solved by bootstrapping virtual storage into the kernel; the kernel is then allocated in pages that are *locked* into main store. The bootstrapping process is outside the layered architecture of the classical kernel, so descriptions in the literature of virtual storage tend to omit the messy details. Once a virtual store has been booted, the storage manager process can operate in the same place as in real-store kernels.

Virtual storage introduces a number of simplifications into a kernel but at the expense of a more complex bootstrap and a more involved storage manager (in particular, it needs to be optimised more carefully, a point discussed in some detail in Chapter 6). Virtual machines also introduce a cleaner separation between the kernel and the rest of the system but imposes the need to switch data between virtual machines (an issue that is omitted from Chapter 6 because there are many solutions).

The introduction of virtual storage and the consequent abstraction of virtual machines appears at first to move away from the classical kernel model. However, as the argument above and the model of Chapter 6 indicate, there is, in fact, no conflict and the classical model can be adapted easily to the virtual storage case. A richer and more radical virtual machine model, say Iliffe's *Basic Language Machine* [17], might turn out to be a different story but one that is outside the scope of the present book and its models.

1.5 Hardware and Its Role in Models

Hardware is one of the reasons for the existence of the kernel. Kernels abstract from the details of individual items of hardware, even processors in the case of portable kernels. Kernels also deal directly with hardware by saving and restoring general-purpose registers on context switches, setting flags and executing ISRs.

The kernel is also where interrupts are handled by ISRs and devices handled by their specific drivers. No model of an operating system kernel is complete without a model (at some level of abstraction) of the hardware on which it is assumed to execute.

In the models below, there is only relatively little material devoted to hardware. Most of this is general and included in Chapter 2. This must be accounted for.

First, consider interrupts. Each processor type has its own way of dealing with interrupts. First, there is the question of vectored or non-vectored interrupts: some processors (the majority) offer vectored interrupts, while others do not. Next, what are the actions performed by the processor when an interrupt occurs? Some processors do very little other than indicate that the interrupt has actually occurred. If the processor uses vectored interrupts, it will execute the code each interrupt vector element associates with its interrupt. Although not modelled below, an interrupt vector would be a mapping between the interrupt number, say, and the code to be executed, and some entries in the vector might be left empty. Some processors save the contents of the general-purpose registers (or a subset of them) in a specific location. This location might be a fixed area of store, an area of store pointed to by a register that is set by the hardware interrupt or it might be on the top of the current stack (it might be none of these).

After the code of an ISR has executed, there must be a return to normal processing. Some processors are designed so that this is implemented as a jump, some implement it as a normal subroutine return, while still others implement it as a special instruction that performs some kind of subroutine return and also sets flags. The advantage of the subroutine return approach is that the saved registers are restored when the ISR terminates—this is a little awkward if a reschedule occurs and a context switch is required, but that is a detail.

There are other properties of interrupts that differentiate processors, the most important of which the prioritised interrupts. It is not possible to consider all the variations. Instead, it is necessary to take an abstract view, as abstract as is consistent with the remainder of the model. The most abstract view is that processors have a way of indicating that an asynchronous hardware event has occurred.

Interrupts are only one aspect of the hardware considerations. The number of general-purpose registers provided by a processor is another. Kernels do not, at least do not *typically*, alter the values in particular registers belonging to the processes it executes (e.g., to return values, as, e.g., in a subroutine call). For this reason, the register set can be modelled as an abstract entity; it consists of a set of registers (the maximum number is assumed known, and it might be 0 as in a stack machine, but not used anywhere other than in the definition of the abstractions) and a pair of operations, one to obtain the registers' values from the hardware and one to set hardware register values from the abstraction.

There is also the issue of whether the processor must be in a special kernel mode when executing the kernel. Kernel mode often enables additional instructions that are not available to user-mode processes.

There are many such issues pertaining to hardware. Most of the time, they are of no interest when engaging in a modelling or high-level specification exercise; they become an issue when refinement is underway. The specification of a low-level scheduler has precious little to do with the exact details of the `rti` instruction's operation. What *is* required is that the specification or model be structured in such a way that, when these details become significant, they can be handled in the most appropriate or convenient way.

The diversity of individual devices connected to a processor also provides a source either of richness or frustration. Where there are no standards, device manufacturers are free to construct the interfaces that are most appropriate to their needs. Where there are standards, there can be more uniformity but there can also be details like requiring a driver to wait n μs before performing the next instruction or to wait m μs before testing a pin again to confirm that it has changed its state.

Again, the precise details of devices are considered a matter of refinement and abstract interfaces are assumed or modelled if required. (The hardware clock and the page-fault mechanism are two cases that are considered in detail below.) In these cases, the refinement argument is supported by device-independent I/O, portable operating systems work over many years and by driver construction techniques such as that used in Linux [25]. The refinement argument is, though, strengthened by the fact that the details of *how* a device interface operates are only the concern to the driver, not to the rest of the kernel; only when refining the driver do the details become important.

Nevertheless, the hardware and its gross behaviour are important to the models. For this reason, a small model of an ideal processor is defined and included in the common structures chapter (Chapter 2). The hardware model includes a single-level interrupt mechanism and the necessary interactions between hardware and kernel software are represented. The real purpose of this model is to capture the *interactions* between hardware and software; this is an aspect of the models that we consider of some importance (indeed, as important as making explicit the above assumptions about hardware abstraction).

1.6 Organisation of this Book

The organisation of this book is summarised in this section. Chapters 2 to 6 contain the main technical material, and the last chapter (Chapter 7) contains a summary of what has been done. It also contains some suggestions about where to go next.

Very briefly, the technical chapters are as follows.

Chapter 2. Common structures. This chapter contains the Z specification of a number of structures that are common to most kernels. These structures include FIFO queues, process tables and semaphores. Also included is a hardware model. This is very simple and quite general and is included just to orient the reader as well as to render explicit our assumptions about the hardware. CCS [21] is used for the operational part of this model. Some relevant propositions are proved in this chapter.

Chapter 3. A simple kernel. This kernel is of the type often found in real-time and embedded systems. It is relatively simple and open. It serves as an introduction to the process of modelling kernels. The focus, as far as formally proved properties are concerned, is the priority queue that is used by this kernel's scheduler.

Chapter 4. The swapping kernel. This is a kernel of the kind often found in mini-computers such as the PDP-11/40 and 44 that did not have virtual storage. It includes IPC (using semaphores), process management and storage management. The system includes a process-swapping mechanism that periodically swaps processes to backing store. The kernel uses interrupts for system calls, as is exemplified by the clock process (the sole example of a device driver). The chapter contains proofs of many properties.

Chapter 5. This is a variation on the kernel modelled in Chapter 4. The difference is that IPC is now implemented as message passing. This requires changes to the system processes, as well as the addition of generic structures for handling interrupts and the context switch. The kernel interface is implemented using message passing. A number of properties are proved.

Chapter 6. The main purpose of this chapter is to show that virtual storage can be included in a kernel model. Virtual storage is today too important to ignore; in the future, it is to be expected that embedded processors will include virtual storage[4]. Many properties are proved.

1.7 Choices and Their Justifications

It is worth explaining some of the choices made in this book.

Originally, the models were written in Z [28]. Unfortunately, a considerable amount of promotion was required. The presence of framing schemata in the specification tended, in our belief, to obscure the details of the models. Object-Z [12, 27] uses a reference-based model that makes promotion a transparent operation.

Chapter 2 still contains a fair amount of pure Z: this is to orient readers who are more familiar with Z than Object-Z and give them some idea of the structures used in the rest of the book. The chapter contains some framing schemata and promoted operations. The reader should be able to see how framing gets in the way of a clear presentation. Chapter 2 also contains some CCS.

Object-Z is an object-oriented specification language. Although the models in this book in no way demand object-oriented specification or implementation, the modularity of Object-Z again seems to make each model's structure clearer since operations can be directly related to the modular structure to which they naturally belong. During the specification in Object-Z, objects were considered more in the light of modules (as in Modula2) or Ada packages. Every effort, however, has been made to conform to Object-Z's semantics, so it could be argued that the specifications are genuinely object-oriented; this is an issue we prefer to ignore.

[4] The StrongArm processor has, for example, included virtual storage support since before the year 2000.

As can be inferred from the comment above, CCS [21] is used in a few places. CCS was chosen over CSP [16], π-calculus [22, 33] or some other process algebra (e.g., [1]) because it expresses everything required of it here in a compact fashion. The *Concurrency Workbench* [8] is available to support work in CCS, as will be seen in Chapter 6. Use of CCS is limited to those places where interactions between component processes must be emphasised or where interactions are the primary issue.

The use of Woodcock *et al.*'s CIRCUS specification [7] language was considered and some considerable work was done in that language. In order to integrate a CIRCUS model with the remainder of the models and to model a full kernel in CIRCUS, it would have been necessary to model message passing and the proof that the model coincided with the one assumed by CIRCUS would have to have been included. Another notation would have tended to distract readers from the main theme of this book, as would the additional equivalence proofs.

It was originally intended to include a chapter on a monitor-based kernel. The use of monitors makes for a clearly structured kernel, but this structure only appears above the IPC layer. Eventually, it turned out that:

1. the chapter added little or nothing to the general argument; and
2. Inclusion of the chapter would have made an already somewhat long book even longer.

For this reason, the chapter was omitted. This is a pity because, as just noted, monitors make for nicely structured concurrent programs and the specification of monitors and monitor-using processes in Object-Z is in itself a rather pleasing entity.

Some readers will be wondering why there are no refinements included in this book. Is this because there have been none completed (for whatever reason, for example because they do not result in appropriate software) or for some other reason? We have almost completed the refinement of two different kernels similar to the swapping kernel (but without the swap space), one based on semaphores and one based on messages. The target for refinement is Ada. These refinements will have been completed by the time this book is published. The reasons for omitting them are that there was no time to include them in this book and that they are rather long (the completed one is more than 100 A4 pages of handwritten notes). It is hoped that the details of these refinements, as well as the code, will be published in due course.

It cannot be stressed enough times that the models presented in this book are *logical* models. The intention is that they should be the subject of reasoning. In order to refine the models to code, some extra work has to be done; for example, some sequential compositions will have to be introduced and predicates rearranged or regrouped. The aim here is to make the constructs as clear as possible, even if this means that the grouping of predicates is not optimal for a refinement attempt.

It is a natural question to ask why temporal logic has not been used in this book. The work by Bevier [2] uses temporal logic. Temporal logic is a natural system for specifying concurrent and parallel programs and systems. The answer is that temporal logic is simply not necessary. Everything can be done in Z or Object-Z. A process algebra (CCS [21]) is used in a few cases to describe interactions between components and to prove behavioural equivalence between interacting processes. The approach adopted here is directly analogous to the use of a sequential programming language to program a kernel: the result might be parallel but the means of achieving it are sequential.

This book concentrates on what is referred to as the "classical" kernel paradigm. The reason for excluding other kernel designs, say those based on events, is that they are not as widely known. In order to demonstrate that formal models are possible for kernels, it would appear wiser to attempt the most widely known paradigm. The classical kernel comes in many different flavours, so the scope for different models is relatively broad.

2

Standard and Generic Components

2.1 Introduction

In this chapter, we introduce some of the more common structures encountered in operating system kernels. Each structure is specified and, frequently, properties of that structure are proved. This provides a formal basis upon which to construct the kernels of this book. Some of the structures are used with minor variations (for example, semaphores will be redefined in the next two chapters), while others are not explicitly used at all (for example, tables).

The reason for explicitly specifying and proving properties of such structures is that they will usually appear *as components* of other structures. For example, the generic table structure, *GENTBL[K,D]*, appears as the process table in all of the following models, with or without some extra components. There are instances of semaphore and message queue tables. As a consequence, properties of these supporting structures might be omitted by accident, even though they are of considerable importance to the overall specification of the system. The purpose of this chapter is to supply those additional proofs.

2.2 Generic Tables

Tables appear in a number of places in the specifications to follow. The process table is one example, as is the queue of alarm requests in the clock driver. Tables are mappings of some kind from a set of keys (e.g., process references) to a set of data items (for example, process descriptors). The state is defined (in Z) as:

$GENTBL\,[K, D]$

$tbl : K \mathbin{\ooalign{$\to$\cr\hidewidth$\shortmid\shortmid$\hidewidth}} D$

$keys : \mathbb{F}\, K$

$keys = \operatorname{dom} tbl$

This is a *generic* schema for obvious reasons. The variable *keys* is the set of domain elements of the mapping *tbl* (i.e., the keys of the table).

The table is initialised by the following operation:

InitGENTBL $[K, D]$

    GENTBL[K, D]'
    ───────────
    keys' = ∅

The set of keys is initialised to empty.

Sometimes, it is useful to determine which keys are in the table. The following schema defines that operation:

TBLContainsKey $[K, D]$

    ΞGENTBL[K, D]
    k? : K
    ───────────
    k? ∈ keys

If a table has been initialised and no other update operations have been performed, then that table contains no keys. This is the point of the following proposition.

Proposition 1.

$$InitGENTBL[K, D] \Rightarrow \neg \exists k : K \bullet TBLContainsKey'[K, D][k/k?]$$

PROOF. The predicate of *InitGENTBL* is $keys' = \varnothing$. The predicate of *TBLContainsKey'* is: $k? \in keys'$. If $keys = \varnothing$, there can be no $k?$ such that is an element of *keys*. □

The following operation adds a key-datum pair to a table. Strictly speaking, if the key is already in the table, an error should be raised. Here, we are just defining the operations, so the error condition is ignored. In any case, a user of this component might want to report a more relevant error than a simple "duplicate key".

AddTBLEntry $[K, D]$

    ΔGENTBL[K, D]
    k? : K
    d? : D
    ───────────
    tbl' = tbl ∪ {k? ↦ d?}

Since the key, $k?$, is assumed not to be in the table, set union can be used rather than domain override ($\oplus$).

Proposition 2. *The conjunction*

$\neg\ TBLContainsKey \wedge AddTBLEntry[K,D][k/k?, d/d?]$

implies $TBLContainsKey[K,D][k/k?]$.

PROOF. The predicate of $TBLContainsKey$ side is:

$$k? \notin keys \wedge \\ tbl' = tbl \cup \{k? \mapsto d?\}$$

Since $keys = \mathrm{dom}\ tbl$, by taking domains:

$$\begin{aligned} &\mathrm{dom}\ tbl' \\ &\quad = \mathrm{dom}(tbl \cup \{k? \mapsto d?\}) \\ &\quad = (\mathrm{dom}\ tbl) \cup \mathrm{dom}(\{k? \mapsto d?\}) \\ &\quad = keys \cup \{k?\} \\ &\quad = keys' \end{aligned}$$

So $k? \in keys'$ □

The next operation is the one that retrieves the datum corresponding to a key. If the key is not present, an error should be raised (or some default value returned); this is ignored for the same reason that was given above. The schema is:

$$\begin{array}{|l} \textit{GetTBLEntry}\ [K,D] \\ \hline GENTBL[K,D] \\ k? : K \\ d! : D \\ \hline d! = tbl(k?) \\ \hline \end{array}$$

The operation to remove a key-datum pair from a table is defined by the following schema. If the key is not present, the table is invariant. This is the point of the second proposition after the schema.

$$\begin{array}{|l} \textit{DelTBLEntry}\ [K,D] \\ \hline GENTBL[K,D] \\ k? : K \\ \hline tbl' = \{k?\} \lhd tbl \\ \hline \end{array}$$

The domain subtraction operator, $\lhd$, is used to remove $k?$ from tbl's domain.

Proposition 3. $DelTBLEntry[K,D] \Rightarrow \neg\ TBLContainsKey'[K,D]$.

PROOF. The right-hand side is $k? \notin keys'$. Since $keys = \mathrm{dom}\, tbl$, the result can be obtained by taking domains:

$$\begin{aligned}
&\mathrm{dom}\, tbl' \\
&\quad = \mathrm{dom}(\{k?\} \ntriangleleft tbl) \\
&\quad = \mathrm{dom}(tbl \setminus \{k?\}) \\
&\quad = (\mathrm{dom}\, tbl) \setminus \{k?\} \\
&\quad = keys \setminus \{k?\} \\
&\quad = keys'
\end{aligned}$$

Therefore $k? \notin keys'$. □

Proposition 4. *If $k \in K$ is not in keys, $DelTBLEntry[K, D][k/k?]$ leaves tbl invariant.*

PROOF. By definition of $\ntriangleleft$. □

Another common operation is overwriting the datum corresponding to a key that is already present in a table. The operation is defined by the following schema:

```
┌─ OverwriteTBLEntry [K, D] ──────────────────────────
│ GENTBL[K, D]
│ k? : K
│ d? : D
├──────────────────
│ tbl' = tbl ⊕ {k? ↦ d?}
└─────────────────────────────────────────────────────
```

The following proposition shows that overwriting is the same as a deletion followed by an addition.

Proposition 5. *If $k \in keys$,*

$$\begin{aligned}
&OverwriteTBLEntry[K, D][k/k?, d/d?] = \\
&\quad (DelTBLEntry[K, D] \mathbin{;} AddTBLEntry[K, D])[k/k?, d/d?]
\end{aligned}$$

PROOF. The composition of $DelTBLEntry \mathbin{;} AddTBLEntry$, when expanded, is:

$$\begin{aligned}
&\exists\, tbl'' : K \twoheadrightarrow\!\!\!\!\!\!\!\;\;+ D \bullet \\
&\quad tbl'' = \{k\} \ntriangleleft tbl \land \\
&\quad tbl' = tbl'' \cup \{k? \mapsto d?\}
\end{aligned}$$

Clearly, $k? \notin \mathrm{dom}(\{k\} \ntriangleleft tbl)$. Equally clearly, $k? \in \mathrm{dom}\{k? \mapsto d?\}$ and is therefore in $\mathrm{dom}\, tbl'$, so $tbl'(k?) = d?$.

The definition of $f \oplus g$ is:

$$(f \oplus g)(x) = \begin{cases} g(x) & : \text{if } x \notin \operatorname{dom} f \\ f(x) & : \text{otherwise.} \end{cases}$$

Setting $f = tbl''$ and $g = \{k? \mapsto d?\}$, it is obvious that:

$$(tbl'' \oplus \{k? \mapsto d?\})(x) = \begin{cases} tbl''(x) & : \text{if } x \neq k? \\ k? \mapsto d?(x) & : \text{if } x = k? \end{cases}$$

The two predicates coincide. □

2.3 Queues and Their Properties

Queues are one of the primary data types used in the specification and implementation of operating system kernels. For this reason, this section contains the basic specification of the queue type, as well as a collection of proofs. The queue type is quite general and is of a FIFO (First-In, First-Out) queue. It is essential that a type as important as the FIFO queue is completely understood and supported by proofs of its major properties.

The queue is generic so it can be instantiated to any element type. The operations specified for this type are the ones that will usually occur in the specifications that follow.

The type that is defined by the following schema is intended to be used for a good many data types within the kernel. After presenting this specification, a version is defined and justified that contains process references (specifically elements of the type *APREF* which is defined below in Sections 3.3 and 4.4) is defined and justified. As will be seen, it has the same operations as the generic queue type that is defined here.

This is the generic FIFO queue type:

```
─ QUEUE [X] ──────────────────────────
│ elts : seq X
└──────────────────────────────────────
```

The queue is represented quite naturally in Z by a sequence of elements of some type (here, the generic type X). Sequences in Z are just partial functions from a subset of $\mathbb{N}$ to another set, here X.

The initialisation operation for $QUEUE[X]$ (recall that the name includes the generic parameter) is as follows:

```
─ InitQUEUE [X] ──────────────────────
│ QUEUE[X]'
├──────────
│ elts' = ⟨ ⟩
└──────────────────────────────────────
```

The schema defines the after state of the operation only. The sequence is set to the empty sequence, $\langle\,\rangle$.

The length of the queue is the number of elements it contains:

LengthOfQUEUE $[X]$

$$\Xi QUEUE[X]$$
$$len! : \mathbb{N}$$

$$len! = \# elts$$

It is necessary to determine whether a queue contains elements (for example, when removing or dequeuing the first element). The following schema defines this test:

EmptyQUEUE $[X]$

$$\Xi QUEUE[X]$$

$$elts = \langle\rangle$$

The *Enqueue*$[X]$ operation adds an element to the end of the queue. It is naturally modelled in Z by the following schema:

Enqueue $[X]$

$$\Delta QUEUE[X]$$
$$x? : X$$

$$elts' = elts \frown \langle x? \rangle$$

To dequeue an element from a queue according to the FIFO scheme, the first element is removed, provided that the queue is not empty. The following schema defines the removal operation only:

RemoveFirst $[X]$

$$\Delta QUEUE[X]$$
$$x! : X$$

$$\langle x! \rangle \frown elts' = elts$$

Equivalently, the *RemoveFirst* operation could be written as:

$$\Delta QUEUE[X]$$
$$x! : X$$

$$x! = head\ elts$$
$$elts' = tail\ elts$$

This form is one that will often be used in proofs below.

Sometimes, it is necessary to test whether an element is present in a queue. The following schema models this test:

$$\begin{array}{|l}
\hline
\textit{IsInQueue}\,[X] \\
\Xi QUEUE \\
x? : X \\
\hline
x? \in \operatorname{ran} elts \\
\hline
\end{array}$$

Occasionally, the index of an element in the queue is required. This operation is modelled by the following schema:

$$\begin{array}{|l}
\hline
\textit{QueueEltIndex}\,[X] \\
\Xi QUEUE[X] \\
x? : X \\
n! : \mathbb{N}_1 \\
\hline
(\exists\, n : \mathbb{N}_1 \mid n \in 1..\#elts \bullet \\
\qquad elts(n) = x? \land n = n!) \\
\hline
\end{array}$$

or $elts(n!) = x?$ for some n (both versions are non-deterministic).

It will frequently be necessary to remove queue elements that are not at the head of the queue. This is modelled by the following schema:

$$\begin{array}{|l}
\hline
\textit{RemoveQueueElt}\,[X] \\
\Delta QUEUE[X] \\
x? : X \\
\hline
\exists\, s, t : \operatorname{seq} X \mid elts = s \frown \langle x? \rangle \frown t \bullet \\
\qquad elts' = s \frown t \\
\hline
\end{array}$$

This operation is a necessary one. In the kernels that appear in this book, the *unready* operation is frequently used. The *unready* operation removes a process from the queue in which it resides. The element to be removed is not, however, the head of the queue. The *unready* operation's core is the *RemoveQueueElt* operation just defined.

Just for completeness, a collection of error types is defined for the generic $QUEUE[X]$ type and the operations defined over it.

$$QERROR ::= \mathsf{emptyqerr} \mid \mathsf{okq}$$

$$\begin{array}{|l}
\hline
\textit{EmptyQError} \\
qerr! : QERROR \\
\hline
qerr! = \mathsf{emptyqerr} \\
\hline
\end{array}$$

$$\begin{array}{|l}
\hline
\textit{QOk} \\
qerr! : QERROR \\
\hline
qerr! = \mathsf{okq} \\
\hline
\end{array}$$

Traditionally, the FIFO queue is equipped with a *Dequeue* operation. This operation would be defined as follows:

$$\begin{aligned}&Dequeue_a[X] \mathrel{\widehat=} \\ &\qquad (\neg\ EmptyQUEUE[X] \land RemoveFirst[z]) \\ &\qquad\qquad \lor\ EmptyQError\end{aligned}$$

The operation must first test the queue to determine that it contains at least one element. If the queue is empty, an error is usually raised. If the queue is not empty, the first element is removed and all is well (denoted by the *QOk* operation).

Meanwhile, for the propositions that follow, the following definition is quite satisfactory:

$$Dequeue[X] \mathrel{\widehat=} RemoveFirst[X]$$

Indeed, this definition permits reasoning about empty queues that would otherwise be complicated by the error schemata (*EmptyQError*).

This is the way in which removal of the queue head is treated in the models that follow. The reason for this is that the emptiness test is performed somewhere else, somewhere that makes better sense for the operation in which *RemoveFirst* occurs. It is, in any case, something of an inconvenience to use the error schemata defined above.

A number of fairly obvious propositions are now proved about $QUEUE[X]$.

This first proposition shows that an enqueue followed immediately by a dequeue produces a queue that is different from the one prior to the operation.

Proposition 6. *If* $Enqueue[X] \mathbin{⨟} Dequeue[X]$, *then* $elts \neq elts'$.

Proof. The predicate of operation *Enqueue* is:

$$elts' = elts \frown \langle x? \rangle$$

While the predicate of operation *Dequeue* is:

$$elts = \langle x! \rangle \frown elts'$$

By the definition of sequential composition (and changing output variable name to avoid confusion):

$$\begin{aligned}&Enqueue \mathbin{⨟} Dequeue \equiv \\ &\qquad \exists\, elts'' : \operatorname{seq} X \bullet \\ &\qquad\qquad elts'' = elts \frown \langle x? \rangle \land \\ &\qquad\qquad elts'' = \langle y! \rangle \frown elts'\end{aligned}$$

The second conjunct can be re-written as:

$$\begin{aligned}&elts' = tail\ elts'' \\ &y! = head\ elts''\end{aligned}$$

So,

$$
\begin{aligned}
&y! \frown ((\mathit{tail}\ \mathit{elts}) \frown \langle x? \rangle) \\
&\qquad = (\mathit{head}\ \mathit{elts}) \frown (\mathit{tail}\ \mathit{elts}) \frown \langle x? \rangle \\
&\qquad = (\mathit{head}\ \mathit{elts}) \frown ((\mathit{tail}\ \mathit{elts}) \frown \langle x? \rangle)
\end{aligned}
$$

which implies that:

$$
\begin{aligned}
\mathit{elts}' &= (\mathit{tail}\ \mathit{elts}) \frown \langle x? \rangle \\
&\neq \mathit{elts}
\end{aligned}
$$

□

Proposition 7. *If* $\#\mathit{elts} \geq 2$*,* $\mathit{RemoveFirst}[X]$ *implies that* $\mathit{tail}(\mathit{tail}\ \mathit{elts}) = \mathit{tail}\ \mathit{elts}'$

Proof. The predicate of the $\mathit{RemoveFirst}[X]$ schema is:

$$
\begin{aligned}
&x! = \mathit{head}\ \mathit{elts} \\
&\mathit{elts}' = \mathit{tail}\ \mathit{elts}
\end{aligned}
$$

For some y:

$$
\begin{aligned}
&\mathit{tail}\ \mathit{tail}\ \mathit{elts} \\
&\qquad = \mathit{tail}(\mathit{tail}\langle x! \rangle \frown \langle y \rangle \frown \mathit{elts}'') \\
&\qquad = \mathit{tail}(\langle y \rangle \frown \mathit{elts}'') \\
&\qquad = \mathit{elts}''
\end{aligned}
$$

□

Proposition 8. *If* $\mathit{elts} \neq \langle\, \rangle$*,* $\mathit{RemoveFirst}[X]$ *implies that:*

$$\mathit{head}\ \mathit{elts} \neq \mathit{head}\ \mathit{elts}'$$

Proof. Let $\mathit{elts} = \langle x \rangle \frown \langle y \rangle \frown \mathit{elts}''$.

$$
\begin{aligned}
&\mathit{elts}' \\
&\qquad = \mathit{tail}\ \mathit{elts} \\
&\qquad = \mathit{tail}(\langle x! \rangle \frown \langle y \rangle \frown \mathit{elts}'') \\
&\qquad = \langle y \rangle \frown \mathit{elts}''
\end{aligned}
$$

Note that, even if $x! = y$, we can consider them to be different *instances* of $x!$. □

Proposition 9. *If, for some element, x, of elts, if $elts(n) = x$ for some n, such that $1 \leq n \leq \#elts$, then $RemoveNext^n[X]$ removes x from the queue, where:*

$$RemoveNext^n \equiv \overbrace{(RemoveNext \fatsemi \ldots \fatsemi RemoveNext)}\ n \text{ times}$$

PROOF. The proof is by induction on the length of the prefix (i.e., the elements $elts(1) \ldots elts(n-1)$). The length is denoted by k.

The predicate of *RemoveNext* is (changing the name of the output variable):

$$elts = \langle y! \rangle \frown elts'$$

Case $k = 0$, then $x = head\ elts$, so *RemoveNext* removes it from the queue. Case $k = n - 1$. So, there are $n - 1$ elements ahead of x in *elts*. By definition of *RemoveNext*, $RemoveNext^{n-1}$ removes them, so $x = head\ elts$. Therefore, $RemoveNext^n$ removes x from *elts*. □

Proposition 10. *If x is an element of elts, and, for some m, such that $1 \leq m \leq \#elts$, $elts(m) = x$, then, assuming no removals from the queue, $Enqueue^n$, for all n, leaves x at the same index.*

PROOF. $Enqueue^n$ is defined as:

$$\overbrace{Enqueue \fatsemi \ldots \fatsemi Enqueue}\ n \text{ times}$$

If $n = 0$, $Enqueue^0$ can be defined as $elts' = elts \frown \langle\rangle$.

It can be assumed without loss of generality that, for some i, $i > 0$, $elts(i) = x$ and $\#elts = i$.

The proof proceeds by induction that $Enqueue^n$, $n \geq 0$, leaves $elts(i) = x$. Case $n = 0$, so $elts(i) = elts'(i) = x$.This is for the reason that $Enqueue^0$ implies that $elts' = elts \frown \langle\rangle = elts$.
Case $n = k - 1$, then $\#elts' = (\#elts) + k - 1$ since

$$Enqueue^{k-1} = elts \frown \langle x_1, \ldots, x_{k-1} \rangle$$

where the elements x_j, $1 \leq j \leq k - 1$ occur after x in $elts'$. The elements in the sequence $(\lambda j : m + 1 \mathinner{..} m + k - 1 \bullet elts(j))$ clearly appear at indices greater than m, so $elts'(m) = elts(m) = x$. □

Corollary 1. *If $\#elts = n$ and $n > 0$, $Dequeue[X]^m \Rightarrow elts' \neq \langle\rangle$, if and only if $m < n$.*

PROOF. Immediate from Proposition 9. □

Corollary 2. *If* $\#elts = n$, $Dequeue[X]^n \Rightarrow elts' = \langle\,\rangle$

PROOF. Immediate from Proposition 9. □

Corollary 3. *If* $elts = \langle\,\rangle$, *then, for all* n *and* m,

$$Enqueue[X]^n \mathbin{\S} Dequeue[X]^m \Rightarrow elts' = \langle\,\rangle$$

if and only if $n = m$.

PROOF. Immediate from Propositions 9 and 10. □

2.4 Hardware Model

As stated above, a hardware model is required. In this section, a simple model is defined. The purpose of this section is to make clear the assumptions about the hardware that are made for the remainder of this book. The section remains somewhat outside the rest of the models because the hardware is rather outside the kernels modelled here.

In this section, the use of Z is replaced by the use of CCS, which is used to model the fundamental behaviour of the hardware, in particular the interrupt structure. There are no proofs to be undertaken: the material is suggestive of a general architecture, not a model of a specific one.

2.4.1 CCS Model

CCS [21] is used to model the hardware. CCS is a well-known process algebra with a small set of operations. It is well-suited to describing the hardware's operations.

The hardware is modelled by a CCS process, HW. The model is not complete and is intended merely to be suggestive of the actions taken by the hardware in response to various signals.

The first action of this process is *start*, which starts the hardware (initialises it, etc.). The hardware then behaves as if it were process HW_1. This process waits for a message. If the message is an interrupt, i_i, it saves the register set (by a hidden action, *saveregs*) and then waits for the signal to restore it; after the *restoreregs* signal has been received, the process iterates. If the message is *setregs*, the hardware loads values (unspecified) into the general-purpose (i.e., programmable) registers; if the message is *getregs*, the hardware returns the register set (by performing an action not shown in the definition of HW_1).

The process is arranged so that neither a *setregs* nor a *getregs* action can be performed while the register set is saved.

The hardware process is defined as:

$$HW = start.HW_1$$
$$\begin{aligned} HW_1 = (i_1.saveregs &+ HW_1 \\ &+setregs.HW_1 \\ &+getregs.HW_1 \\ &+restoreregs.HW_1) \setminus saveregs \end{aligned}$$

The following pair of processes are intended to model the behaviour of hardware when an interrupt occurs. The process Int_i represents the ith interrupt. When it receives its internal interrupt signal, i_i, it signals that the Interrupt Service Routine (ISR) corresponding to this interrupt should be executed; this is done by sending the $\overline{runisr}_i$ message. The interrupt process then recurs, ready to accept another interrupt signal. The second process, ISR_i, is intended roughly to model the actions of the ISR corresponding to interrupt i. When the ISR receives the signal to execute ($runisr_i$), it performs the *service* action and then instructs the hardware to restore the register set to the way it was before the interrupt occurred. The ISR process then recurs, so that it can accept another interrupt.

$$Int_i = \bar{i}_1.\overline{runisr}_1.Int_i$$
$$ISR_i = runisr.service.restoreregs.ISR_i$$

The hardware and interrupt subsystem can be thought of as the following (parallel) composition of processes:

$$H = HW \mid \Pi_{i \in I}(Int_i \mid ISR_i)$$

The next process models the *interrupt mask*. The interrupt mask determines whether interrupts are signalled or not (it is modelled in this book by the *Lock* Object-Z class).

$$IntMask = on.IntMask(1)$$
$$\begin{aligned} IntMask(v) = &\ off.IntMask(0) \\ &+on.IntMask(1) \\ &+stat.\overline{istat}(n).IntMask(n) \end{aligned}$$

The interrupt mask enables the hardware model to be extended so that interrupts can be enabled and disabled under programmer control. Integration of the interrupt mask and the process P is left as an exercise for the interested reader.

The model works as follows. Initially, the *IntMask* accepts an *on* event to initialise the mask. Initialisation enables interrupts and takes the state of the mask as its parameter (the value in parentheses). After this, the mask offers three possible actions: *off* to disable interrupts, *on* to enable them and *stat* to enquire about the state of the mask. When *IntMask* engages in an *off* action, it disables interrupts (denoted by the 0 parameter). Alternatively, it can engage in an *on* action. If interrupts are currently disabled, the *on* action re-enables them; otherwise, it does nothing. Finally, some other component (say, some software) can enquire as to the state of the interrupt mask by engaging in the third possible action, *stat* (status). The *IntMask* process then returns the current status (denoted by n) via an $\overline{istat}$ (*i*nterrupt *stat*us) action; enquiry does not affect the state of the mask. This is indicated by the recursion on the same value as that communicated by the $\overline{istat}$ action.

This is a single-level interrupt scheme. Some processors have a multi-level one. At the level of detail required in this book, the differences between single- and multi-level interrupt schemes are not significant, so a single-level scheme is assumed for simplicity.

Purely for interest, a multi-level interrupt mask, *MLIMask*, can be defined as follows. First, the mask is initialised by participating in an *allon* (all on) action:

$$MLIMask = allon.MLIMask(S)$$

Here, the parameter S denotes the set of all interrupt levels. The mask now behaves as follows:

$$\begin{aligned} MLIMask(S) = {} & off(i).MLIMask(S \setminus \{i\}) \\ & + on(i).MLIMask(S \cup \{i\}) \\ & + ison(i).\overline{istat}(i \in S).MLIMask(S) \\ & + offm(I).MLIMask(S \setminus I) \\ & + onm(I).MLIMask(S \cup I) \end{aligned}$$

where I denotes a set of interrupt levels and i is an individual level.

2.4.2 Registers

The processor contains a set of general-purpose registersas well as a set of more specialised ones: stack register, instruction pointer and status register (sometimes called the "status word"). It is assumed that each register is one *PSU* wide.

The model of the registers is rather minimal. There is not a lot that can be proved about it.

It is assumed that the hardware is not a stack machine (i.e., a single-address machine, that is). If a stack machine were the target, the registers would not strictly be required. Actually, many stack machines do have the odd off-stack register just as an optimisation.

The number of general-purpose registers is given by:

$$numregs : \mathbb{N}_1$$

Note that no value is given. This is a partial specification (it is, in any case, impossible to assign a value to *numregs* without knowing which processor is being used).

The register names form the following set:

$GENREG == \{\mathsf{r}_0, \ldots, \mathsf{r}_{numregs-1}\}$

The contents of this set are of no further interest to us because the register set will be manipulated as a complete entity.

The register set is defined as a function from register (index) to the value it contains:

$GENREGSET == GENREG \rightarrow PSU$

The status register contains a value. That value is of the following type. It is assumed to be of the same size (in bits) as an element of PSU.

$[STATUSWD]$

This will be an enumeration, for example: overflow, division_by_zero, carry_set.

The register state is defined by the following schema:

$$
\begin{array}{l}
\textit{HWREGISTERS} \\
\hline
hwregset : GENREGSET \\
hwstack : PSTACK \\
hwip : \mathbb{N} \\
hwstatwd : STATUSWD \\
\hline
\end{array}
$$

The general register set is *hwregset*, the stack is in *hwstack*, the instruction pointer (program counter) is *hwip* and the status word is denoted by *hwstatwd*.

The following defines the zero elements for PSU and $STATUSWD$:

$$
\begin{array}{|l}
0_{PSU} : PSU \\
clear : STATUSWD
\end{array}
$$

The registers are initialised when the hardware starts up. This initialisation is modelled by the following operation:

$$
\begin{array}{l}
\textit{InitHWREGISTERS} \\
\hline
HWREGISTERS' \\
\hline
(\forall r : GENREGSET \bullet \\
\quad r \in hwregset' \Rightarrow hwregset'(r) = 0_{PSU}) \\
hwip' = 0 \\
hwstack' = EmptyStack \\
hwstatwd' = clear \\
\hline
\end{array}
$$

This schema does not appear in any of the kernels because it will have been referenced (executed) before any kernel initialisation operations are executed. It is included for completeness.

2.4.3 Interrupt Flag

The interrupt flag is of crucial importance in the models that follow. The flag is of a type containing two values (they could be *true* and *false* or 0 and 1—symbolic values are used instead for easier interpretation of often complex schemata):

$$INTERRUPTSTATUS ::= \mathsf{inton} \mid \mathsf{intoff}$$

The value inton represents the hardware state in which interrupts are enabled. The value intoff denotes the fact that interrupts have been disabled.

The interrupt flag itself is defined as:

INTERRUPTFLAG

$$iflag : INTERRUPTSTATUS$$

When the hardware starts up, it will execute an operation similar to that denoted by the following schema:

InitINTERRUPTFLAG

$$INTERRUPTFLAG'$$

$$iflag' = \mathsf{inton}$$

This schema is similar to the register-initialisation schema. It is assumed that the hardware executes it before the kernel bootstrap starts executing. This will be the only time we see this schema.

There are three operations associated with the interrupt flag. Two are under program control: one disables and one enables interrupts. The remaining operation raises the interrupt and performs operations such as saving the current register state and transferring control to an ISR.

The operation to disable interrupts is modelled here as:

DisableInterrupts

$$\Delta INTERRUPTFLAG$$

$$iflag' = \mathsf{intoff}$$

The operation to enable interrupts is:

EnableInterrupts

$$\Delta INTERRUPTFLAG$$

$$iflag' = \mathsf{inton}$$

Interrupts are disabled and then enabled again to ensure that no interrupts occur during the execution of a piece of code. They are used as a kind of low-level mutual exclusion mechanism.

It is usual to define a couple of operations, named *Lock* and *Unlock*, to perform the disabling and enabling of interrupts. These operations are usually defined as assembly language macros. The names are used because they are better mnemonics. They are defined as:

$Lock \mathrel{\widehat{=}} DisableInterrupts$

and:

$Unlock \mathrel{\widehat{=}} EnableInterrupts$

2.4.4 Timer Interrupts

Most processors have a hardware clock that generates interrupts at a regular rate (e.g., typically 60Hz, the US mains supply frequency). Timer interrupts are used to implement process alarms (sleep periods—the term "alarm" is used in this book by analogy with "alarm clock"). A process suspends itself for a specified period of time. When that time, as measured by the hardware clock, has expired, the process is resumed (by giving it an "alarm call"). A piece of code, which will be called the *clock driver* in this book, is responsible for (among other things) suspending processes requesting alarms and for resuming them when the timer has expired.

This subsection is concerned with the general operation of the clock driver and with clock interrupts. The clock will be used in a number of places in the kernels that follow and it will be re-modelled in various forms. The purpose of the current section is just to orient the reader and to show that such a low-level model can be produced in Z (later in Object-Z) in a fashion that is relatively clear and, what is more, in a form that allows a number of properties to be proved.

The hardware clock is associated with the interrupt number:

| *clockintno* : *INTNO*

Time is modelled as a subset of the naturals:

$TIMEVAL == \mathbb{N}$

Here, time is expressed in terms of uninterpreted units called "ticks" (assumed to occur at regular intervals, say every 1/60 second).

The clock is just a register that contains the current time, expressed in some units:

```
┌─ CLOCK ──────────────────────────
│ timenow : TIMEVAL
└──────────────────────────────────
```

The hardware initialises the clock on startup. (The clock can also be reset on some processors.)

$$
\begin{array}{|l}
\hline
\textit{InitCLOCK} \\
\textit{CLOCK}' \\
\hline
\textit{timenow}' = 0 \\
\hline
\end{array}
$$

The length of the clock tick often needs to be converted into some other unit. For example, a 60Hz "tick" might be converted into seconds.

$$
\begin{array}{|l}
\textit{ticklength} : \textit{TIMEVAL}
\end{array}
$$

The clock updates itself on every hardware "tick":

$$
\begin{array}{|l}
\hline
\textit{UpdateCLOCKOnTick} \\
\Delta \textit{CLOCK} \\
\hline
\textit{timenow}' = \textit{timenow} + \textit{ticklength} \\
\hline
\end{array}
$$

When the current time is required, the following operation is used:

$$
\begin{array}{|l}
\hline
\textit{TimeNow} \\
\Xi \textit{CLOCK} \\
\textit{now}! : \textit{TIMEVAL} \\
\hline
\textit{now}! = \textit{timenow} \\
\hline
\end{array}
$$

When a process needs to set an alarm, it sends the clock driver a message of the following type:

$$\textit{TIMERRQ} == \textit{PREF} \times \textit{TIMEVAL}$$

The message contains the identifier of the requesting process (here, of type *PREF*, the most general process reference type) plus the time by which it expects to receive the alarm.

The following axioms define functions to access elements of *TIMERRQ* (which are obvious and merit no comment):

$$
\begin{array}{|l}
\textit{timerrq\_pid} : \textit{TIMERRQ} \rightarrow \textit{PREF} \\
\textit{timerrq\_time} : \textit{TIMERRQ} \rightarrow \textit{TIMEVAL} \\
\hline
\forall t : \textit{TIMERRQ} \bullet \\
\quad \textit{timerrq\_pid}(t) = \textit{fst}\ t \\
\quad \textit{timerrq\_time}(t) = \textit{snd}\ t
\end{array}
$$

The queue is represented as a finite set of requests. An instance of $QUEUE[X]$ could be used but, as will be seen, searching might have to be performed to find the process to wake and the actual arrival time of requests in the queue is not of any particular importance, so a more abstract view of the queue, as a collection, is used instead.

The request queue is defined as:

TIMERRQQUEUE
$telts : \mathbb{F}\ TIMERRQ$

The request queue is initialised by the following operation. It can be called at any time the kernel is running, say on a warm reboot.

InitTIMERRQQUEUE
$TIMERRQQUEUE'$

$telts' = \varnothing$

The following schema defines a predicate that is true when the request queue is empty:

EmptyTIMERRQQUEUE
$\Xi TIMERRQQUEUE$

$telts = \varnothing$

The following three schemata define operations that add and remove requests:

EnqueueTIMERRQ
$\Delta TIMERRQQUEUE$
$tr? : TIMERRQ$

$telts' = telts \cup \{tr?\}$

RemoveFirstTIMERRQ
$\Delta TIMERRQQUEUE$
$tr! : TIMERRQ$

$\{tr!\} \cup telts' = telts$

This operation removes the first element of the queue. It is a non-deterministic operation.

RemoveTIMERRQQueueElt
$\Delta TIMERRQQUEUE$
$tr? : TIMERRQ$

$tr? \in telts$
$telts' = telts \setminus \{tr?\}$

This schema defines an operation that removes an arbitrary element of the request queue.

The following schema defines a combination of a clock and a request queue. The instance of *CLOCK* is intended to be a register holding a copy of the hardware clock's current value. The idea is that the clock driver copies the hardware clock's value so that the driver can refer to it without needing to access the hardware.

$$TIMER \mathrel{\widehat{=}} TIMERRQQUEUE \land CLOCK$$

This expands into:

$$\begin{array}{|l}
telts : \mathbb{F}\ TIMERRQ \\
timenow : TIMEVAL \\
\hline
\end{array}$$

The timer is initialised by the obvious operation:

$$\begin{array}{l}
TIMERInit \mathrel{\widehat{=}} \\
\quad InitCLOCK \land \\
\quad TIMERInit
\end{array}$$

This expands into:

$$\begin{array}{|l}
TIMER' \\
CLOCK' \\
\hline
timenow' = 0 \\
telts' = \varnothing \\
\hline
\end{array}$$

The following condition must always hold:

Proposition 11. *At any time, now:*

$$\begin{array}{l}
\forall tr : TIMERRQ \bullet \\
\quad tr \in telts \Rightarrow timerrq_time(tr) > now
\end{array}$$

Proposition 12. *At any time, now:*

$$\begin{array}{l}
\neg\ \exists tr : TIMERRQ \bullet \\
\quad tr \in telts \land timerrq_time(tr) \leq now
\end{array}$$

Both of these propositions are consequences of Proposition 92 (p. 173). Their proofs are omitted.

In order to unblock those processes whose alarms have gone off, the following schema is used. It returns a set of requests whose time component specifies a time that is now in the past:

$$\begin{array}{|l}
\multicolumn{1}{l}{__ \mathit{TimerRequestsNowActive} ____} \\
\Delta \mathit{TIMER} \\
\mathit{trqset!} : \mathbb{F}\ \mathit{TIMERRQ} \\
\hline
\mathit{trqset!} = \{ \mathit{trq} : \mathit{TIMERRQ} \mid \mathit{trq} \in \mathit{telts} \land \mathit{timerrq_time}(\mathit{trq}) \leq \mathit{timenow} \bullet \mathit{trq} \} \\
\mathit{telts}' = \mathit{telts} \setminus \mathit{trqset!} \\
\hline
\end{array}$$

This is the basis of a CLOCK process:

$$\begin{array}{l}
\mathit{OnTimerInterrupt} \mathrel{\widehat{=}} \\
\quad (\mathit{UpdateCLOCKOnTick} \mathbin{;} \\
\quad ((\mathit{TimerRequestsNowActive}[\mathit{trqset}/\mathit{trqset!}] \land \\
\qquad (\forall \mathit{trq} : \mathit{TIMERRQ} \mid \mathit{trq} \in \mathit{trqset} \land \mathit{timerrq_pid}(\mathit{trq}) \in \mathit{known_procs} \bullet \\
\qquad\quad (\exists p : \mathit{PREF}; \mid p = \mathit{timerrq_pid}(\mathit{trq}) \bullet \\
\qquad\qquad \mathit{MakeReadypq}[p/\mathit{pid}?]))) \setminus \{\mathit{trqset}\}))
\end{array}$$

The operation works as follows. First, the clock is updated by one tick. Then, those processes whose alarms have gone off (expired) are found in and removed from the set of waiting processes. Each one of these processes is put into the ready queue ($\mathit{MakeReadypq}[p/\mathit{pid}?]$).

The basic operation executed by a process when requesting an alarm is the following:

$$\begin{array}{l}
\mathit{WaitForTimerInterrupt} \mathrel{\widehat{=}} \\
\quad ((([\mathit{CURRENTPROCESS}; \mathit{time}? : \mathit{TIMEVAL}; \mathit{trq} : \mathit{TIMERRQ} \mid \\
\qquad\quad \mathit{trq} = (\mathit{currentp}, \mathit{time}?)] \land \\
\qquad \mathit{Lock} \land \mathit{EnqueueTIMERRQ}[\mathit{trq}/\mathit{tr}?]) \setminus \{\mathit{trq}\} \land \\
\qquad \mathit{MakeUnready}[\mathit{currentp}/\mathit{pid}?] \land \\
\qquad \mathit{SwitchFullContextOut}[\mathit{currentp}/\mathit{pid}?] \land \\
\qquad \mathit{SCHEDULENEXT}) \mathbin{;} \\
\qquad \mathit{Unlock}
\end{array}$$

In Chapter 4, some properties of the clock process and its alarm mechanism will be proved.

2.4.5 Process Time Quanta

In some of the kernels to follow, user processes are scheduled using a pre-emptive method. Pre-emption is implemented in part using time quanta. Each *user* process (system processes are not allocated time quanta and cannot be pre-empted) is allocated a time quantum, a value of type *TIMEVAL*. On each hardware clock "tick", the time quantum is decremented. When the quantum reaches some threshold value, the process is suspended. When that same process is executed the next time, it is assigned a new quantum.

To begin, the type of *time_quantum* is defined:

$$\mathit{time\_quantum} : \mathit{TIMEVAL}$$

For the purpose of this book, every user process uses the same values for initialisation and threshold.

The following schema retrieves the value of a process' time quantum from the process table.

$$
\begin{array}{|l}
\hline
\textit{ProcessQuantum} \\
\hline
\Xi PROCESSES \\
pid? : PREF \\
timeq! : TIMEVAL \\
\hline
timeq! = pquants(pid?) \\
\hline
\end{array}
$$

The next schema defines an operation that sets the initial value for its time quantum:

$$
\begin{array}{|l}
\hline
\textit{SetInitialProcessQuantum} \\
\hline
\Delta PROCESSES \\
pid? : PREF \\
time\_quant? : TIMEVAL \\
\hline
pquants' = pquants \cup \{pid? \mapsto time\_quant?\} \\
\hline
\end{array}
$$

When a process' time quantum is to be reset, the following operation does the work:

$$
\begin{array}{l}
ResetProcessTimeQuantum \mathrel{\widehat=} \\
\quad (\exists q : TIMEVAL \mid q = time\_quantum \bullet \\
\qquad UpdateProcessQuantum[time\_quantum/timeq?])
\end{array}
$$

The following schema models an operation that sets the current value of its time quantum in its process descriptor:

$$
\begin{array}{|l}
\hline
\textit{UpdateProcessQuantum} \\
\hline
\Delta PROCESSES \\
pid? : PREF \\
timeq? : TIMEVAL \\
\hline
pquants' = pquants \oplus \{pid? \mapsto timeq?\} \\
\hline
\end{array}
$$

This operation can be used when the process is interrupted or when a higher-priority process must be scheduled.

There is a storage location that holds the current process' time quantum while it executes:

$$
\begin{array}{|l}
\hline
\textit{SetCurrentProcessQuantum} \\
\hline
\Delta CURRENTPROCESSpq \\
timequant? : TIMEVAL \\
\hline
tq' = timequant? \\
\hline
\end{array}
$$

The quantum is updated by:

$$
\begin{array}{|l}
\hline
\textit{UpdateCurrentProcessQuantum} \\
\Delta CURRENTPROCESSpq \\
now? : TIMEVAL \\
\hline
tq' = tq - now? \\
\hline
\end{array}
$$

This schema defines a predicate that is satisfied when the current process' time quantum has expired:

$$
\begin{array}{|l}
\hline
\textit{CurrentProcessQuantumHasExpired} \\
\Xi CURRENTPROCESSpq \\
\hline
tq \leq 0 \\
\hline
\end{array}
$$

The current process quantum is read from the storage location by the next schema:

$$
\begin{array}{|l}
\hline
\textit{CurrentProcessQuantum} \\
\Xi CURRENTPROCESSpq \\
tquant! : TIMEVAL \\
\hline
tquant! = tq \\
\hline
\end{array}
$$

On each hardware clock tick, the current process' time quantum is updated by the following operation:

$$
\begin{array}{l}
UpdateCurrentQuantumOnTimerClick \mathrel{\widehat=} \\
\quad (TimeNow[now/now!] \land \\
\quad UpdateCurrentProcessQuantum[now/now?]) \setminus \{now\}
\end{array}
$$

This operation is already represented in the last line of *OnTimerInterrupt*.

When a process is blocked, the following are required:

$$
\begin{array}{l}
SaveCurrentProcessQuantum \mathrel{\widehat=} \\
\quad (CurrentProcessQuantum[tquant/tquant!] \land \\
\qquad UpdateProcessQuantum[tquant/timeq?]) \setminus \{tquant\}
\end{array}
$$

This expands into:

$$
\begin{array}{|l}
\hline
\Xi CURRENTPROCESSpq \\
\Delta PROCESSES \\
\hline
(\exists\, tquant : TIMEVAL \bullet \\
\quad tquant = tq \land \\
\quad pquants' = pquants \oplus \{pid? \mapsto tquant?\}) \\
\hline
\end{array}
$$

On each clock tick, the CLOCK process executes the following operation:

$$
\begin{aligned}
&\mathit{SuspendOnExhaustedQuantum} \mathrel{\widehat=} \\
&\quad (\mathit{CurrentProcessQuantumHasExpired} \land \\
&\qquad \mathit{ResetProcessTimeQuantum} \land \\
&\qquad (\mathit{SuspendCurrent} \mathbin{;} \mathit{SCHEDULENEXTn})) \\
&\quad \lor (\mathit{UpdateCurrentProcessQuantum} \land \\
&\qquad \mathit{ContinueCurrent})
\end{aligned}
$$

$$
\begin{aligned}
&\mathit{SetNewCurrentProcessQuantum} \mathrel{\widehat=} \\
&\quad (\mathit{ProcessQuantum}[\mathit{tquant}/\mathit{timeq}!] \land \\
&\qquad \mathit{SetCurrentProcessQuantum}[\mathit{tquant}/\mathit{timequant}?]) \setminus \{\mathit{tquant}\}
\end{aligned}
$$

This expands into:

$$
\begin{array}{|l}
\hline
\Xi \mathit{PROCESSES} \\
\Delta \mathit{CURRENTPROCESSpq} \\
\mathit{pid}? : \mathit{PREF} \\
\hline
(\exists\, \mathit{tquant} : \mathit{TIMEVAL} \bullet \\
\quad \mathit{tquant} = \mathit{pquants}(\mathit{pid}?) \land \\
\quad \mathit{tq}' = \mathit{tquant}) \\
\hline
\end{array}
$$

It simplifies to $tq' = pquants(pid?)$.

2.5 Processes and the Process Table

This section deals with a representation of processes and the process table. Each process is represented by an entry in the process table; the entry is a *process descriptor*. The process descriptor contains a large amount of information about the state of the process it represents; the actual contents of the process descriptor depend upon the kernel, its design and its purpose (e.g., a real-time kernel might contain more information about priorities and time than one for an interactive system as well as the hardware).

The purpose of this section is not to define the canonical process descriptor and process table for the kernels in this book (which, in any case, differ among themselves), nor to define the canonical structure for the process table (the one here is somewhat different from those that follow). Instead, it is intended as a general definition of these structures and as a place where general properties can be identified and proved.

As will become clear from the kernel models that follow, the process table in this section differs somewhat from the others in this book. In particular, the process table here is modelled as a collection of mappings, while the others are more obviously "tables". The reasons for this difference are many. The most important, for present purposes, are:

- The current model is at a higher level than the others.
- The current model separates the different attributes of the process descriptor into individual mappings.

Some kernels (e.g., some versions of Unix) use the representation used here. The representation of this section has a slight advantage over the standard table representation: for fast real time, it is possible to access components of the process descriptor simultaneously—this might also be of utility in a kernel running on a multi-processor system.

The section begins with a set of definitions required to support the definition of the process descriptor and the process table.

In particular, there is a limit to the number of processes that can be present in the system. There is one process descriptor for each process, so this represents the size of the process table.

$$maxprocs : \mathbb{N}_1$$

A type for referring to processes must be defined:

$$PREF == 0 \mathinner{..} maxprocs$$

The null and idle processes must be defined:

$$\begin{array}{l} IdleProcRef : PREF \\ NullProcRef : PREF \\ \hline NullProcRef = 0 \\ IdleProcRef = maxprocs \end{array}$$

where *NullProcRef* is the "name" of no process and *IdleProcRef* is the "name" of the idle process.

It is possible to define a set of "real" process names, that is process identifiers that represent actual processes. An "actual" process can be defined as a process associated with code that does something useful. The null process has no code. The idle process consists of a empty infinite loop.

Given this definition, the set, *REALPROCS* can be defined as:

$$REALPROCS == PREF \setminus \{NullProcRef, IdleProcRef\}$$

That is, $REALPROCS == 1 \mathinner{..} (maxprocs - 1)$. Another, but less useful, set of identifiers can also be defined:

$$IREALPROCS == PREF \setminus \{NullProcRef\}$$

Writing out the definitions, this is $IREALPROCS == 1 \mathinner{..} maxprocs$. These additional types will not be used in this specification but might be of some use in refinement.

Hardware devices are assigned an identifying number:

$DEVICEID == \mathbb{N}$

Each process has a state in addition to that denoted by the PSW.

$$
\begin{aligned}
PROCSTATUS ::= \; & \mathsf{pstnew} \\
| \; & \mathsf{pstrunning} \\
| \; & \mathsf{pstready} \\
| \; & \mathsf{pstwaiting} \\
| \; & \mathsf{pstswappedout} \\
| \; & \mathsf{pstzombie} \\
| \; & \mathsf{pstterm}
\end{aligned}
$$

Processes come in three kinds:

$$
\begin{aligned}
PROCESSKIND ::= \; & \mathsf{ptsysproc} \\
| \; & \mathsf{ptuserproc} \\
| \; & \mathsf{ptdevproc}
\end{aligned}
$$

These kinds are system, user and device processes.

The code and data areas of a process' main-store image need to be represented:

$[PCODE, PDATA]$

For the time being, we can ignore *PCODE* and *PDATA*. Their elements are structured in a way that will only be relevant during refinement; similarly, the *PSTACK* type also has elements whose structures can, for the most part, be ignored. (The structure of elements of type *PSTACK* is only really of relevance to interrupt service routines and the mechanisms that invoke them—typically they push a subset of the current register set onto the stack.)

The *process descriptor* (sometimes called the *process record*) is defined by the following schema; together all process descriptors in the system form the *process table*. It is the primary data structure for recording important information about processes. The information includes a representation of the process' state, which is retained while the process is not executing. On a context switch, the state (primarily, hardware registers, IP and stack) is copied into the process descriptor for storage until the process is next executed. When next selected to run, the state is copied back into registers. The process descriptor does hold other information about the process: data about the storage areas it occupies, message queues, priority information and a symbolic representation of its current state (in this book, an element of type *PROCSTATUS*).

In this chapter, process descriptors are represented as sets of mappings from process identifiers to the various attribute types. This representation finds a natural representation as a collection of arrays. An alternative that is commonly encountered in working systems, is to collect the information about each process into a record or structure; all process descriptors are then implemented as an array of these records. The record implementation has the advantage that all relevant information about a process is held in one data structure. The main disadvantage is that the record has to be accessed as an entity. In the array-based implementation (the one adopted in this chapter, i.e.), individual components are accessed separately. The advantage to the separate-access approach is seen when locking is considered: when one component array is being accessed under a lock, the others remain available to be locked.

In this representation, the process table is implicitly defined as the mapping from process reference (*PREF*) to attribute value:

PROCESSES

$$
\begin{array}{l}
pstatus : PREF \ffun PROCSTATUS \\
pkinds : PREF \ffun PROCESSKIND \\
pprios : PREF \ffun PRIO \\
pregs : PREF \ffun GENREGSET \\
pstacks : PREF \ffun PSTACK \\
pstatwds : PREF \ffun STATUSWD \\
pcode : PREF \ffun PCODE \\
pdata : PREF \ffun PDATA \\
pips : PREF \ffun \mathbb{N} \\
known\_procs : \mathbb{F}\, PREF
\end{array}
$$

$$
\begin{array}{l}
NullProcRef \notin known\_procs \\
known\_procs = \mathrm{dom}\, pstatus \\
\mathrm{dom}\, pstatus = \mathrm{dom}\, pkinds \\
\mathrm{dom}\, pkinds = \mathrm{dom}\, pprios \\
\mathrm{dom}\, pprios = \mathrm{dom}\, pregs \\
\mathrm{dom}\, pregs = \mathrm{dom}\, pstacks \\
\mathrm{dom}\, pstacks = \mathrm{dom}\, pstackwds \\
\mathrm{dom}\, pstackwds = \mathrm{dom}\, pcode \\
\mathrm{dom}\, pcode = \mathrm{dom}\, pdata \\
\mathrm{dom}\, pdata = \mathrm{dom}\, pips \\
known\_uids = \mathrm{ran}\, pips
\end{array}
$$

The conjunct, *NullProcRef*, is added to the predicate because it is required that *NullProcRef* actually refer to the null process. It should never be the case that the null process appears in the process table.

It is possible to exclude *IdleProcRef* from the process table in a manner identical to *NullProcRef*. In the case of *IdleProcRef*, matters are less clear. The idle process *is* a real process: it has code but probably no data or stack areas. It is quite possible to have an idle process without representing it in the process table. The idle process is only executed whenever the ready queue is empty, so a small piece of code can do all the necessary work. To some, this will appear an *ad hoc* solution; to others, it will appear totally natural.

At the moment, the idle process will be represented in the process table, even though it requires an additional slot (this is, after all, a specification, not an implementation).

When refinement is performed, the inclusion of *IdleProcRef* and the exclusion of *NullProcRef* are of some importance. They determine the range of possible values for the domains of the components of process descriptors. In other words, their inclusion and exclusion determine what a "real process" can be; this is reflected in the type to which *PREF* refines: *REALPROCS* or *IREALPROCS*. (A hidden goal of the refinement process is to represent *NullProcRef* as, for example, a null pointer.)

Proposition 13. *NullProcRef does not refer to a "real" process.*

Definition 1 . *A "real" process must be interpreted as one that has code and other attributes (stack, data, status, instruction pointer and so on). Alternatively, a "real" process is one that can be allocated either by the kernel or as a user process.*

More technically, a "real" process is one whose parameters are represented in the process table and, hence, whose identifier is an element of known_procs.

Note that this definition is neutral with respect to the idle process. Some systems might regard it as "real" and include an operation to create the idle process. Other systems, MINIX [30] for example, regard the idle process as a pseudo-process that is implemented as just a piece of kernel code that is executed when there is nothing else to do; as in other systems built using this assumption, the idle process is not represented by an entry in the process table.

The above definition could be extended to include the idle process, of course.

PROOF. The components of the process description, *pstate*, *pkind*, *pstack*, *pregs*, etc., all have identical domains by the first part of the invariant of *PROCESSES*. That is:

$$\begin{array}{l}\mathrm{dom}\, pstatus = \mathrm{dom}\, pkind \land \\ \mathrm{dom}\, pkinds = \mathrm{dom}\, pprios \land \\ \mathrm{dom}\, pprios = \mathrm{dom}\, pregs \land \\ \mathrm{dom}\, pregs = \mathrm{dom}\, pstacks \land \\ \mathrm{dom}\, pstacks = \mathrm{dom}\, pips \land \\ \mathrm{dom}\, pcode = \mathrm{dom}\, pstacks \land \\ \mathrm{dom}\, pdata = \mathrm{dom}\, pcode \land \\ \mathrm{dom}\, pips = \mathrm{dom}\, pstatus\end{array}$$

Furthermore, the domains are all identical to *known_procs* since $\mathrm{dom}\, pstatus = known\_procs$. Since $NullProcRef \notin known\_procs$, it follows that $NullProcRef \notin \mathrm{dom}\, pregs$ (for example). By Definition 1, the null process is not a "real" process. □

The process table is initialised by the following operation:

$$
\begin{array}{|l}
\hline
\textit{InitPROCESSES} \\
\textit{PROCESSES}' \\
\hline
\textit{known\_procs}' = \varnothing \\
\hline
\end{array}
$$

A process is removed from the process table by the operation modelled by the following schema:

$$
\begin{array}{|l}
\hline
\textit{DelProcess} \\
\Delta \textit{PROCESSES} \\
\textit{pid}? : \textit{PREF} \\
\hline
\textit{pstatus}' = \{\textit{pid}?\} \lhd \textit{pstatus} \\
\textit{pkinds}' = \{\textit{pid}?\} \lhd \textit{pkinds} \\
\textit{pprios}' = \{\textit{pid}?\} \lhd \textit{pprios} \\
\textit{pregs}' = \{\textit{pid}?\} \lhd \textit{pregs} \\
\textit{pstacks}' = \{\textit{pid}?\} \lhd \textit{pstacks} \\
\textit{pips}' = \{\textit{pid}?\} \lhd \textit{pips} \\
\hline
\end{array}
$$

or, more simply: $\textit{pid}? \notin \textit{known\_procs}'$.

Proposition 14. *DelProcess*$[p/\textit{pid}?]$ *implies that* $p \notin \textit{known\_prcs}'$.

PROOF. Since all domains are identical, take, for example, the case of *pregs*:

$$\textit{pregs}' = \{p\} \lhd \textit{pregs}$$

Taking domains and using the identity $\operatorname{dom} \textit{pregs} = \textit{known\_procs}$:

$$
\begin{aligned}
&\operatorname{dom} \textit{pregs}' \\
&\qquad = \textit{known\_procs}' \\
&\qquad = \operatorname{dom}(\{p\} \lhd \textit{pregs}) \\
&\qquad = \operatorname{dom}(\textit{pregs} \setminus \{p\})
\end{aligned}
$$

Therefore, $p \notin \textit{known\_procs}'$.

The same reasoning can be applied to all similar functions in *PROCESSES*. □

A process is added to the process table by the *AddProcess* operation:

$$
\begin{array}{|l}
\hline
\textit{AddProcess} \\
\Delta \textit{PROCESSES} \\
\textit{pid}? : \textit{PREF} \\
\textit{knd}? : \textit{PROCESSKIND} \\
\textit{status}? : \textit{PROCSTATUS} \\
\textit{stat}? : \textit{STATUSWD} \\
\textit{regs}? : \textit{GENREGSET} \\
\end{array}
$$

$stk? : PSTACK$
$prio? : PRIO$
$ip? : \mathbb{N}$

$pstatus' = pstatus \cup \{pid? \mapsto status?\}$
$pkinds' = pkinds \cup \{pid? \mapsto knd?\}$
$pprios' = pprios \cup \{pid? \mapsto prio?\}$
$pregs' = pregs \cup \{pid? \mapsto regs?\}$
$pstatwds' = pstatwds \cup \{pid? \mapsto stat?\}$
$pips' = pips \cup \{pid? \mapsto ip?\}$
$pstacks' = pstacks \cup \{pid? \mapsto stk?\}$

Proposition 15. *($AddProcess[p/pid?, \ldots] \mathbin{;} DelProcess[p/pid?]$) is the identity on the process table.*

PROOF. This proposition states that the effect of adding a process and immediately deleting it leaves the process table invariant.

Since *PROCESSES* is rather large, only a part will be considered in detail.

The composition can be written as:

$$
\begin{aligned}
&\exists\, pstacks'' : PREF \twoheadrightarrow PSTACK \bullet \\
&\quad pstacks'' = pstacks \cup \{p? \mapsto \mathsf{stk}\} \wedge \\
&\quad pstacks' = \{p?\} \mathbin{⩤} pstacks''
\end{aligned}
$$

which simplifies to:

$$pstacks' = \{p?\} \mathbin{⩤} (pstacks \cup \{p? \mapsto \mathsf{stk}\})$$

So:

$$
\begin{aligned}
&\mathrm{dom}(\{p?\} \mathbin{⩤} (pstacks \cup \{p? \mapsto \mathsf{stk}\})) \\
&\quad = (\mathrm{dom}\, pstacks \cup \mathrm{dom}\{p? \mapsto \mathsf{stk}\}) \setminus \{p?\} \\
&\quad = ((\mathrm{dom}\, pstacks) \cup \{p?\}) \setminus \{p?\}
\end{aligned}
$$

Therefore:

$$
\begin{aligned}
\mathrm{dom}\, pstacks' &= \mathrm{dom}(pstacks \cup \{p?\}) \setminus \{p?\} \\
&= \mathrm{dom}\, pstacks
\end{aligned}
$$

□

The priority of a process is returned by the following operation:

ProcessPriority

$\Xi PROCESSES$
$pid? : PREF$
$prio! : PRIO$

$prio! = pprios(pid?)$

The kind of process is returned by:

KindOfProcess

$\Xi PROCESSES$
$pid? : PREF$
$knd! : PROCESSKIND$

$knd! = pkinds(pid?)$

A process' current status is retrieved from the process table by the following operation:

StatusOfProcess

$PROCESSES$
$pid? : PREF$
$ps! : PROCSTATUS$

$ps! = pstatus(pid?)$

InitialiseProcessStatus

$\Delta PROCESSES$
$pid? : PREF$
$pstat? : PROCSTATUS$

$pstatus' = pstatus \cup \{pid? \mapsto pstat?\}$

Process status changes frequently during its execution. The following operation alters the status:

UpdateProcessStatus

$\Delta PROCESSES$
$pid? : PREF$
$pstat? : PROCSTATUS$

$pstatus' = pstatus \oplus \{pid? \mapsto pstat\}$

The following operations set the process status to designated values as and when required:

$$
\begin{array}{l}
SetProcessStatusToNew \mathrel{\widehat=} \\
\quad ([pstat : PROCSTATUS \mid pstat = \mathsf{pstnew}] \land \\
\qquad UpdateProcessStatus[pstat/pstat?]) \setminus \{pstat\}
\end{array}
$$

This operation is called when a process has been created but not added to the ready queue.

When a process enters the ready queue, the following operation changes its status to reflect the fact:

$$\begin{array}{l}
SetProcessStatusToReady \mathrel{\widehat=} \\
\quad ([pstat : PROCSTATUS \mid pstat = \mathsf{pstready}] \land \\
\qquad UpdateProcessStatus[pstat/pstat?]) \setminus \{pstat\}
\end{array}$$

The *SetProcessStatusToRunning* operation should be called when a process begins execution:

$$\begin{array}{l}
SetProcessStatusToRunning \mathrel{\widehat=} \\
\quad ([pstat : PROCSTATUS \mid pstat = \mathsf{pstrunning}] \land \\
\qquad UpdateProcessStatus[pstat/pstat?]) \setminus \{pstat\}
\end{array}$$

When a process is suspended for whatever reason, the following operation is called to set its status to pstwaiting:

$$\begin{array}{l}
SetProcessStatusToWaiting \mathrel{\widehat=} \\
\quad ([pstat : PROCSTATUS \mid pstat = \mathsf{pstwaiting}] \land \\
\qquad UpdateProcessStatus[pstat/pstat?]) \setminus \{pstat\}
\end{array}$$

In the second kernel below (Chapter 4), processes can be swapped out to disk space. The status of such a process is set by the following schema. As with all of these schemata, the variable *pstat* represents the new state.

$$\begin{array}{l}
SetProcessStatusToZombie \mathrel{\widehat=} \\
\quad ([pstat : PROCSTATUS \mid pstat = \mathsf{pstzombie}] \land \\
\qquad UpdateProcessStatus[pstat/pstat?]) \setminus \{pstat\}
\end{array}$$

This schema is used when a process terminates but has not yet released its resources:

$$\begin{array}{l}
SetProcessStatusToTerminated \mathrel{\widehat=} \\
\quad ([pstat : PROCSTATUS \mid pstat = \mathsf{pstterm}] \land \\
\qquad UpdateProcessStatus[pstat/pstat?]) \setminus \{pstat\}
\end{array}$$

For many purposes, it is necessary to know whether a given process reference denotes a process that is in the process table. The following schema defines that test:

$$\begin{array}{|l}
\multicolumn{1}{l}{KnownProcess} \\
\hline
\Xi PROCESSES \\
pid? : PREF \\
\hline
pid? \in known_procs \\
\hline
\end{array}$$

It is not possible to allocate processes indefinitely. The following operation determines whether new processes can be allocated.

$$\begin{array}{|l}
\multicolumn{1}{l}{CanAllocateProcess} \\
\hline
PROCESSES \\
\hline
known_procs \subset PREF \\
\hline
\end{array}$$

The identifier of the next new process is generated by the following (relatively abstract) schema.

$$
\begin{array}{|l}
\textit{NextPREF} \\
\hline
\textit{PROCESSES} \\
\textit{pid}! : \textit{PREF} \\
\hline
(\exists\, p : \textit{PREF} \mid p \in (\textit{PREF} \setminus \textit{known\_procs}) \bullet \\
\quad p \neq \textit{NullProcRef} \wedge p \neq \textit{IdleProcRef} \wedge \\
\quad \textit{pid}! = p) \\
\hline
\end{array}
$$

or:

$$\textit{pid}! \in \{p : \textit{PREF} \bullet p \notin \textit{known\_procs}\}$$

The way names are allocated to new processes is as follows. There is a set of all possible process references, *PREF*. If a process' identifier is not in *known_procs*, the set of known processes (i.e., the names of all processes that are currently in the system—the domain of all attribute mappings), it can be allocated. Allocation is, here, the addition of a process reference to *known_procs*.

If all processes have been allocated, the following schema's predicate is satisfied.

$$
\begin{array}{|l}
\textit{ProcessesFullyAllocated} \\
\hline
\textit{PROCESSES} \\
\hline
\textit{known\_procs} = \textit{PREF} \setminus \{\textit{NullProcRef}, \textit{IdleProcRef}\} \\
\hline
\end{array}
$$

Note that neither *IdleProcRef*, nor *NullProcRef* represent real processes that can be allocated and deallocated in the usual way. Indeed, *IdleProcRef* denotes the idle process that runs whenever there is nothing else to do; it is already defined within the kernel. The constant *NullProcRef* denotes the null process and is only used for initialisation or in cases of error.

The following is just a useful synonym:

$$\textit{CannotAllocateProcess} \mathrel{\widehat{=}} \textit{ProcessesFullyAllocated}$$

Proposition 16. *For any process, p, such that $p \in \textit{known\_procs}$:*

$$\textit{DelProcess} \Rightarrow \neg\, \textit{ProcessesFullyAllocated}'$$

PROOF. By Proposition 14, the operation *DelProcess* applied to a process, p, implies that $p \notin \textit{known\_procs}'$.

Note, first of all, that p cannot be one of *NullProcRef* or *IdleProcRef*.

The definition of *CanAllocatePREF* is $\textit{known\_procs} \subset \textit{PREF}$, which implies that there is some subset $S : \mathbb{F}\,\textit{PREF}$ s.t., $\textit{known\_procs} \cup S = \textit{PREF}$; (strictly speaking:

$$known\_procs \cup S = PREF \setminus \{NullProcRef, IdleProcRef\}$$

so *CanAllocatePREF* implies that if $S \neq \varnothing$, then $p \in S$ will be the next *PREF* to be allocated, so $known\_procs \cup \{p\} \cup (S \setminus \{p\}) = PREF$. If $S = \varnothing$, clearly $known\_procs = PREF$. Therefore $PREF \setminus S = known\_procs$.

In the case of deletion, $known\_procs' = known\_procs \setminus \{p?\}$, so:

$$\begin{aligned} &known\_procs' \\ &\quad = known\_procs \setminus \{p\} \\ &\quad = (PREF \setminus S) \cup \{p\} \\ &\quad = PREF \setminus (S \cup \{p\}) \end{aligned}$$

For $PREF = known\_procs$, it is impossible that $p \in S$. Therefore, it can be concluded that the predicate of $\neg$ *ProcessesFullyAllocated′* does not hold. □

It might be useful to know whether there are any processes in the system. The following schema provides that ability:

```
┌─ NoProcessesInSystem ──────────────────────
│ ΞPROCESSES
├──────────────
│ known_procs = ∅
└────────────────────────────────────────────
```

Proposition 17. $AddProcess \Rightarrow pid \in known\_procs'$.

PROOF. For *AddProcess*, consider the case $pstacks' = pstacks \cup \{p? \mapsto \mathsf{stk}\}$. Since $\operatorname{dom} pstacks = known\_procs$ and $\operatorname{dom} pstacks' = known\_procs'$, it follows that: $\operatorname{dom} pstacks = known\_procs$ and:

$$\begin{aligned} \operatorname{dom} pstacs &= \operatorname{dom}(pstacks \cup \{p? \mapsto \mathsf{stk}\}) \\ &= (\operatorname{dom} pstacks) \cup (\operatorname{dom}\{p? \mapsto \mathsf{stk}\}) \\ &= \operatorname{dom} pstacks \cup \{p?\} \\ &= known\_procs' \end{aligned}$$

□

Proposition 18.

$$\neg\ NoProcessesInSystem \wedge (NextPREF \wedge AddProcess)^n \wedge 0 < n \leq maxprocs \Rightarrow \\ \neg\ ProcessesFullyAllocated$$

PROOF. The proposition statement expands to:

$$\begin{aligned} &known\_procs = \varnothing \wedge \\ &\quad (NextPref \wedge AddProcess)^n \wedge \\ &\quad 0 < n \leq maxprocs \end{aligned}$$

It has already been established (Proposition 17) that

$$AddProcess[p/pid?, \ldots] \Rightarrow p \in known\_procs' \qquad (2.1)$$

so:

$$(NextPref[p/pid!] \land AddProcess[p/pid?]) \Rightarrow p \in known\_procs$$

Writing the available identifiers as:

$$A = (PREF \setminus \{NullProcRef, IdleProcRef\}) \setminus known\_procs$$

We write A for the set of available identifiers before $NextPREF \land AddProcess$ and A' for that afterwards. Then $\#A$ is the cardinality of A, and so $\#A' = \#A - 1$. This is justified by:

$$\begin{aligned} &A' \\ &\quad = (PREF \setminus \{NullProcRef, IdleProcRef\}) \setminus known\_procs' \\ &\quad = (PREF \setminus \{NullProcRef, IdleProcRef\}) \setminus (known\_procs \cup \{p\}) \end{aligned}$$

where p is the newly allocated identifier.

Consequently, for $(NextPREF[p_m] \land AddProcess[p_m/pid?, \ldots])^m$ and for some $m, 0 < m < maxprocs - 1$

$$\begin{aligned} &A' \\ &\quad = (PREF \setminus \{NullProcRef, IdleProcRef\}) \setminus known\_procs' \\ &\quad = (PREF \setminus \{NullProcRef, IdleProcRef\}) \setminus (known\_procs \cup \{p_1, \ldots, p_{m-1}\}) \end{aligned}$$

Therefore, for $m = maxprocs$,

$$\begin{aligned} &A' \\ &\quad = (PREF \setminus \{NullProcRef, IdleProcRef\}) \setminus known\_procs' \\ &\quad = (PREF \setminus \{NullProcRef, IdleProcRef\}) \setminus known\_procs \cup \{p_1, \ldots, p_m\} \end{aligned}$$

Since the interval $1 \mathinner{\ldotp\ldotp} maxprocs$ contains exactly m elements:

$$(PREF \setminus \{NullProcRef, IdleProcRef\}) \setminus known\_procs' = \varnothing$$

□

It should be noted that if the idle process is not regarded as a "real" process, the statement of the proposition should be restricted to $\land\ 0 < n < maxprocs$.

Proposition 19. *The operations AddProcess and DelProcess are inverse operations. That is, for any* p, $AddProcess[p/pid?] \mathbin{;} DelProcess[p/pid?]$ *implies that* $\#known\_procs = \#known\_procs'$.

PROOF. This follows immediately by induction from Proposition 15. □

2.6 Context Switch

Context switches occur when a process is swapped on or off the processor. This section outlines a scheme for modelling context switching.

Basically, a context switch involves the transfer of hardware and other state information from or to the process descriptor. Of the registers, the most important is the instruction pointer. Context switches are expensive because they copy the contents of all hardware registers into the current process descriptor. They occur when the scheduler determines that another process should be allocated to the processor. They are also required for the specification of semaphores.

There are two main operations involved in context switching: one to copy state data from the process descriptor to the hardware and one to copy data in the opposite direction. The schemata defined in this section are included as an illustration. They will be redefined with slight variations when required.

The *SaveAllHWRegisters* operation copies the contents of the registers used by a process into its process descriptor. The operation is complemented by *RestoreAllHWRegisters*, which reads the process descriptor and copies items from it to the hardware's general-purpose registers. In the representation below, the instruction pointer is the last register to be set from the process descriptor.

SaveAllHWRegisters

$\Delta PROCESSES$
$HWREGISTERS$
$pid? : PREF$

$(\forall r : GENREG \bullet$
$\quad pregs'(pid?)(r) = hwregset(r))$
$pstacks'(pid?) = hwstack$
$pstatwds'(pid?) = hwstatwd$
$pips'(pid?) = hwip$

$SwitchContextOut \mathrel{\widehat=} SaveAllHWRegisters$

RestoreAllHWRegisters

$PROCESSES$
$HWREGISTERS'$
$pid? : PREF$

$hwstack' = pstacks(pid?)$
$hwstatwd' = pstatwds(pid?)$
$hwip' = pips(pid?)$
$(\forall r : GENREG \bullet$
$\quad hwregset'(r) = pregs(pid?)(r))$

$SwitchContextIn \mathrel{\widehat=} RestoreAllHWRegisters$

Sometimes, for example when an interrupt occurs, a *partial* context switch can occur. Partial context switches only save part of the data normally switched by a context switch. Although the detailed workings of the hardware interrupt subsystem are mostly ignored in this book, it is interesting, just as an orienting exercise, to include the following schemata.

A partial context switch is described by the following two schemata:

$$
\begin{array}{l}
\textbf{SaveHWGeneralRegisters} \\
\hline
\Delta PROCESSES \\
HWREGISTERS \\
pid? : PREF \\
\hline
(\forall r : GENREG \bullet \\
\quad pregs'(pid?)(r) = hwregset(r))
\end{array}
$$

$SavePartialContext \mathrel{\widehat=} SaveHWGeneralRegisters$

$$
\begin{array}{l}
\textbf{RestoreHWGeneralRegisters} \\
\hline
\Delta PROCESSES \\
\Xi HWREGISTERS \\
pid? : PREF \\
\hline
\forall r : GENREG \bullet \\
\quad hwregset(r) = pregs'(pid?)(r)
\end{array}
$$

$RestorePartialContext \mathrel{\widehat=} RestoreHWGeneralRegisters$

2.7 Current Process and Ready Queue

This section presents a simple model of the operation of the kernel's scheduler. It uses a simple FIFO queue to hold the processes that are ready to execute; this is *readyq*. The identifier of the process currently executing is stored in *currentp*.

$$
\begin{array}{l}
\textbf{CURRENTPROCESS} \\
\hline
currentp : PREF \\
readyq : ProcQueue \\
\hline
\end{array}
$$

Here, *ProcQueue* can either be regarded as an instantiation of the generic *QUEUE* type or of a new type.

$$
\begin{array}{|l}
\hline
\textit{MakeCurrent} \\
\Delta CURRENTPROCESS \\
pid? : PREF \\
\hline
currentp' = pid? \\
\hline
\end{array}
$$

Because this is Z, a framing schema is used to promote operations on the process queue:

$$
\begin{array}{|l}
\hline
\Phi CURRENTPROCESS_q \\
\Delta CURRENTPROCESS \\
\Delta ProcQUEUE \\
\hline
readyq = \theta ProcQUEUE \\
readyq' = \theta ProcQUEUE' \\
\hline
\end{array}
$$

$$
\begin{array}{|l}
\hline
\Psi CURRENTPROCESS_q \\
CURRENTPROCESS' \\
ProcQUEUE' \\
\hline
readyq' = \theta ProcQUEUE' \\
\hline
\end{array}
$$

The scheduler is initialised by the following operation:

$$
\begin{array}{l}
InitCURRENTPROCESS \mathrel{\widehat=} \\
\quad \Psi CURRENTPROCESS_q \land \\
\qquad InitProcQueue \land \\
\qquad [CURRENTPROCESS' \mid currentp' = NullProcRef]
\end{array}
$$

A process is added to *readyq*, the queue of processes ready to execute, by the *MakeReady* operation. It is defined as:

$$
\begin{array}{l}
MakeReady \mathrel{\widehat=} \\
\quad \Phi CURRENTPROCESS_q \land \\
\qquad EnqueueProc[readyq/procs, readyq'/procs']
\end{array}
$$

It is often necessary for a process whose execution was interrupted or blocked by some operation, immediately to be resumed. The following does this:

$$
\begin{array}{|l}
\hline
\textit{ContinueCurrent} \\
\Xi CURRENTPROCESS \\
\hline
currentp' = currentp \\
readyq' = readyq \\
\hline
\end{array}
$$

We can now prove some propositions about the scheduler.

Proposition 20. $p = currentp \land RunNext \Rightarrow currentp' \neq p$.

The currently executing process can suspend itself by a call to the following operation:

$$
\begin{array}{l}
SuspendCurrent \mathrel{\widehat=} \\
\quad SwitchContextOut[currentp/pid?] \mathbin{;} \\
\quad\quad (\Phi CURRENTPROCESS_q \land \\
\quad\quad EnqueueProc[currentp/p?])
\end{array}
$$

$$
\begin{array}{l}
DequeueProc \mathrel{\widehat=} \\
\quad (\neg\ EmptyProcQueue[PREF] \land RemoveFirst[PREF] \land ProcQOk) \\
\quad\quad \lor EmptyProcQueue[PREF]
\end{array}
$$

There is no need for a dequeue operation that raises an error: should the process queue ever become empty, the idle process will be executed. Therefore, the following will be adequate for current needs:

$$
\begin{array}{l}
DequeueProc \mathrel{\widehat=} \\
\quad (EmptyQueue[PREF] \land MakeNextIdle) \\
\quad\quad \lor RemoveFirst[PREF]
\end{array}
$$

where:

$$MakeNextIdle \mathrel{\widehat=} [x! : PREF \mid x! = IdleProcRef]$$

The core of this little scheduler is the *SCHEDULENEXT* operation. It is defined as:

$$
\begin{array}{l}
SCHEDULENEXT \mathrel{\widehat=} \\
\quad ((\Phi CURRENTPROCESS_q \land \\
\quad\quad DequeueProc[p/x!] \land \\
\quad\quad SetProcessStatusToRunning[p/pid?] \land \\
\quad\quad MakeCurrent[p/pid?]) \mathbin{;} \\
\quad SwitchContextIn[p/pid?]) \setminus \{p\}
\end{array}
$$

3

A Simple Kernel

Scimus, et veniam petimusque damusque vicissim.
– Horace, Ars Poetica, 9

3.1 Introduction

This chapter contains the first of the kernel models in this book. The kernel modelled here is deliberately simple but is still useful. It is intended to be a kernel for an embedded or small real-time system and to be a specification of a workable system—one aim of the exercise is to produce a specification that *could* be refined to working code. (As noted in Chapter 1, the kernels in this book have been revised somewhat and are being refined to code as a concurrent activity by the author.)

The model defined in this chapter is intended as an existence proof. It is intended to show that it is possible to model an operating system kernel, albeit a small one, using purely formal models. The model is simple, as is the kernel—more extensive kernels will be modelled in the next two chapters, so readers will find increasingly complex kernels that deal with some of the issues left unresolved by the current one (for example, properties of semaphores in the next chapter).

The current effort is also intended as an orienting exercise for the reader. The style of the models is the same in this chapter as in the following ones. As will be seen, there are structures that are common (at least in high-level terms) and this chapter introduces them in their actual context.

This chapter uses Object-Z rather than Z.

3.2 Requirements

The operating system to be modelled in this chapter is intended to be suitable for real-time processing, possibly in an embedded context. The kernel should be as small. The kernel should also be portable and, as such, there is no need to specify any ISRs or the clock and associated driver and ISR. Devices and the uses to which the clock is to be put are considered matters that depend upon the particular instantiation of the kernel (e.g., some kernels might not use drivers, or the clock might do more than just record the time and wake sleeping processes).

The kernel must implement a priority-based scheduler. Initially, all priorities are to be fixed. The priority of a process is defined before it is loaded and assigned to it via a parameter; that parameter is to be used to set the process' priority in the scheduler.

The kernel is not to contain any storage-management modules. All storage is to be allocated statically, offline, as part of the kernel configuration process.

The kernel is, basically, to implement the process abstraction, a scheduler and IPC. The IPC is to be relatively rich and must include:

- semaphores (and shared memory);
- mailboxes and asynchronous messages.

All shared memory must be allocated statically when process storage is allocated.

The kernel is to be statically linked with the user processes that run on it. The memory map of the system is used to define where the various processes reside and where the shared storage is. Primitives are to be provided to:

- create processes and enter them into the scheduler's queue;
- terminate a process and release its process descriptor, together with any semaphores and message queues that it owns.

There are operations, moreover, to perform the following operations:

- suspend a running process;
- create and dispose of IPC structures;
- perform IPC operations.

In addition, the kernel will support an operation that permits a process to alter its priority.

3.3 Primary Types

This section contains the definitions of the basic types used by this model.

Processes must be referenced. The basic reference type is the following:

$[PREF]$

As noted elsewhere, it is necessary to define constants to denote the null and idle processes. The types are:

$$NullProcRef, IdleProcRef : PREF$$

Two more process reference types can now be defined. The first excludes the null process reference, while the second excludes both the null and idle process references:

$$IPREF == PREF \setminus \{NullProcRef\}$$
$$APREF == IPREF \setminus \{IdleProcRef\}$$

Without loss of generality, these types can be given a more concrete representation. First, it is necessary to define the maximum number of processes that the kernel can contain:

$$maxprocs : \mathbb{N}$$

(This is, in fact, the size of the process table or the number of process descriptors in the table.)

Next, the types and values of the constants denoting the null and idle processes are defined:

$$NullProcRef : PREF$$
$$IdleProcRef : IPREF$$

$$NullProcRef = 0$$
$$IdleProcRef = maxprocs$$

They are defined so that they form the extrema of the *IPREF* type.

The *PREF* type can now be defined as:

$$PREF == NullProcRef \mathrel{..} IdleProcRef$$

The above definitions of *IPREF* and *APREF* still hold. For example, writing constants out:

$$APREF == 1 \mathrel{..} maxprocs - 1$$

These definitions will, *inter alia*, make process identifier generation simpler and easier to understand.

Each process is in one and exactly one state at any time during its existence. The following type defines the names of these states (the pst prefix just denotes "process table"):

$$PROCSTATUS ::= \mathsf{pstnew} \mid \mathsf{pstrunning} \mid \mathsf{pstready} \mid \mathsf{pstwaiting} \mid \mathsf{pstterm}$$

The states can be understood as follows. When a process is newly created but not added to the ready queue, it has state pstnew. When a process is ready to execute (is resident in the ready queue), it has state pstready. When a process is executed, its state is pstrunning. Processes block or are suspended for a variety of reasons (e.g., when they are waiting for a device). While a process is waiting (for whatever reason), it is in state pstwaiting. Finally, when a process terminates, it enters the pstterm state.

In this kernel, each process has a stack and code and data areas. The process descriptor records the address and size of each of these areas. In addition, the process descriptor records the pointer to the top of the stack. The relevant types are defined as the following atomic types:

$[PSTACK, PCODE, PDATA]$

Of these, *PSTACK* is the only one that is used much in this model. It is assumed that the process state consists partly of the state of its current stack and that there is hardware support for the stack, so there is a stack register. For the purposes of this model, it is assumed that values of type *PSTACK* are atomic values that can be assigned to registers.

The other types, *PCODE* and *PDATA*, are only used in the current model to represent values stored in each process' process descriptor. If they were expanded, they could be used for error checking; we ignore this possibility, however.

Finally, the type representing process priorities is defined:

$PRIO == \mathbb{Z}$

The interpretation is that the higher the priority, the greater the magnitude. Therefore, the priorities -1, 20 and 0 are ordered (highest to lowest) as: -1, 0 and 20. There are no bounds placed on priorities, so it is always possible to produce a priority that is lower than any other. In an implementation, there would be a minimum priority equal to the greatest integer representable by the hardware ($2^{32} - 1$ on a 32-bit machine); conversely, the highest possible priority would be the most negative integer representable by the hardware (-2^{32} on a 32-bit machine).

3.4 Basic Abstractions

This section is concerned with the definition of the basic constructs used to define the kernel. Three of these abstractions, *ProcessQueue*, *HardwareRegisters* and *Lock* have appeared before, so they will be presented without comment. The reader is warned again that the model in this chapter is written in Object-Z and not in Z. For this reason, the constructs just listed are represented as Object-Z classes and methods.

The Object-Z version of the *ProcessQueue* is modelled as follows. With the exception of the constructs required for Object-Z, the model is exactly the same as the Z version:

ProcessQueue

$\upharpoonright(INIT, IsEmpty, Enqueue, RemoveFirst,$
$QueueFront, RemoveElement)$

$elts : \text{iseq}\, APREF$

INIT

$elts' = \langle\, \rangle$

IsEmpty

$elts = \langle\, \rangle$

Enqueue

$\Delta(elts)$
$x? : APREF$

$elts' = elts \frown \langle x? \rangle$

RemoveFirst

$\Delta(elts)$
$x! : APREF$

$x! = head\ elts$
$elts' = tail\ elts$

QueueFront

$x! : APREF$

$x! = head\ elts$

RemoveElement

$\Delta(elts)$
$x? : APREF$

$$(\exists\, s_1, s_2 : \text{iseq}\, APREF \bullet s_1 \frown \langle x? \rangle \frown s_2 = elts \wedge s_1 \frown s_2 = elts')$$

The class exports the following operations, in addition to the initialisation (*Init*) operation: *IsEmpty*, *Enqueue*, *RemoveFirst*, *QueueFront* and *RemoveElement*.

In a similar fashion, the hardware register class is composed of operations that are identical to those defined in the last chapter.

HardwareRegisters

$\upharpoonright(SetGPRegs, GetGPRegs, GetStackReg, SetStackReg,$
$SetIntsOff, SetIntsOn, GetIP, SetIP$
$GetStatWd, SetStatWd)$

$hwgenregs : GENREGSET$
$hwstack : PSTACK$
$hwstatwd : STATUSWD$
$hwip : \mathbb{N}$

INIT

$hwgenregs.INIT$
$hwstack' = 0$
$hwstatwd' = 0_S$
$hwip' = 0$

SetGPRegs

$\Delta(hwgenregs)$
$regs? : GENREGSET$

$hwgenregs' = regs?$

GetGPRegs

$regs! : GENREGSET$

$regs! = hwgenregs$

GetStackReg

$stk! : PSTACK$

$stk = hwstack$

SetStackReg

$stk? : PSTACK$

$hwstack' = stk?$

GetIP

$ip! : \mathbb{N}$

$ip! = hwip$

SetIP

$ip? : \mathbb{N}$

$hwip' = ip?$

GetStatWd

$stwd! : STATUSWD$

$hwstatwd' = stwd?$

SetStatWd

$stwd? : STATUSWD$

$stwd! = hwstatwd$

SetIntsOff

$intflg' = \mathsf{intoff}$

SetIntsOn

$intflg' = \mathsf{inton}$

The lock class is as follows. It exports the *Lock* and *Unlock* operations, as well as an initialisation operation. The class differs very slightly from the specification of the previous chapter. In the Z specification, the operations worked directly on the interrupt able/disable flag. Here, the class takes a reference to the hardware registers as its only initialisation parameter. The *Lock* and *Unlock* operations are defined in terms of the reference to the hardware. The net effect is that the *Lock* class must be instantiated before it is used.

Lock

$\upharpoonright(INIT, Lock, Unlock)$

$hw : HardwareRegisters$

Assume that registers have been initialised.

INIT

$hwrgs? : HardwareRegisters$

$hw' = hw?$

$Lock \mathrel{\widehat{=}} hw.SetIntsOff$

$Unlock \mathrel{\widehat{=}} hw.SetIntsOn$

The lowest level of the kernel uses locking to ensure mutual exclusion. Above this, semaphores are used. Semaphores are modelled by the following class, which exports an initialisation operation (*INIT*), *Wait* (*P* in Dijkstra's terminology) and *Signal* (*V*). The class contains a process queue (*waiters*) to hold its waiting processes; the *waiters* queue is unrelated to any other queue, a fact denoted by the $_{©}$ subscript. The other semaphore component is the counter, *scnt*, which has type $\mathbb{Z}$. (The initialisation value, *initval*, is also retained.)

The semaphore has to cause processes to be scheduled and suspended. It needs to access the scheduler (via the reference *sched*) and to the process table (via *ptab*). Semaphores also need to switch contexts when they block waiting processes, so a reference to the *Context* class is required to provide access to the context-switching operations.

Semaphores work by updating the counter (*scnt*) as an atomic operation. To do this, the semaphore uses the *Lock* class to exclude all processes except the calling one from the counter. In the model, the lock operations are placed around more operations than simply the counter update. This is to make the specification easier to read.

Here, the decrement-based model for semaphores is adopted (see, for example, [26]). The decrement-based model has certain advantages from the implementation viewpoint. In particular, the sign of the counter is used as the basis for the major decisions in the two operations.

Although semaphores are of considerable interest, the reader will see that we do not prove any of their properties in this chapter. Indeed, the only proofs in this chapter relate to priority queues, because priority queues in this form do not appear in any of the following chapters—semaphores are used in the next chapter. The interested reader will have to wait until the next chapter for proofs of the basic properties of semaphores.

Semaphore

$\upharpoonright(INIT, Wait, Signal)$

$waiters : ProcessQueue_{©}$
$scnt, initval : \mathbb{Z}$
$ptab : ProcessTable$
$sched : LowLevelScheduler$
$ctxt : Context$
$lck : Lock$

INIT

$iv? : \mathbb{Z}$
$pt? : ProcessTable$
$sch? : LowLevelScheduler$
$ct? : Context$
$lk? : Lock$

$initval' = iv? \land scnt' = iv? \land ptab' = pt?$
$sched' = sch? \land ctxt' = ct? \land lck' = lk?$
$waiters.Init$

$$
\begin{array}{l}
NegativeSemaCount \mathrel{\widehat=} scnt < 0\\
NonPositiveSemaCount \mathrel{\widehat=} scnt \leq 0\\
IncSemaCount \mathrel{\widehat=} scnt' = scnt + 1\\
DecSemaCount \mathrel{\widehat=} scnt' = scnt - 1\\
Wait \mathrel{\widehat=}\\
\quad lck.Lock ⨟\\
\qquad (DecSEMACnt ⨟\\
\qquad\quad (NegativeSEMACount \land\\
\qquad\qquad waiters.Enqueue[currentp/x?] \land\\
\qquad\qquad (\exists cpd : ProcessDescr \bullet\\
\qquad\qquad\quad ptab.DescrOfProcess[currentp/pid?, cpd/pd!] \land\\
\qquad\qquad\quad cpd.SetProcessStatusToWaiting \land\\
\qquad\qquad ctxt.SwitchContextOut) \land shed.MakeUnready[currentp/pid?]\\
\qquad\qquad \land sched.RunNextProcess)\\
\quad \lor sched.ContinueCurrent) ⨟\\
\quad lck.Unlock\\
Signal \mathrel{\widehat=}\\
\quad lck.Lock ⨟\\
\qquad (IncSEMACnt ⨟\\
\qquad\quad (NonPositiveSEMACount \land\\
\qquad\qquad waiters.RemoveFirstProc[cand/x!] \land\\
\qquad\qquad\quad (\exists cpd : ProcessDescr \bullet\\
\qquad\qquad\qquad ptab.DescrOfProcess[cand/pid?, cpd/pd!] \land\\
\qquad\qquad\qquad cpd.SetProcessStatusToReady) \land\\
\qquad\qquad\qquad sched.MakeReady[cand/pid?]) \setminus \{cand\}\\
\qquad\quad \lor sched.ContinueCurrent) ⨟\\
\quad lck.Unlock
\end{array}
$$

At this point, it is worth making an observation or two about the use of locks in this book. When writing code that manipulates the interrupt flag, it is always wise to make the period during which interrupts are disabled as short as possible so that new interrupts are not missed. In the models in this book, there are operations entirely bracketed by locking and unlocking operations, thus giving the impression that interrupts *must* be disabled for relatively long periods of time: these operations are high-level specifications that have been written with clarity as a goal. When refining the specifications, use should be made of the two following propositional calculus theorems:

$$p \land q \Leftrightarrow q \land p$$

and:

$$p \land q \Leftrightarrow p \land q \land p$$

These theorems (and their corollaries) are of use in distributing the lock/unlock conjuncts through the rest of the operation, thus adjusting the regions over which interrupts are disabled.

In this model, the process descriptor is a record-like structure stored in a table (an array of records). The process descriptor stores the priority (*prio*), registers (*regs*), status word (*statwd*) and the instruction pointer (*ip*), as well as the process' current state (*status*). Furthermore, it also contains a pointer to the process' stack (*stack*) and to its data and code areas (*data* and *code*, respectively). It is assumed that the stack, code and data are stored in one contiguous region of main store of size *memsize* and pointed to by *mem*. The type *MEMDESC* can be considered, for the time being, as simply a pointer into main store.

The process descriptor is modelled by the next class. It exports an initialisation operation (*INIT*), together with the following operations: *FullContext* (to extract the process' context from the record) and *SetFullContext* (to save the context in the descriptor). In addition, there are operations to store and update the process' priority and to update its state record (*status*) and return and set its storage descriptor (this is required by the storage-allocation operations). Operations are also provided to read and set the process' priority.

The reader might care to compare this definition with the much more complex one required in the next chapter (Section 4.4). Although the record-based approach to process descriptors is common and easy to work with (it turns the process table into an array of records), the approach has some disadvantages, the most interesting of which relates to contention. If more than one process needs to access a process descriptor at the same time, it is necessary to protect it in some way, by a lock in the uni-processor case. However, locking prevents access to *all* of the components of the descriptor (and possibly all of the process table). An alternative implementation must therefore be sought; for example, representing the components of the records by mappings from *IPREF* to their type (e.g., the process' stack by $IPREF \mathrel{+\!\!\!\!\!\twoheadrightarrow} PSTACK$).

ProcessDescr

$\upharpoonright(INIT, FullContext, SetFullContext, Priority, SetPriority,$
$\quad SetProcessStatusToNew, SetProcessStatusToTerminated,$
$\quad SetProcessStatusToReady, SetProcessStatusToWaiting,$
$\quad StoreSize, SetStoreDescr)$

$prio : PRIO;\ status : PROCSTATUS;\ regs : GENREGSET$
$statwd : STATUSWD;\ ip : \mathbb{N};\ stack : PSTACK$
$data : PDATA;\ code : PCODE;\ mem : MEMDESC$
$memsize : \mathbb{N}$

$INIT \mathrel{\widehat=} \ldots$

$Priority \mathrel{\widehat=} \ldots$

$SetPriority \mathrel{\widehat=} \ldots$

$ProcessStatus \mathrel{\widehat=} \ldots$

$SetProcessStatusToNew \mathrel{\widehat=} \ldots$

$SetProcessStatusToTerminate \mathrel{\widehat=} \ldots$

$SetProcessStatusToReady \mathrel{\widehat=} \ldots$

$SetProcessStatusToRunning \mathrel{\widehat=} \ldots$

$SetProcessStatusToWaiting \mathrel{\widehat=} \ldots$

$StoreSize \mathrel{\widehat=} \ldots$

$StoreDescr \mathrel{\widehat=} \ldots$

$SetStoreDescr \mathrel{\widehat=} \ldots$

$FullContext \mathrel{\widehat=} \ldots$

$SetFullContex \mathrel{\widehat=} \ldots$

INIT

$pr? : PRIO$
$stat? : PROCSTATUS$
$pstack? : PSTACK$
$pdata? : PDATA$
$pcode? : PCODE$
$mem? : MEMDESC$
$msz? : \mathbb{N}$

$prio' = pr?$
$status' = stat?$
$waitingtype' = none$
$kind' = knd?$
$regs.INIT$
$statwd' = 0_S$
$ip' = 0$
$data' = pdata?$
$code' = pcode?$
$stack' = pstack?$
$mem' = mem?$
$memsize' = msz?$

Priority
$$pr! : PRIO$$

$$pr! = prio$$

SetPriority
$$\Delta(prio)$$
$$pr? : PRIO$$

$$prio' = pr?$$

ProcessStatus
$$st! : PROCSTATUS$$

$$st! = status$$

SetProcessStatusToNew
$$\Delta(status)$$

$$status' = \mathsf{pstnew}$$

SetProcessStatusToTerminated
$$\Delta(status)$$

$$status' = \mathsf{pstterm}$$

SetProcessStatusToReady
$$\Delta(status)$$

$$status' = \mathsf{pstready}$$

SetProcessStatusToRunning
$$\Delta(status)$$

$$status' = \mathsf{pstrunning}$$

SetProcessStatusToWaiting
$$\Delta(status)$$

$$status' = \mathsf{pstwaiting}$$

StoreSize

$$memsz! : \mathbb{N}$$

$$memsize = memsz!$$

StoreDescr

$$memdescr! : MEMDESC$$

$$memdescr! = mem$$

SetStoreDescr

$$\Delta(pmem, pmemsize)$$
$$newmem? : MEMDESC$$

$$mem' = newmem?$$
$$memsize' = hole\_size(newmem?)$$

FullContext

$$pregs! : GENREGSET$$
$$pip! : \mathbb{N}$$
$$ptq! : TIME$$
$$pstatwd! : STATUSWD$$
$$pstack! : PSTACK$$

$$pregs! = regs$$
$$pip! = ip$$
$$pstatwd! = statwd$$
$$pstack! = stack$$

SetFullContext

$$pregs? : GENREGSET$$
$$pip? : \mathbb{N}$$
$$pstatwd? : STATUSWD$$
$$pstack? : PSTACK$$

$$regs' = pregs?$$
$$ip' = pip?$$
$$statwd' = pstatwd?$$
$$stack' = pstack?$$

The process table is defined next. Its main structure, *procs*, is a finite mapping between *IPREF* and *ProcessDescr*. It is a table or array indexed by elements of *IPREF* and whose elements are objects of type *ProcessDescr*. The class also has state variable *known_procs* to record the elements in the domain of *procs*; it is a record of the identifiers of those processes currently in the system. The variable *freeids* is a set of actual process identifiers that represent those process references not currently referring to processes in the system. The idea is that the identifier of a process is its index in the process table.

The kernel only allocates "actual" processes; that is, processes other than the null and idle processes. For this reason, *freeids* is a set of type *APREF*. The *procs* mapping (table) is of type *IPREF*, the reason for this being that the idle process is represented by a process descriptor that is allocated in the process table when the kernel is initialised.

Apart from its initialisation operation (again, *INIT*), the process table exports operations to create the idle process (*CreateIdleProcess*) and to add and delete process descriptors (*AddProcess* and *DelProcess*, respectively), as well as an operation to return the descriptor of a process (*DescrOfProcess*).

The operation to create the idle process could be defined in a higher layer of the model. Since the idle process owns no resources and executes a piece of code that will be supplied with the kernel (and whose address can, therefore, be made available at kernel initialisation time), it seems reasonable to make idle process creation a process table operation.

ProcessTable

$$\upharpoonright(INIT, CreateIdleProcess, AddProcess, DelProcess, DescrOfProcess)$$

$$
\begin{aligned}
&procs : IPREF \twoheadrightarrow\!\!\!\!\!\!\shortmid\;\; ProcessDescr\\
&known\_procs : \mathbb{F}\, IPREF\\
&freeids : \mathbb{F}\, APREF
\end{aligned}
$$

INIT

$$
\begin{aligned}
&known\_procs' = \{IdleProcRef\}\\
&freeids' = 1 \mathrel{..} maxprocs - 1\\
&(\exists\, ipd : ProcessDescr \bullet createIdleProcess)
\end{aligned}
$$

CreateIdleProcess

$$
\begin{aligned}
&(\exists\, pr : PRIO;\ stat : PROCSTATUS;\ stwd : STATUSWD;\\
&\qquad emptymem : MEMDESC;\ stkdesc : MEMDESC;\\
&\qquad memsz : \mathbb{N};\ ipd : ProcessDescr \bullet\\
&\quad stat = \mathsf{pstready} \land prio = pr \land stwd = 0_s\\
&\quad \land\ emptymem = (0,0) \land stkdesc = (0,20) \land memsz = 0\\
&\quad \land\ ipd.INIT[stat/stat?, knd/knd?, schdlv/slev?, tq/tq?,\\
&\qquad\qquad stkdesc/pstack?, emptymem/pdata?,\\
&\qquad\qquad emptymem/pcode?, emptymem/mem?, memsz/msz?]\\
&\quad procs' = procs \oplus \{IdleProcRef \mapsto ipd\})
\end{aligned}
$$

AddProcess

$\Delta(procs)$
$pid? : APREF$
$pd? : ProcessDescr$

$procs' = procs \oplus \{pid? \mapsto pd?\}$

DelProcess

$\Delta(procs)$
$pid? : APREF$

$procs' = \{pid?\} \lhd procs$

DescrOfProcess

$pid? : IPREF$
$pd! : ProcessDescr$

$pd! = procs(pid?)$

The *Context* class implements the context-switching operations. It is just an encapsulation of the operations described in the previous chapter. It is, in any case, relatively simple. The reader should note the comments in the class definition. The operations defined in this class are extended by *SwapIn* and *SwapOut*—they are defined for convenience.

Context

$\upharpoonright(INIT, SaveState, RestoreState, SwapIn, SwapOut)$

$ptab : ProcessTable$
$sched : LowLevelScheduler$
$hw : HardwareRegisters$

INIT

$ptb? : ProcessTable$
$shd? : LowLevelScheduler$
$hwregs? : HardwareRegisters$

$ptab' = ProcessTable$
$sched' = LowLevelScheduler$
$hw' = hwregs?$

$$
\begin{array}{l}
\textit{SaveState} \\
\hline
(\exists cp : IPREF \bullet \\
\quad sched.CurrentProcess[cp/cp!] \\
\quad (\exists pd : ProcessDescr \bullet \\
\quad\quad ptab.DescrOfProcess[cp/pid?, pd/pd!] \\
\quad\quad \land (\exists regs : GENREGSET;\ stk : PSTACK;\ ip : \mathbb{N}; \\
\quad\quad\quad\quad\quad stat : STATUSWD \bullet \\
\quad\quad\quad hw.GetGPRegs[regs/regs!] \\
\quad\quad\quad \land hw.GetStackReg[stk/stk!] \\
\quad\quad\quad \land hw.GetIP[ip/ip!] \\
\quad\quad\quad \land hw.GetStatWd[stat/stwd!] \\
\quad\quad\quad \land pd.SetFullContext[regs/pregs?, ip/pip?, stat/pstatwd?, \\
\quad\quad\quad\quad\quad stk/pstack?])))) \\
\hline
\end{array}
$$

$$
\begin{array}{l}
\textit{RestoreState} \\
\hline
(\exists cp : IPREF \bullet \\
\quad sched.CurrentProcess[cp/cp!] \\
\quad \land (\exists pd : ProcessDescr \bullet \\
\quad\quad ptab.DescrOfProcess[cp/pid?, pd/pd!] \\
\quad\quad \land (\exists regs : GENREGSET;\ stk : PSTACK;\ ip : \mathbb{N}; \\
\quad\quad\quad\quad\quad stat : STATUSWD \bullet \\
\quad\quad\quad\quad pd.FullContext[regs/pregs!, ip/pip!, stat/pstatwd!, \\
\quad\quad\quad\quad\quad stk/pstack!] \\
\quad\quad\quad\quad \land hw.SetGPRegs[regs/regs?] \\
\quad\quad\quad\quad \land hw.SetStackReg[stk/stk?] \\
\quad\quad\quad\quad \land hw.SetStatWd[stat/stwd?] \\
\quad\quad\quad\quad \land hw.SetIP[ip/ip?]))) \\
\hline
\end{array}
$$

$$
\begin{array}{l}
SwapOut \mathrel{\widehat=} \\
\quad (\exists cp : IPREF;\ pd : ProcessDescr \bullet \\
\quad\quad sched.CurrentProcess[cp/cp!] \\
\quad\quad \land ptab.DescrOfProcess[pd/pd!] \\
\quad\quad \land pd.SetProcessStatusToWaiting \\
\quad\quad \land SaveState \fatsemi sched.MakeUnready[currentp/pid?] \\
\quad\quad \land sched.ScheduleNext)
\end{array}
$$

$$
\begin{array}{l}
SwapIn \mathrel{\widehat=} \\
\quad (\exists cp : IPREF;\ pd : ProcessDescr \bullet \\
\quad\quad sched.CurrentProcess[cp/cp!] \\
\quad\quad \land pd.SetProcessStatusToRunning \\
\quad\quad \land RestoreState)
\end{array}
$$

$$
SwitchContext \mathrel{\widehat=} SwapOut \fatsemi SwapIn
$$

3.5 Priority Queue

This kernel uses a priority queue as the core of its scheduler. The *PRIO* type is equivalent to the integers, so the priorities cannot be arranged as broad classes as they are in the kernel modelled in the next chapter (where there are three priority classes, each modelled by a separate queue). This kernel does not make assumptions about how the priority bands are defined, so a representation has to be chosen to reflects this. The representation is a sequence of process references.

Three relations are required for the definition of priorities. They are the usual $\leq$, $\geq$ and $=$ operations. The subscript is used just to differentiate them from the corresponding relations over the integers.

$$
\begin{array}{l}
\_ \leq_P \_ : PRIO \leftrightarrow PRIO \\
\_ =_P \_ : PRIO \leftrightarrow PRIO \\
\_ \geq_P \_ : PRIO \leftrightarrow PRIO \\
\hline
\forall p_1, p_2 : PRIO \bullet \\
\quad p_1 \leq_P p_2 \Leftrightarrow p_1 \leq p_2 \\
\quad p_1 =_P p_2 \Leftrightarrow p_1 = p_2 \\
\quad p_1 \geq_P p_2 \Leftrightarrow p_1 \geq p_2
\end{array}
$$

The following derived relations are defined in the obvious fashion:

$$
\begin{array}{l}
\_ <_P \_ : PRIO \leftrightarrow PRIO \\
\_ >_P \_ : PRIO \leftrightarrow PRIO \\
\hline
\forall p_1, p_2 : PRIO \bullet \\
\quad p_1 <_P p_2 \Leftrightarrow (p_1 \leq_P p_2) \land \neg (p_1 =_P p_2) \\
\qquad p_1 >_P p_2 \Leftrightarrow (p_1 \geq_P p_2) \land \neg (p_1 =_P p_2)
\end{array}
$$

For completeness, the definitions of these relations are given, even though they should be obvious. Moreover, the $<_P$ relation is not used in this book.

A class defining the process priority queue (or queue of processes ordered by priority) is as follows:

$$
\begin{array}{l}
\textbf{PROCPRIOQUEUE} \\
\upharpoonright (INIT, EnqueuePROCPRIOQUEUE, \\
\quad NextFromPROCPRIOQUEUE, IsInPROCPRIOQUEUE, \\
\quad IsEmptyPROCPRIOQUEUE, PrioOfProcInPROCPRIOQUEUE, \\
\quad RemoveProcPrioQueueElem) \\
\\
\quad qprio : PREF \twoheadrightarrow\!\!\!\!\!\!\!\!\;\shortmid\;\; PRIO \\
\quad procs : \mathrm{iseq}\, PREF \\
\hline
\quad \mathrm{dom}\, qprio = \mathrm{ran}\, procs \\
\quad \forall p_1, p_2 : PREF \bullet \\
\qquad p_1 \in \mathrm{ran}\, procs \land p_2 \in \mathrm{ran}\, procs \land qprio(p_1) \leq_P qprio(p_2) \Rightarrow \\
\qquad\quad (\exists i_1, i_2 : 1 \ldots \#procs \bullet i_1 \leq i_2 \land procs(i_1) = p_1 \land procs(i_2) = p_2)
\end{array}
$$

$$\begin{array}{l}
\underline{\quad INIT \quad} \\
PROCPRIOQUEUE' \\
\hline
procs' = \langle\,\rangle
\end{array}$$

$EnqueuePROCPRIOQUEUE \mathrel{\widehat=} \ldots$

$NextFromPROCPRIOQUEUE \mathrel{\widehat=} \ldots$

$IsInPROCPRIOQUEUE \mathrel{\widehat=} \ldots$

$IsEmptyPROCPRIOQUEUE \mathrel{\widehat=} \ldots$

$PrioOfProcInPROCPRIOQUEUE \mathrel{\widehat=} \ldots$

$RemovePrioQueueElem \mathrel{\widehat=} \ldots$

$reorderProcPrioQueue \mathrel{\widehat=} \ldots$

The queue's state consists of a finite mapping from process identifiers to their associated priority (the priority mapping) and a queue of processes. The invariant states that if the priority of one process is less than that of another, the first process should precede the second.

The class exports an initialisation operation (*INIT*), an enqueue and removal operations. There is an emptiness test and a test to determine whether a given process is in the queue. A removal operation, as well as a reordering operation, is also provided (the removal operation is used when re-prioritising processes). The final operation that is exported returns the priority of a process that is in the queue.

The reader will have noted that the priority record in the priority queue duplicates that in the process table. What would happen in refinement is that the two would be identified.

The enqueue operation is:

$$\begin{array}{l}
\underline{\quad EnqueuePROCPRIOQUEUE \quad} \\
\Delta(qprio, procs) \\
pid? : PREF \\
pprio? : PRIO \\
\hline
qprio' = qprio \cup \{pid? \mapsto pprio?\} \\
(procs = \langle\,\rangle \land procs' = \langle pid? \rangle) \\
\lor\ (procs \neq \langle\,\rangle \\
\qquad \land\ ((qprio(pid?) \leq_P qprio(head\ procs)) \land procs' = \langle pid? \rangle \frown procs) \\
\qquad \lor\ ((qprio(pid?) >_P qprio(last\ procs)) \land procs' = procs \frown \langle pid? \rangle) \\
\qquad \lor\ (\exists\, s_1, s_2 : \mathrm{iseq}\, PREF \mid s_1 \frown s_2 = procs \bullet \\
\qquad\qquad ((qprio(last\ s_1) \leq_P qprio(head\ s_2)) \land \\
\qquad\qquad\qquad procs' = s_1 \frown (\langle pid? \rangle \frown s_2)))
\end{array}$$

The operation uses the priority of the process in determining where the process is to be inserted in the queue. If the new process' priority is greater (has a lower value) than the first element of the queue, the new process is prepended to the queue; conversely, if the priority is lower (has a greater value) than the last element, it is appended to the queue. Otherwise, the priority is somewhere between these two values and a search of the queue is performed (by the existential quantifier) for the insertion point.

Proposition 21. *The predicate of the EnqueuePROCPRIOQUEUE schema satisfies the invariant of the PROCPRIOQUEUE schema that defines the state.*

PROOF. Let I denote the invariant:

$$\begin{array}{l}
\mathrm{dom}\, qprio = \mathrm{ran}\, procs \\
\forall p_1, p_2 : PREF \bullet \\
\quad p_1 \in \mathrm{ran}\, procs \land p_2 \in \mathrm{ran}\, procs \land qprio(p_1) \leq_P qprio(p_2) \Rightarrow \\
\quad\quad (\exists i_1, i_2 : 1 \mathinner{..} \#procs \bullet \\
\quad\quad\quad i_1 \leq i_2 \land procs(i_1) = p_1 \land procs(i_2) = p_2)
\end{array}$$

Case 1. $procs = \langle\rangle \land proc' = \langle p \rangle \Rightarrow I$. Since $qprio(p) \leq_P qpiro(p)$, for all p. Clearly, in the case of $procs'$, $i_1 \leq i_2$ (since $i_1 = i_2$).
Case 2. $procs \neq \langle\rangle$. Let $p_s = \{p : PREF \mid p \in \mathrm{ran}\, procs' \bullet qprio(p)\}$. There are three cases to consider:

i. p is at the head of $procs'$—$qprio(p) \leq_P min\ p_s$;
ii. p is the last of $procs'$—$qprio(p) >_P max\ p_s$;
iii. p appears in the middle of the sequence—i.e., $min\ p_s \leq_P (p_q prio(p) \leq_P max\ p_s$.

Case 2i. Immediate.
Case 2ii. Immediate.
Case 2iii. Assume that there are two increasing sequences, s_1 and s_2, of $PREF$s s.t. $s_1 \frown s_2 = procs$. Then, if $qprio(last\ s_1) \leq_P qprio(p) \leq_P qprio(head\ s_2)$, $procs' = s_1 \frown \langle p \rangle \frown s_2$. By induction, s_1 and s_2 satisfy I, therefore $s_1 \frown \langle p \rangle \frown s_2$ satisfies I. □

Proposition 22. *If a process, p, has a priority, p_r, such that, for any priority queue, the value of p_r is less than all of its elements, then p is the head of the queue.*

PROOF. By the ordering $\leq_P$, p_r is less than all the elements of $procs$, i.e., $p_r \leq_P min\ p_s$, where $p_s = \{p : PREF \mid p \in \mathrm{ran}\, procs \bullet qprio(p)\}$, as above. By the invariant, $\langle p_r \rangle \frown procs = procs'$. □

Proposition 23. *If a process, p, has a priority, p_r, such that, for any priority queue, the value of p_r is greater than all of its elements, then p is the last element of the queue.*

When a process has executed, the next element has to be selected from the priority queue. The operation to do this is:

NextFromPROCPRIOQUEUE

$\Delta(procs, qprio)$
$pid! : PREF$

$procs = \langle pid! \rangle \frown procs'$
$qprio' = \{pid!\} \ntriangleleft qprio$

The priority is updated as well as the sequence of queues.

Proposition 24. *If an element, p, has been removed from a priority queue and the priority of p is p_r, then $p_r \leq head\ procs'$.*

PROOF. If $\langle p_r \rangle \frown procs = procs'$, then $p_r > p_s$ (where p_s is as in Proposition 22). By the invariant, I,

$$qprio(head\ procs) \leq_P qprio(head(tail\ procs))$$
$$\Rightarrow qprio(head(tail\ procs)) = min\ p_s$$

Therefore, $qprio(head(tail\ procs)) \geq_P qprio(head\ procs)$. □

The following pair of schemata define predicates. The first, *IsInPROCPRIOQUEUE*, determines whether the process, *pid?*, is in the queue of processes. The second is true when the process queue is empty.

IsInPROCPRIOQUEUE

$pid? : PREF$

$pid? \in \mathrm{ran}\ procs$

IsEmptyPROCPRIOQUEUE

$procs = \langle \rangle$

The following schema defines an operation to return the priority of the process denoted by *pid?*:

PrioOfProcInPROCPRIOQUEUE

$pid? : PREF$
$pprio! : PRIO$

$pprio! = qprio(pid?)$

Proposition 25. *For any pair of processes, p_1 and p_2, that are both in the same priority queue, if $p_{r,1} < p_{r,2}$ (where $p_{r,i}$ denotes the priority of process 1 or 2), then p_1 occurs before p_2 in the queue; otherwise, the order in which they occur is reversed.*

PROOF. This proposition is an immediate consequence of I. □

The *unready* operation, among others, requires that processes be removed from the priority queue. The operation to do this is modelled by the following schema:

$$\begin{array}{|l}
\hline
\textit{RemovePrioQueueElem} \\
\Delta(procs, qprio) \\
pid? : PREF \\
\hline
procs' = procs \rhd \{pid?\} \\
qprio' = \{pid?\} \lhd qprio \\
\hline
\end{array}$$

There is the case in which a process' priority is re-calculated at some time. Such a re-evaluation will affect the order in which the processes occur in the queue. For the following, as for all process priority queue operations, it is assumed that the priority value is established by a mechanism that is not described by schemata in this section and that it is supplied to the operation reordering the queue upon re-calculation.

It should be noted that the case in which more than one process' priority is re-calculated at a time does not cause any problems: if n processes have their priority re-calculated, n iterations of the *reorderProcPrioQueue* are required. Usually, however, re-calculations occur one at a time.

The following schema represents the desired operation:

$$\begin{array}{l}
reorderProcPrioQueue \mathrel{\widehat=} \\
\quad RemovePrioQueueElem \fatsemi EnqueuePROCPRIOQUEUE[newprio?/pprio?]
\end{array}$$

After substituting $qprio'$ for $qprio''$ in the appropriate places, this expands into:

$$\begin{array}{|l}
\hline
\textit{reorderProcPrioQueue} \\
\Delta(procs, qprio) \\
pid? : PREF \\
newprio? : PRIO \\
\hline
\exists\, procs'' : \mathrm{iseq}\, PREF;\ qprio'' : PREF \pinj PRIO \bullet \\
\qquad procs'' = procs \rhd \{pid?\} \\
\qquad \land\ qprio'' = \{pid?\} \lhd qprio \\
\qquad \land\ qprio' = qprio'' \cup \{pid? \mapsto newprio?\}
\end{array}$$

$$
\begin{array}{l}
\land ((procs'' = \langle\rangle \land procs' = \langle pid? \rangle) \lor (procs'' = \langle\rangle \\
\qquad \land ((qprio'(pid?) \leq_P qprio'(head\, procs'')) \\
\qquad\qquad \land procs' = \langle pid? \rangle \frown procs'') \\
\qquad \lor ((qprio'(pid?) >_P qprio'(last\, procs'')) \\
\qquad\qquad \land procs' = procs'' \frown \langle pid \rangle) \\
\qquad \lor (\exists s_1, s_2 : \mathrm{iseq}\, PREF \mid s_1 \frown s_2 = procs'' \bullet \\
\qquad\qquad (qprio'(last\, s_1) \leq_P qprio'(pid?) \leq_P qprio'(head\, s_2)) \\
\qquad\qquad\qquad \land procs' = s_1 \frown \langle pid? \rangle \frown s_2)))
\end{array}
$$

This operation must, clearly, respect *PROCPRIOQUEUE*'s invariant, so we have the following:

Proposition 26. *ReorderProcPrioQueue respects PROCPRIOQUEUE's invariant.*

PROOF. By Proposition 21, *EnqueuePROCPRIOQUEUE* respects the invariant. It remains, therefore, to show that *ReorderProcPrioQueue* does.

Recall that *ReorderProcPrioQueue* is defined as:

$$
\begin{array}{l}
\Delta(procs, qprio) \\
p? : PREF \\
\hline
procs' = procs \rhd \{p?\} \\
qprio' = \{p?\} \lhd qprio
\end{array}
$$

There are three cases to consider:

i. $p? = head\, procs$: this is clear;
ii. $p? = last\, procs$: this is clear;
iii. $procs = s_1 \frown \langle p? \rangle \frown s_2$: this is proved by induction.

It is clear that the composition implies the invariant. □

Proposition 27. *If a process, p, is removed from a priority queue, P, its priority, p_r, recalculated, and returned to P to form P', then:*

1. *If the length of P is exactly 1, $P' = P$.*
2. *If the length of P is exactly 2, then p is either the first or last element of P'.*

PROOF. There are two cases.
1. If $p \in P$ and $\#P = 1$, then $P = \langle p \rangle$. This case is covered by Proposition 37 (first case).
2. If $P = \langle p_1, p_2 \rangle$, then it follows that if $p = p_1$, $p \leq_P p_2$; otherwise, if $p = p_2$, $p_1 \leq_P p$. The result then follows by Proposition 21. □

Proposition 28. *If a process, p, is removed from a priority queue, P, its priority, p_r, recalculated and is reinserted into P to form P′, provided that P has more than one element, exactly one of the following holds:*

1. *If the new value of p_r is less than the old one, p will appear closer to the head of P′ than in P.*
2. *If the new value of p_r is greater than the old one, p will appear closer to the end of P′ than P.*
3. *If the new value of p_r is the same as the old one, one of the two previous conditions will hold.*

PROOF. By Proposition 21. The interesting case is case 3, whose proof follows from $\leq_P$. □

Proposition 29. *If a process, p, has priority, p_r, is in a priority queue at the nth position, if there are no processes of higher priority inserted into that queue, and if the priority of p is not recomputed, process p will have the highest priority after n processes have been removed from the queue.*

PROOF. This is just an instance of the corresponding proposition for FIFO queues. □

Proposition 30. *If a process, p, is at the nth position in a priority queue and if m processes, each with priority higher than p, are added to the queue, p will then occupy position $n + m$ in the queue, provided that the priority of p is not recomputed.*

PROOF. Let $procs = s_1 \frown \langle p \rangle \frown s_2$ with $\#s_1 = n$. By Proposition 21 (case 2iii), $s_1' = s_{1,1} \frown s_m \frown s_{1,2}$, where $\#s_m = m$, and s_n are sequences of new processes, where $s_= s_{1,1} \frown s_{1,2}$. (More simply but less generally assume $s_1' = s_1 \frown s_m$.)

If $\#s_1 = n$ and $\#s_m = m$, and $s_1 = s_{1,1} \frown s_{1,2}$ and $\#s_{1,1} + \#s_{1,2} = n$, then $\#(s_{1,1} \frown s_m \frown s_{1,2}) = n + m$. Since $qprio(p)$ remains constant, the proposition is proved. □

3.6 Current Process and Prioritised Ready Queue

This is just a redefinition of *CURRENTPROCESS* so that the *ready* is replaced by a priority queue. The result is still called *CURRENTPROCESS*.

Structural theorems carry over from *PROCPRIOQUEUE*.

The class *CURRENTPROCESS* represents the scheduler in this kernel. The class contains a variable holding the currently executing process, *currentp*, as well as the queue of ready processes (*readyq*, an instance of the

PROCPRIOQUEUE class). The class exports two schemata that manipulate *currentp*: *CurrentProcess* (which returns the value of *currentp*) and *MakeCurrent* (which sets the value of *currentp*). The operation *MakeReady* adds a process to the scheduler's queue; to do this, the priority of the process must be supplied. The *ContinueCurrent* is defined: it is, in fact, the identity relation and serves only to continue the execution of the current process. The *SuspendCurrent* operation removes the currently executing process but does not swap in another process. The schema for *RunNextProcess* defines an operation that takes the next element from the ready queue, *ready*, and sets *currentp* so that the new process can be executed. The operation defined by the schema called *SCHEDULENEXT* is the primary interface to the scheduler.

Clearly, the scheduler needs to perform context operations. It is also clear that it should use locking to ensure exclusive access to data structures.

CURRENTPROCESS

$\upharpoonright(INIT, CurrentProcess, MakeCurrent,$
$\quad MakeReady, ContinueCurrent, SuspendCurrent,$
$\quad RunNextProcess, SCHEDULENEXT)$

$currentp : PREF$
$readyqp : PROCPRIOQUEUE$
$ctxt : Context$
$lck : Lock$

INIT

$ct? : Context$
$lk? : Lock$

$readyqp.INIT$
$currentp' = NullProcRef$
$lck' = lk?$
$ctxt' = ct?$

$CurrentProcess \mathrel{\widehat=} \ldots$

$MakeCurrent \mathrel{\widehat=} \ldots$

$MakeReady \mathrel{\widehat=} \ldots$

$MakeUnready \mathrel{\widehat=} \ldots$

$isCurrentProc \mathrel{\widehat=} \ldots$

$reloadCurrent \mathrel{\widehat=} \ldots$

$ContinueCurrent \mathrel{\widehat=} \ldots$

$SuspendCurrent \mathrel{\widehat=} \ldots$

$RunNextProcess \mathrel{\widehat=} \ldots$

$selectIdleProcess \mathrel{\widehat=} \ldots$

$$SCHEDULENEXT \mathrel{\widehat=} \ldots$$

CurrentProcess

$$cp! : PREF$$

$$cp! = currentp$$

MakeCurrent

$$\Delta(currentp)$$
$$pid? : PREF$$

$$currentp' = pid?$$

The *MakeReady* schema is defined as follows:

$$\begin{array}{l} MakeReady \mathrel{\widehat=} \\ \quad (\exists\, pd : ProcessDescr \bullet \\ \qquad ptab.DescrOfProcess[pd/pd!] \\ \qquad \land\ pd.ProcessPriority[prio/prio!] \\ \qquad \land\ pd.SetProcessStatusToReady \\ \qquad \land\ readyq.EnqueuePROCPRIOQUEUE[prio/pprio?]) \setminus \{prio\} \end{array}$$

The *MakeReady* operation inserts a new process into the queue. As can be seen, the priority of the process must be used to compute the point where the process is to be inserted.

As can be seen, *ContinueCurrent* is the identity:

reloadCurrent

$$currentp' = currentp$$
$$readyqp' = readyqp$$

$$\begin{array}{l} ContinueCurrent \mathrel{\widehat=} \\ \quad reloadCurrent \\ \quad \land\ ctxt.RestoreState \end{array}$$

The *MakeUnready* operation is defined as:

$$\begin{array}{l} MakeUnready \mathrel{\widehat=} \\ \quad lck.Lock \fatsemi \\ \qquad ((IsCurrentProc\ \land \\ \qquad\quad (ctxt.SaveState \fatsemi RunNextProcess) \fatsemi \\ \qquad lck.Unlock) \\ \qquad \lor\ RemovePrioQueueElem) \fatsemi \\ \qquad\quad lck.Unlock \end{array}$$

where:

$$
\begin{array}{|l}
\hline
\textit{isCurrentProc} \\
\Xi CURRENTPROCESS \\
p? : PREF \\
\hline
p? = currentp \\
\hline
\end{array}
$$

This operation is used to remove a process from the *readyq*. It can be used, for example, when a process has to wait on a device queue.

In this kernel, it is possible for a process to suspend itself. It does so by calling the following operation:

$$
\begin{array}{l}
SuspendCurrent \mathrel{\widehat=} \\
\quad lck.Lock \fatsemi \\
\quad ((ctxt.SaveState \\
\quad \land (\exists pd : ProcessDescr \bullet \\
\qquad ptab.DesrcOfProcess[currentp/pid?, pd/pd!] \\
\qquad \land ((pd.ProcessPriority[prio/prio!] \\
\qquad\quad \land pd.SetProcessStatusToWaiting) \\
\qquad \land readyqp.EnqueuePROCPRIOQUEUE[currentp/x?, prio/prio?]) \\
\qquad\qquad\qquad \setminus \{prio\})) \fatsemi \\
\quad RunNextProcess) \fatsemi \\
\quad lck.Unlock
\end{array}
$$

The *RunNextProcess* operation calls the scheduler and executes the next process. Note that it assumes that the context of the previously executing process has been saved *before* this operation executes.

$$
\begin{array}{l}
RunNextProcess \mathrel{\widehat=} \\
\quad SCHEDULENEXT \fatsemi ctxt.RestoreState
\end{array}
$$

The primary interface to the scheduler is the following operation:

$$
\begin{array}{l}
SCHEDULENEXT \mathrel{\widehat=} \\
\quad lck.Lock \fatsemi \\
\quad ((\exists pd : ProcessDescr \bullet \\
\qquad ptab.DescrOfProcess[p/pid?, pd/pd!] \\
\qquad \land (\lnot\ readyqp.IsEmptyPROCPRIOQUEUE \\
\qquad\quad \land readyqp.NextFromPROCPRIOQUEUE[p/pid!] \\
\qquad\quad \land readyqp.MakeCurrent[p/pid?] \\
\qquad\quad \land pd.SetProcessStatusToRunning[p/pid?]) \\
\qquad \lor selectIdleProcess)) \fatsemi \\
\quad lck.Unlock
\end{array}
$$

The auxiliary operation, *selectIdleProcess*, is defined as follows:

$$\begin{array}{|l} \hline \textit{selectIdleProcess} \\ \Delta(currentp) \\ \hline currentp' = IdleProcRef \\ \hline \end{array}$$

This operation is required to ensure that the idle process runs when there is nothing to execute in *readyq*.

Proposition 31. *If* $currentp = p$, *for some process reference,* $p : APREF$, *then* $RunNextProcess$ *implies that* $currentp' \neq p$.

PROOF. The proof divides into two cases.
Case 1. The ready queue is empty. Therefore, by $SCHEDULENEXT$:

$$readyqp.IsEmptyPROCPRIOQUEUE \wedge selectIdleProcess$$

and $currentp' = IdleProcRef$ follows immediately.
Case 2. The ready queue is not empty. The following conjunction occurs in the predicate of schema $SCHEDULENEXT$:

$$readyqp.NextFromPROCPRIOQUEUE \wedge readyqp.MakeCurrent[p/pid?]$$

The operation $NextFromPROCPRIOQUEUE$ removes the first element from the ready queue. This will, in general, be different from the current value of *currentp*. □

3.7 Messages and Semaphore Tables

Semaphores have already been defined. This kernel also requires asynchronous message queues (mailboxes), according to the requirements. This section contains the outline specification for mailboxes.

First, the message type is defined. Messages are composed of data (modelled by the atomic type $MSGDATA$) and a record of the message's source ($MSGSRC$).

$$[MSGDATA]$$

A message can be sent by any "actual" process or by a hardware device.

$$MSGSRC == APREF \cup \{\mathsf{hardware}\}$$

Putting these components together, we obtain $MBOXMSG$, the type of messages:

$$MBOXMSG == MSGSRC \times MSGDATA$$

The following (obvious) functions are defined to assist in manipulating messages:

$$
\begin{array}{|l}
msgsender : MBOXMSG \rightarrow MSGSRC \\
msgdata : MBOXMSG \rightarrow MSGDATA \\
\hline
\forall m : MBOXMSG \bullet \\
\quad msgsender(m) = fst\ m \\
\quad msgdata(m) = snd\ m
\end{array}
$$

The class that actually defines the mailbox type is as follows. The mailbox is defined in the obvious way as a queue of messages.

The class exports operations to add a message (*PostMessage*) and obtain the next message from the mailbox (*NextMessage*) and an operation that determines whether there are messages in the mailbox (*HaveMessages*). The initialisation operation just clears the queue of messages.

$$
\begin{array}{|l}
\textit{Mailbox} \\
\hline
\upharpoonright(INIT, PostMessage, HaveMessages, NextMessage) \\
\\
\quad msgs : QUEUE[MBOXMSG] \\
\quad lck : Lock \\
\\
\quad \textit{INIT} \\
\quad l? : LOCK \\
\quad \hline
\quad msgs.INIT \\
\quad lck' = l? \\
\\
\quad \textit{PostMessage} \\
\quad m? : MBOXMSG \\
\quad \hline
\quad lck.Lock \fatsemi (msgs.Enqueue[m?/x?] \fatsemi lck.Unlock) \\
\\
\quad \textit{HaveMessages} \\
\quad lck.Lock \fatsemi \neg\ msgs.IsEmpty \fatsemi lck.Unlock \\
\\
\quad \textit{NextMessage} \\
\quad m! : MBOXMSG \\
\quad \hline
\quad lck.Lock \fatsemi msgs.RemoveFirst[m!/x!] \fatsemi lck.Unlock
\end{array}
$$

Each process can have no more than one mailbox in this kernel (it does not need more than one). This suggests that:

- An instance of *GENTBL* can be used to define and implement a central table of mailboxes.

- *APREF* can be used for the key type in this table.

The table type requires the following operations: initialise, create a mailbox for a process, delete a process' mailbox and retrieve the mailbox upon demand. The type will have, in addition, to ensure mutual exclusion so that, should two processes attempt, say, to create a mailbox at the same time, only one will succeed.

The definition of the mailbox table is relatively simple and is, in any case, similar to the *SemaphoreTable* type, which will be defined next.

Processes can also have one or more semaphores to provide synchronisation on their shared data. The semaphore table contains all the semaphores in the system that are available for use by processes. The table is an instance of the generic $GENTBL[K, D]$ class. Locking is used to ensure mutual exclusion.

The index type, *SEMAID*, is of little relevance. Processes will use values of this type to refer to semaphores. It is defined as an atomic type:

$[SEMAID]$

Semaphore identifiers can be assumed to be generated by:

$$
\begin{array}{l}
\textit{GenSemaId} \\
\hline
sid! : SEMAID \\
\hline
\ldots
\end{array}
$$

An infinite number of semaphore identifiers is assumed. It is also assumed that no identifier is generated twice.

$$
\begin{array}{l}
\textit{SemaphoreTable} \\
\hline
\upharpoonright(INIT, NewSemaphore, DelSemaphore, GetSemaphore) \\[1em]
\quad lck : Lock \\
\quad stbl : GENTBL[SEMAID, Semaphore] \\[1em]
\quad \underline{INIT} \\
\quad l? : Lock \\
\quad \hline
\quad lck' = l? \land stbl.INIT \\[1em]
\quad NewSemaphore \mathrel{\widehat=} \\
\qquad lck.Lock \fatsemi \\
\qquad (\exists s : Semaphore; sid : SEMAID \bullet \\
\qquad\quad s.INIT \\
\qquad\quad \land GenSemaId[sid/sid!] \\
\qquad\quad \land stbl.AddTBLEntry[sid/k?, s/d?] \\
\qquad \fatsemi lck.Unlock
\end{array}
$$

$DelSemaphore \mathrel{\widehat=} lck.Lock \mathbin{;} stbl.DelTBLEntry[sid?/k?] \mathbin{;} lck.Unlock$

$GetSemaphore \mathrel{\widehat=} lck.Lock \mathbin{;} stbl.GetTBLEntry[sid/k?, s!/d!] \mathbin{;} lck.Unlock$

3.8 Process Creation and Destruction

In this kernel, processes are created statically, assigned a static priority and linked to the kernel. This has the implication that process creation and termination primitives are not really required. However, it *is* necessary to communicate some basic parameters about each process to the kernel. This can be done using the following operations (defined, for convenience, inside an object called *UserLibrary*).

The operations defined below should be easily understood: their names state what they should do.

UserLibrary

$\upharpoonright(INIT, CreateProcess, TerminateProcess, Suspend)$

$procid : IPREF$
$ptab : ProcessTable$
$sched : Scheduler$

INIT

$ptb? : ProcessTable$
$schd? : Scheduler$

$ptab' = ptb?$
$sched' = schd?$

CreateProcess

$pprio? : PRIO$
$stkd? : PSTACK$
$datad? : PDATA$
$cdd? : PCODE$
$allocin? : MEMDESC$
$totmemsz? : \mathbb{N}$
$pid! : PREF$

$\exists pd : ProcessDescr; stat : PROCSTATUS \mid stat = \mathsf{pstnew} \bullet$
$\quad pd.INIT[pprio?/pr?, stat/stat?, stkd?/pstack?,$
$\qquad datad?/pdata?, cdd?/pcode?, allocin?/mem?,$
$\qquad totmemsz?/msz?]$
$\quad \land ptab.AddProcess[pd/pd?, pid!/p!] \land sched.MakeReady[pid!/pid?]$
$\quad \land procid' = pid!$

$TerminateProcess \mathrel{\widehat{=}}$
$\quad sched.MakeUnready[procid/pid?] \land$
$\quad ptab.DelProcess[procid/pid?]$

$Suspend \mathrel{\widehat{=}} sched.SuspendCurrent$

3.9 Concluding Remarks

The kernel modelled in this chapter is extremely simple. It is not so simple that it cannot be used. Indeed, it is of a complexity not far from that of the tiny kernels used for embedded and some real-time systems. The μC/OS [18] is a good example of such a kernel.

The kernel is minimal in the sense that it contains no facilities for performing device-specific operations and contains no clock process. If the kernel were to be used in reality, these operations would have to be modelled and implemented. This would not be a particularly difficult operation, for all the necessary operations have been provided by the above model.

The kernel is also open as far as security is concerned. This is an area that requires further development.

The above kernel is also a fairly static affair. Processes are statically linked to the kernel via a library of routines (basically the interface routines and the small library defined at the end of the last section). Processes are free to change their priority and to create new processes when they are running but this is subject to the constraint that all processes must always be resident in main store.

Main store itself is partitioned statically as a configuration operation when user code is linked to the kernel. There are no storage-management operations in this kernel, so the process creation primitives are, really, of limited use.

Even though the kernel is so limited, it is reasonable useful. In the next chapter, a larger kernel, one that can perform storage management (it performs process swapping) is presented and discussed.

Perhaps the most significant aspect of this chapter's model is that it acts as an existence proof. It *is* possible to define a formal model of an operating system kernel and to prove some of its properties. In the next chapter, the model becomes more involved and more properties are proved. As noted in the introduction, the general method of this book is to increase complexity and to prove more properties (when appropriate) as the book progresses.

4

A Swapping Kernel

4.1 Introduction

The last chapter presented the model of a simple kernel. The kernel is similar to those found in embedded and small real-time systems such as μC/OS [18]. The model of the last chapter serves as an *existence proof*: the formal modelling of an operating system can be performed. The purpose of this chapter is to expand upon this by presenting a model of the kernel with much more functionality. It is a kernel of about the same complexity as minicomputer (and some mainframe) operating systems of the 1970s and 1980s, and a large proportion of its functionality can be found in present-day kernels.

Since the requirements for this kernel are listed in Section 4.2, they will not be repeated here. Instead, it will be noted that the kernel was greatly influenced by the Linux [4] and, particularly, MINIX [30] kernels. Some might object that this kernel is really too simple and that it is a poor example. To this, it must be replied that the model presented below is a high-level description of the functions and interactions of the kernel, a kernel that does not contain file systems or browsers as integral components[1]. The lack of complexity is merely superficial, as the kernel contains models of all the operations required of it in Section 4.2.

4.2 Requirements

The kernel specified in this chapter should have a layered design. In order, the layers should be:

1. the hardware;
2. Interrupt Service Routines;
3. the process abstraction;

[1] As far as the author is concerned, feature creep leading to kernel bloat!

4. Inter-Process Communications;
5. system processes;
6. a scheduler to be based on a priority scheme. Instead of arranging matters as in the previous kernel, three broad bands are to be used: one each for device processes, one for system processes and one for user processes. User processes have the lowest priority; device processes have the highest. Within each priority band, a round-robin scheme should be used.

The system processes should include, at a minimum, the following:

- a clock to support alarms of various kinds;
- a storage-management mechanism. Each process should be composed of a contiguous segment of main store that is allocated when the process is created.

In addition, a swapping mechanism for storing active processes in a reserved disk space is to be used. When a process has been swapped out for a sufficiently long time, it should be returned to main store so that it can continue execution. If there is insufficient store free when a process is created, it should be allocated on disk, later to be swapped into main store. These mechanisms apply *only* to user-level processes.

The organisation of the kernel is shown in Figure 4.1. This figure, which first appeared as Figure 1.1, shows the organisation of the kernel in terms of layers: hardware appears at the bottom and the user's programs at the top. The model displays all the characteristics of the classical operating system kernel, hence the duplication of the figure[2].

4.3 Common Structures

In this section, some structures common to many of the models in this book are defined. With the exception of the hardware model that immediately follows (Section 4.3.1), the definitions are in Z rather than Object-Z. This is because many readers will be more familiar with Z, so this chapter serves as a gentle introduction to the remainder of this book. The use of Z in some cases leads to framing and promotion. It was decided to include these so that the reader can gain some idea of what a full specification in Z might look like (and, thereby, see why Object-Z was eventually preferred).

4.3.1 Hardware

Operating system kernels operate the system's hardware. This subsection contains a number of definitions that will be used when defining the rest of the kernel. The specifications of this subsection are loose for the reason that it

[2] This is certainly *not* to claim that this kernel is the paradigm upon which all should follow.

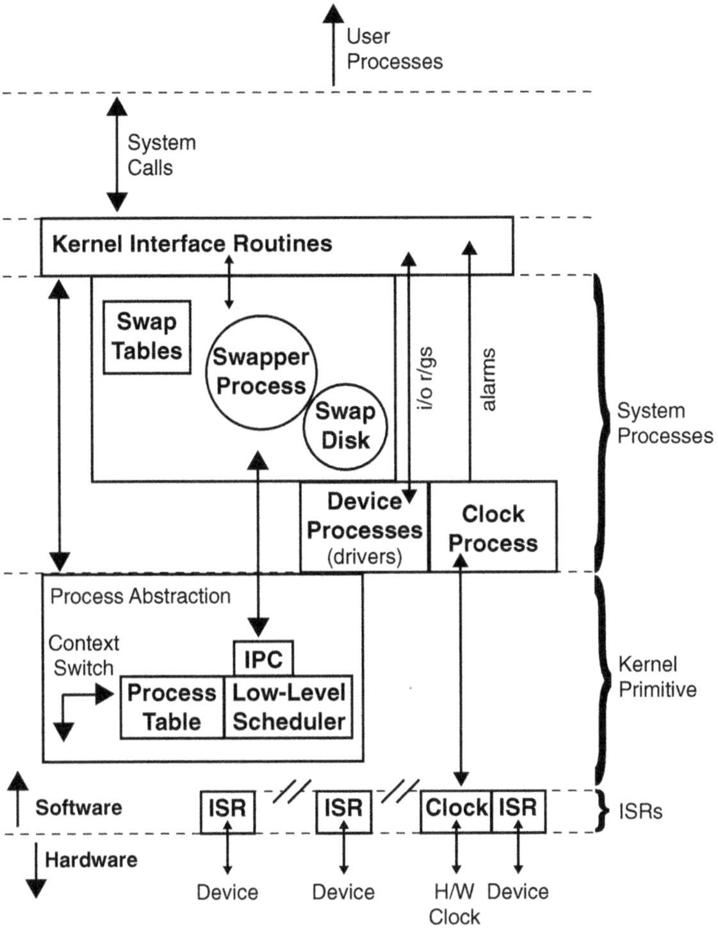

Fig. 4.1. *The layer-by-layer organisation of the kernel.*

is not possible, *a priori*, to determine anything other than the most general properties of the actual hardware upon which a kernel runs. If the kernel specified in this chapter were refined to code, the hardware-related parts would, at some stage, need to be made more precise. Without target hardware in mind, this cannot be done. On the other hand, the looseness of the specification of this subsection shows how little one need assume about the hardware in order to construct a working kernel; this is good news for portability and for abstraction.

The specifications begin with a type that represents time. Time is important to the hardware and, although this chapter is not about a real-time kernel, the current time is important to a number of kernel components: for example, the swapper process and the scheduler. The scheduler uses *time slicing* to implement pre-emption on user processes.

$TIME == \mathbb{N}$

Time is defined as an infinite, discrete type of atomic elements. The elements (in one-one correspondence with the naturals) can be events or arbitrary ticks of the hardware. For the purposes of this specification, the actual denotation of elements of this type is left unspecified.

Next, there is the runtime stack. Each process maintains a stack and there is a slot in the process descriptor of each process for a stack. The type is defined as:

$[PSTACK]$

For the time being, there is no need to ask what constitutes a stack. It is, therefore, left as an atomic type. (A refinement of this specification would add structure to this type; for example, a pointer to the start of the storage area reserved for the stack, the size of the reserved area and a pointer to the top of the stack.) In order to initialise various data structures, as well as the (model of the) hardware registers, a null value is needed for stacks. It is defined as:

$$NullStack : PSTACK$$

It is assumed that the kernel will run on hardware with one or more registers. The actual number of registers is defined by the following constant:

$$maxgreg : \mathbb{N}$$

There are no assumptions made about the value of *maxgreg*. For the purposes of the next definition, it can be said that *maxgreg* should be assigned to a value that is one less than the actual number of registers:

$GENREG == 0 \mathrel{..} maxgreg$

This is the type defining the indices used to refer to the actual hardware registers that appear in the register set. The register set is defined as the following Object-Z class:

GENREGSET

$\upharpoonright(regs)$

$$regs : GENREG \rightarrow PSU$$

INIT

$$\forall i : GENREG \bullet \\ \quad regs'(i) = 0$$

The general registers are modelled as a function from the index set to the value in each register. Another way to view this definition is as an array—this is what is really wanted. This specification does not include operations to access and update individual registers. At present, there is no need for individual register access in the more general kernel specification, so they are not included (they are simple enough to define, in any case).

The general registers that are accessible to programmers come next. This is what is usually known as the processor's *register set*. The registers include the general registers, the stack register, the program counter and the status register; a flag controlling interrupts is also included.

HardwareRegisters

$$\upharpoonright(SetGPRegs, GetGPRegs, GetStackReg, SetStackReg, SetIntsOff, SetIntsOn, GetIP, SetIP, GetStatWd, SetStatWd)$$

$$hwgenregs : GENREGSET$$
$$hwstack : PSTACK$$
$$hwstatwd : STATUSWD$$
$$hwip : \mathbb{N}$$

INIT

$$hwgenregs.INIT$$
$$hwstack' = 0$$
$$hwstatwd' = 0_S$$
$$hwip' = 0$$

$$SetGPRegs \mathrel{\widehat{=}} \ldots$$
$$GetGPRegs \mathrel{\widehat{=}} \ldots$$
$$GetStackReg \mathrel{\widehat{=}} \ldots$$
$$SetStackReg \mathrel{\widehat{=}} \ldots$$
$$GetIP \mathrel{\widehat{=}} \ldots$$
$$SetIP \mathrel{\widehat{=}} \ldots$$
$$GetStatWd \mathrel{\widehat{=}} \ldots$$
$$SetIntsOff \mathrel{\widehat{=}} \ldots$$
$$SetIntsOn$$

Operations to read and set each register are defined now.

SetGPRegs

$$\Delta(hwgenregs)$$
$$regs? : GENREGSET$$

$$hwgenregs' = regs?$$

$$\begin{array}{|l}\hline \textit{GetGPRegs} \\ regs! : GENREGSET \\ \hline regs! = hwgenregs \\ \hline\end{array}$$

$$\begin{array}{|l}\hline \textit{GetStackReg} \\ stk! : PSTACK \\ \hline stk = hwstack \\ \hline\end{array}$$

$$\begin{array}{|l}\hline \textit{SetStackReg} \\ stk? : PSTACK \\ \hline hwstack' = stk? \\ \hline\end{array}$$

The *program counter* or *instruction pointer* can be read and set. The operations now follow.

$$\begin{array}{|l}\hline \textit{GetIP} \\ ip! : \mathbb{N} \\ \hline ip! = hwip \\ \hline\end{array}$$

$$\begin{array}{|l}\hline \textit{SetIP} \\ ip? : \mathbb{N} \\ \hline hwip' = ip? \\ \hline\end{array}$$

There is usually a status register provided by the processor. This register contains a collection of flags representing conditions such as result non-zero, result zero and so on. The operation to read the contents of this register is defined by the next schema:

$$\begin{array}{|l}\hline \textit{GetStatWd} \\ stwd! : STATUSWD \\ \hline hwstatwd' = stwd? \\ \hline\end{array}$$

The initialisation operation uses the constant value 0_S as a means to ensure that it is well-typed. The status register is always initialised by the hardware, not the software.

Interrupts play a large part in the current kernel. They are used as the basic mechanism for mutual exclusion. At the bottom of the kernel, they are used for this purpose, as well as in the definition of semaphores. There are two operations: one to switch off and one to switch on the interrupt notification mechanism. It is assumed that all interrupts can be disabled by the following schema:

$$\textit{SetIntsOff}$$
$$\textit{intflg}' = \mathsf{intoff}$$

Interrupts can be enabled by the following operation:

$$\textit{SetIntsOn}$$
$$\textit{intflg}' = \mathsf{inton}$$

Below, these operations will be aliased and referred to as the "locking" operations.

The following is left undefined. It is only included for reasons of completeness. The swapping procedure can involve the relocation of code and data; without relocation registers, it would not be possible to access and update data or to access code properly. The details of the relocation mechanism are of no interest to us in this specification.

$$\textit{UpdateRelocationRegisters}$$
$$\ldots$$

4.3.2 Queues

Queues are ubiquitous. In this specification, a distinction is made between general queues (of messages, requests and so on) and queues of process references. The former type is modelled using the generic *QUEUE[X]* type, while the other is modelled using the *ProcessQueue* type.

Type *QUEUE[X]* is the generic type encountered in Chapter 2. It is exactly the same as the queue used to define semaphores there.

$$\textit{QUEUE}\,[X]$$
$$\upharpoonright(\textit{INIT}, \textit{IsEmpty}, \textit{Enqueue}, \textit{RemoveNext}, \textit{QueueFront}, \textit{RemoveElement})$$

$$\textit{elts} : \operatorname{seq} X$$

$$\textit{INIT}$$
$$\textit{elts}' = \langle\,\rangle$$

$$\textit{IsEmpty} \mathrel{\widehat=} \ldots$$
$$\textit{Enqueue} \mathrel{\widehat=} \ldots$$
$$\textit{RemoveNext} \mathrel{\widehat=} \ldots$$
$$\textit{QueueFront} \mathrel{\widehat=} \ldots$$
$$\textit{RemoveElement} \mathrel{\widehat=} \ldots$$

Enqueue

$$\Delta(elts)$$
$$x? : X$$

$$elts' = elts \frown \langle x? \rangle$$

IsEmpty

$$elts = \langle\, \rangle$$

RemoveNext

$$\Delta(elts)x! : X$$

$$elts = \langle x! \rangle \frown elts'$$

QueueFront

$$x! : X$$

$$x! = head\ elts$$

RemoveElement

$$\Delta(elts)$$
$$x? : X$$

$$(\exists s_1, s_2 : \mathrm{seq}\, X \bullet elts = s_1 \frown \langle x? \rangle \frown s_2 \land elts' = s_1 \frown s_2)$$

4.3.3 Process Queue

The *ProcessQueue* type is used to model queues of processes. This type differs from the *QUEUE[X]* type in one critical respect: it is defined in terms of an *injective* sequence and not a simple sequence. Semantically, the distinction is that the range of an injective sequence can contain no duplicates (an injective sequence is an injective mapping from a subset of the naturals to the range set).

The reason for adopting an injective sequence is that if a process occurs in a queue, it can only occur once. It would have been tiresome to prove this invariant each time an enqueue operation was performed.

The actual details of the type definition are given immediately below. The class has the same operations as that of *QUEUE[X]*; in fact, it defines an extra one, which will be discussed after all the operations have been defined. The only significant difference between the two (apart from the fact that one is generic and the other is not) is that the underlying sequence type is different.

ProcessQueue

$\upharpoonright(INIT, IsEmpty, Enqueue, RemoveNext, QueueFront, Catenate, RemoveFirst)$

$\_ \otimes \_ : ProcessQueue \times ProcessQueue \rightarrow ProcessQueue$

$\forall q_1, q_2 : ProcessQueue \bullet$
$q_1 \otimes q_2 = q_1.elts \frown q_2.elts$

$elts : \text{iseq } X$

INIT

$elts' = \langle\rangle$

$IsEmpty \mathrel{\widehat=} \ldots$

$Enqueue \mathrel{\widehat=} \ldots$

$RemoveNext \mathrel{\widehat=} \ldots$

$QueueFront \mathrel{\widehat=} \ldots$

$RemoveFirst \mathrel{\widehat=} \ldots$

$Catenate \mathrel{\widehat=} \ldots$

As can be seen from the class definition, the methods are the same as for *QUEUE[X]*. The operations that follow are also the same. The properties of this type are the same as those proved in Chapter 2 for the general queue type; they are not repeated here.

It should be noted that the operation, $\otimes$, defined at the start of this class, cannot be exported. In order for the operation to be performed as an exportable operation, it has to be implemented as an operation, here with the rather unpleasant name *Catenate*. This restriction must be imposed because the $\otimes$ operation acts upon the internal representation of the queue, which is assumed hidden by the class construct.

Enqueue

$\Delta(elts)$
$x? : X$

$elts' = elts \frown \langle x? \rangle$

IsEmpty

$elts = \langle\rangle$

RemoveNext

$$\Delta(elts) x! : X$$

$$elts = \langle x! \rangle \frown elts'$$

QueueFront

$$x! : X$$

$$x! = head\ elts$$

RemoveFirst

$$\Delta(elts)$$

$$elts' = tail\ elts$$

RemoveElement

$$\Delta(elts)$$
$$x? : X$$

$$(\exists s_1, s_2 : \text{iseq } X \bullet elts = s_1 \frown \langle x? \rangle \frown s_2 \land elts' = s_1 \frown s_2)$$

The extra operation added to *ProcessQueue* is the *Catenate* (concatenate) operation. It is defined in terms of the operation $\otimes$ defined at the start of the *ProcessQueue* class. It is defined in order to export $\otimes$.

Catenate

$$q_1?, q_2? : ProcessQueue$$
$$q! : ProcessQueue$$

$$q! = q_1? \otimes q_2?$$

The operation concatenates two instances of *ProcessQueue* to produce another instance.

Because *ProcessQueue* is based on injective sequences, the following property is immediate (the proof is omitted):

Proposition 32. *The formulæ:*

$$\forall q_1, q_2 : ProcessQueue \bullet q_1 \otimes q_2$$

or:

$$\forall q_1, q_2 : ProcessQueue \bullet Catenate[q_1/q_1?, q_2/q_2?]$$

yield a queue with no duplicate elements.

In addition to this property, $\otimes$ (and, hence, *Catenate*) has the same properties as $\frown$, the concatenation operation for finite sequences.

4.3.4 Synchronisation and IPC

In this subsection, the synchronisation and Inter-Process Communication primitives employed by this kernel are specified.

The primary synchronisation primitive is the *lock*. There are two operations: *Lock* and *Unlock*. The requirement is that the operations between a *Lock* and the corresponding *Unlock* cannot be interrupted; in any case, they are executed by a *single* thread of control. The critical issue is that execution of the locked code cannot be interrupted. Locks will be used in defining semaphores below.

Locks are defined in terms of interrupts. This is a standard technique for implementing critical regions in OS kernels. It is quick to apply and easy to implement. The only problem is that the programmer has to remember to unlock a lock. (This can be solved by the judicious definition of a macro.) Locks must be used to implement other, higher-level synchronisation and IPC primitives if the hardware does not support a mutual-exclusion instruction like test-and-set; even when the hardware does support such an instruction, locks are often more appropriate for the reasons just given.

Lock

$\upharpoonright(Init, Lock, Unlock)$

$hw : HardwareRegisters$

Assume that registers have been initialised.

INIT

$hwrgs? : HardwareRegisters$

$hw' = hw?$

$Lock \mathrel{\widehat=} hw.SetIntsOff$

$Unlock \mathrel{\widehat=} hw.SetIntsOn$

Because this is an object-oriented specification, an instance of the hardware has to be passed to the *Lock* class. This is the purpose of the *INIT* operation. The other two operations perform locking and unlocking. The locking operation disables interrupts, while the unlocking one enables them again. Clearly, when interrupts are disabled, the thread of control executing the *lock* operation has sole access to the hardware and to the contents of the store; it cannot be interrupted. This permits the safe manipulation of shared data structures without the use of higher-level operations such as semaphore operations.

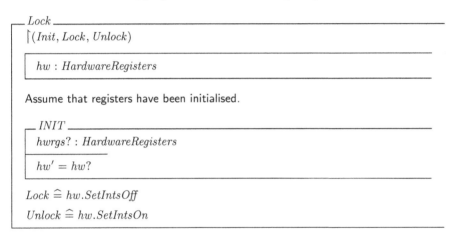

The reader should be aware that the locking method suffers from the problem that any process manipulation must be done by hand. Semaphores and messages are implemented in such a way that processes are automatically handled in the correct ways. Locks are defined at a lower level in the kernel (at a level below that at which the process abstraction has been defined).

Next, we come to the semaphores. The semaphore abstraction is defined in exactly the same way as in Chapter 2. As was noted there, the semaphore defined in this book can be used as a binary as well as a counting semaphore. The class is as follows.

Semaphore

$\upharpoonright(Init, Wait, Signal)$

$waiters : ProcessQueue_{©}$
$scnt, inital : \mathbb{Z}$
$ptab : ProcessTable$
$sched : LowLevelScheduler$
$ctxt : Context$
$lck : Lock$

INIT

$iv? : \mathbb{Z}$
$pt? : ProcessTable$
$sch? : LowLevelScheduler$
$ct? : Context$
$lk? : Lock$

$inital' = iv? \land scnt' = iv? \land ptab' = pt?$
$sched' = sch? \land ctxt' = ct? \land lck' = lk?$
$waiters.Init$

$NegativeSemaCount \mathrel{\widehat=} scnt < 0$

$NonPositiveSemaCount \mathrel{\widehat=} scnt \leq 0$

$IncSemaCount \mathrel{\widehat=} scnt' = scnt + 1$

$DecSemaCount \mathrel{\widehat=} scnt' = scnt - 1$

$Wait \mathrel{\widehat=} \ldots$

$Signal \mathrel{\widehat=} \ldots$

The *Wait* (or *P*) operation is defined first (some results will be proved before *Signal* is defined):

$$
\begin{array}{l}
Wait \mathrel{\widehat=} \\
\quad lck.Lock \mathbin{⨟} \\
\quad (DecSemaCnt \mathbin{⨟} \\
\qquad\qquad (NegativeSemaCount \land ctxt.SaveState \\
\qquad\qquad\quad \land\ waiters.Enqueue[currentp/x?] \land
\end{array}
$$

$$
\begin{array}{l}
\qquad\qquad (\exists\, cpd : ProcessDescr \bullet \\
\qquad\qquad\qquad ptab.DescrOfProcess[currentp/pid?, cpd/pd!] \wedge \\
\qquad\qquad\qquad cpd.SetProcessStatusToWaiting) \wedge \\
\qquad\qquad sched.MakeUnready[currentp/pid?] \wedge sched.ScheduleNext) \\
\quad \vee\ sched.ContinueCurrent) \fatsemi \\
\quad lck.Unlock
\end{array}
$$

(This is exactly the same as in Chapter 2 but is now written in Object-Z.)

Assuming a fair scheduling mechanism (e.g., round-robin) and the fairness of the queue abstraction, the major properties of the semaphore can be proved. The properties of *Wait* and *Signal* are almost symmetrical but the differences between them mean that properties of *Wait* cannot be used in the reverse direction to prove properties of *Signal.*

Lemma 1. $scnt < 0 \Rightarrow elts' = elts \frown \langle caller? \rangle$.

PROOF. Writing out the definition of *Wait*:

$$
\begin{array}{l}
scnt' = scnt - 1 \wedge \\
(scnt' < 0 \wedge \\
\qquad waiters.Enqueue[caller?/x?] \wedge \ldots
\end{array}
$$

This expands into:

$$
\begin{array}{l}
scnt' = scnt - 1 \wedge \\
\qquad scnt' < 0 \wedge \\
\qquad elts' = elts \frown \langle caller? \rangle \wedge \ldots
\end{array}
$$

Clearly, the caller's process is enqueued on the semaphore's queue of waiting processes. □

Lemma 2. $scnt < 0 \Rightarrow currentp \neq currentp'$.

PROOF. The proof of this reduces to the proof that $ScheduleNext \fatsemi RunNext$ implies $currentp \neq currentp'$. □

Lemma 3. $scnt \geq 0 \Rightarrow currentp = currentp'$.

PROOF. $scnt' \geq 0$ implies that the caller is resumed. □

Lemma 4. $scnt \geq 0 \Rightarrow elts' = elts$

PROOF. $scnt \geq 0$ implies $\neg\ NegativeSemaCount$ (i.e., $scnt \geq 0 \Rightarrow \neg\ (scnt < 0)$, so $currentp = currentp'$. Furthermore, $waiters.elts = waiters.elts'$. □

Lemma 5. $|\, scnt \,| - initval = \#elts$

PROOF. A semaphore can be initialised with an arbitrary integral value denoting the number of processes that can be simultaneously in the critical section. If this number is k, if there are more than k calls on *Wait*, say l calls, then $l - k$ of those calls will be blocked.

For the time being, and without loss of generality, let $k = 1$. By the definition of the semaphore type, $scnt' = k$ upon initialisation. (This is a binary semaphore.) Let this be denoted by $scnt_I$.

If m processes simultaneously wait on the same semaphore before the original process has performed a *Signal* operation and leave the critical section, by definition of *Wait*, for each process, *scnt* will be decremented by 1. The value of *scnt* will therefore be decremented by m. Therefore, $Wait^m \Rightarrow scnt' = scnt_I - (m + 1)$. Meanwhile, let $elts_I = \langle\,\rangle$ (the initialisation value of *elts*). It is clear that $Wait \vdash scnt' = scnt - 1$, so $Wait \vdash scnt = 0$. The enqueue operation is performed only when $scnt < 0$, and $Wait^m \vdash \#elts' = \#elts_I + m$, while $scnt = scnt_I + m = 1 + m$, so $\#elts = |\, m \,| + 1 - 1 = m = \#elts$. □

The last proof was written in terms of the initialisation values $scnt_I$ and $elts_I$. These are values that the semaphore attains whenever there are no processes in its waiting queues and no processes in its critical section—a quiesence or inactive semaphore; it is also the state of a semaphore immediately prior to use.

Lemma 6. *Wait implies:*

$$\begin{array}{l} Wait \vdash \forall p : APREF \mid p \in \operatorname{ran} elts \bullet \\ \quad \exists pd : ProcessDescr;\ s : PROCSTATUS \bullet \\ \quad\quad ptab.DescrOfProcess[p/pid?, pd/pd!] \\ \quad\quad \land\ pd.ProcessStatus[s/stat!] \\ \quad\quad \land\ s = \mathsf{pstwaiting} \end{array}$$

if $scnt' < 0$.

PROOF. The inner existential simplifies to:

$$ptab.procs(p).status = \mathsf{pstwaiting}$$

By the definition of *Wait*, if $scnt' < 0$, *Enqueue* is applied. By the definition of *Enqueue*, $elts' = elts \frown \langle p \rangle$, so $last\ elts' = \langle p \rangle$. *SetProcessStatus* sets p's status to pstwaiting.

However, the only way in which processes can enter *waiters.elts* is via a call to *Enqueue*, so *every* process in *waiters.elts* is in the pstwaiting state. □

Lemma 7. *If* $scnt \leq 0$, *then there are scnt elements in the semaphore's waiting queue. The status of each such element is* pstwaiting.

PROOF. By a previous Lemma (Lemma 5), $|\,scnt| = \#elts$. By Lemma 6 and induction, the result is immediate. □

The *Signal* (or V) operation is defined as:

$$
\begin{array}{l}
Signal \mathrel{\widehat=} \\
\quad lck.Lock \fatsemi \\
\quad\quad (IncSemaCnt \fatsemi \\
\quad\quad\quad (NonPositiveSemaCount \land \\
\quad\quad\quad\quad waiters.RemoveFirst[cand/x!] \land \\
\quad\quad\quad\quad (\exists\, cpd : ProcessDescr \bullet \\
\quad\quad\quad\quad\quad ptab.DescrOfProcess[cand/pid?, cpd/pd!] \land \\
\quad\quad\quad\quad\quad cpd.SetProcessStatusToReady) \land \\
\quad\quad\quad\quad sched.MakeReady[cand/pid?]) \setminus \{cand\} \\
\quad\quad\quad \lor\ sched.ContinueCurrent) \fatsemi \\
\quad lck.Unlock
\end{array}
$$

Lemma 8. $scnt \leq 0 \Rightarrow elts = \langle p \rangle \frown elts'$

PROOF. By the predicate of *Signal*:

$$
\begin{array}{l}
waiters.RemoveFirst[cand/x!] \\
\quad = waiters.elts = \langle cand \rangle \frown waiters.elts'
\end{array}
$$

That is, $elts = \langle cand \rangle \frown elts'$. □

Lemma 9. $scnt \leq 0 \Rightarrow \#elts' < \#elts$

PROOF. Again, if $scnt < 0$, $waiters.elts = \langle cand \rangle \frown waiters.elts'$. Rewriting (omitting class names), $elts = \langle cand \rangle \frown elts'$ is obtained. Taking sizes on both sides yields:

$$
\begin{array}{l}
\#elts = \#(\langle cand \rangle \frown elts') \\
\quad = \#\langle cand \rangle + \#elts' \\
\quad = 1 + \#elts'
\end{array}
$$

□

Lemma 10. $scnt \leq 0 \Rightarrow MakeReady$.

PROOF. By the predicate, $scnt \leq 0$ implies *MakeReady*. □

Lemma 11. $scnt \leq 0 \Rightarrow currentp' \neq currentp$

PROOF. In this case, *MakeReady* is the only operation related to scheduling. The effects of *MakeReady* do not include updating *currentp*, because nothing else in the predicate of *Signal* affects the scheduler's data variables. □

Lemma 12. $scnt > 0 \Rightarrow currentp' = currentp$.

PROOF. $scnt \leq 0$ is true when *NonPositiveSemaCount* is not true. From this, it can be inferred that $sched.ContinueCurrent \vdash currentp = currentp'$. □

Proposition 33. $Wait^n \fatsemi Signal^m \vdash elts = \langle\,\rangle$ *iff* $n = m$.

PROOF. Assume that *scnt* was initialised to 1. Assume that exactly one process is already in the critical section.

By Lemma 5, the next n callers to *Wait* extend *elts* by n elements. Conversely, by another lemma (Lemma 9), a call on *Signal* removes one element from *elts*, so n calls remove n elements. It is the case that $Dequeue^m \fatsemi Enqueue^n \vdash elts' = \langle\,\rangle$ iff $n = m$ (by Corollary 3). The process tha initially entered the critical region (calling *Wait*, therefore) must eventually exit it (assuming that meanwhile the process does not terminate abnormally). Therefore, while one process is inside the critical section, and there are n processes waiting to enter it, there must be m calls to *Signal* to restore the semaphore to its initial state; alternatively, for all processes to exit the critical section. □

Corollary 4. *If a semaphore is initialised to* k *and* $elts = \langle\,\rangle$,

$$Wait^{n+k} \fatsemi Signal^{m+k} \vdash elts' = \langle\,\rangle \text{ iff } n = m$$

PROOF. Immediate from previous results. □

Corollary 5. $Wait \fatsemi Signal$ *pairs are fair in the sense that any process performing a Wait operation will eventually enter the critical section and, by performing a* Signal, *will exit it also. When all Waits are matched by corresponding Signals, it is the case that* $\#elts = \#elts'$.

PROOF. For this proof, it is important to observe that the semaphore's operations are defined in terms of sequential compositions. The intermediate state of *elts* is denoted $elts''$, and the before and after states are written using the normal Z convention.

For $\#elts'' = \#elts + 1$, $scnt < 0$, so a process is in the critical section and the callers must have been enqueued.

Conversely, for $\#elts'' = \#elts' + 1$, $scnt \leq 0$, so a process must already have been made ready and, therefore, removed from $elts''$.

If $scnt = -k$ when a process is enqueued by *Wait*, there must be exactly k calls to *Signal* before the process can next be made ready. □

These results establish the correctness of the semaphore specification. By the assumption that the underlying scheduler is fair, the fairness of the semaphore is also established.

In fact, the scheduler used by this kernel is slightly more complex than a simple round-robin one. It is organised as a priority queue with three priorities, the lowest of which employs timesliced pre-emption. The lower two priorities use what amounts to a round-robin scheme (they contain system processes that either run to completion or suspend in a natural fashion—when they are readied, they are enqueued in FIFO fashion). This structure complicates the proof of fairness; because it is a fair scheme, by assumption the results above remain valid.

4.4 Process Management

This section contains the specification of those kernel components that implement the process abstraction.

It is first necessary to have some way of designating or referring to processes. With this defined, it will be possible to refer to individual processes without having to introduce references to all of the apparatus that implements the process abstraction. To do this, three types and two constants are defined.

The first type is:

$[PREF]$

This is the basic *P*rocess *R*eference type. There are two constants of this type:

$| \; NullProcRef, IdleProcRef : PREF$

The constant *NullProcRef* denotes the null process. Normally, the null process should never appear in any data structure; when it does, a significant error has occurred and the system should halt. For the reason that this model is typed, it is possible to prove that the null process has this property. This fact reduces the occurrence of the null process to an error in refinement and/or transcription or to some unforeseen event that should cause immediate termination of the kernel.

The second constant, *IdleProcRef*, denotes the *idle process*. This is the process that should be run when there is nothing else to do; it merely executes an infinite loop containing no instructions, absorbing cpu cycles until an external interrupt occurs or a new process is created and made ready for execution. It might be encoded in a programming language as:

```
while true do skip od;
```

If the scheduler's queue (queues in the present case) ever become empty (i.e., there are no processes ready to execute), the idle process is executed. This process is a way of ensuring that the scheduler always has something to do.

With these constants and *PREF* defined, it is possible to define two more types:

$$IPREF == PREF \setminus \{NullProcRef\}$$
$$APREF == IPREF \setminus \{IdleProcRef\}$$

The type *IPREF* is the type of all process references except the null process reference (*NullProcRef*). It is a type used by the scheduler. (It is also a useful type to have when refining the process management and scheduler specifications.) The type *APREF* is the type of *actual* processes. Type *APREF* contains the identifiers of all those processes that actually exist. (The idle process is, in some systems, a virtual process in the sense that it is implemented as a piece of kernel code.) In any case, the idle process cannot appear in semaphore or device queues; it can *only* appear in the scheduler and in a few other special places, as will be seen. The *APREF* type is used, then, to denote processes that can wait for devices, wait on semaphores, request alarms from the clock, and so on. System processes, except the idle process, and all user processes are denoted by an element of *APREF*.

Without loss of generality, the types *PREF*, *IPREF* and *APREF* can be given a more concrete representation:

$$maxprocs : \mathbb{N}$$

There must be an *a priori* limit to the number of processes that the system can handle. This constant denotes that limit. The constant is used to limit the number of entries in the process table.

The constants *NullProcRef* and *IdleProcRef* can be defined (again) as constants:

$$NullProcRef : PREF$$
$$IdleProcRef : IPREF$$

$$NullProcRef = 0$$
$$IdleProcRef = maxprocs$$

This representation is intended to bracket the actual process references so that simple tests can be performed to determine legality.

The following definitions are of the two other process reference types in view of the values denoting the null and idle processes:

$$IPREF == 1 \mathrel{..} maxprocs$$
$$APREF == 1 \mathrel{..} maxprocs - 1$$

These definitions will, in addition, make process identifier generation simpler and easier to understand.

The state of each process must be recorded in the corresponding process table entry. The next type denotes the possible states a process can be in. It contains designations for the obvious states (the prefix pst denotes *Process STate*; constants will usually be printed in sans serif font):

- pstnew: the state of a newly created process that has not yet been readied (not all of its necessary resources have yet been allocated);
- pstrunning: the state of a process that is currently executing;
- pstready: the state of a process that can run but is not currently executing;
- pstwaiting: the state of a process that is waiting for a device, semaphore, etc.
- pstterm: the state of a process that has terminated and is waiting to be deleted (it might still own resources that are to be deallocated).

In addition, processes can have the following additional states:

- pstswappedout: the state of a process whose stored image (code, data and stack) is currently on the swapper disk and not in main store;
- pstzombie: the state of a process that is waiting to terminate but is prevented from doing so because not all its children have terminated, (The children still require at least the code segment of the parent to remain accessible so that they can continue execution.)

(The use of these last two states will become clearer when the swapper and the zombie scheme are modelled in Section 4.4 of this chapter.)

It is an invariant of this kernel that each process can be in *exactly one* of the above states at any one time.

The type representing process states is Z free type and is defined as follows:

$$
\begin{aligned}
PROCSTATUS ::= &\ \mathsf{pstnew} \\
\mid &\ \mathsf{pstrunning} \\
\mid &\ \mathsf{pstready} \\
\mid &\ \mathsf{pstwaiting} \\
\mid &\ \mathsf{pstswappedout} \\
\mid &\ \mathsf{pstzombie} \\
\mid &\ \mathsf{pstterm}
\end{aligned}
$$

Processes fall into natural kinds: system, user and device processes. The kind of process matters as far as resource allocation, scheduling and swapping are concerned. The process kind is an attribute that is set when each process is created; it is constant thereafter. In the type declaration that follows, the pt prefix is denotes *Process Table*:

$$
\begin{aligned}
PROCESSKIND ::= &\ \mathsf{ptsysproc} \\
\mid &\ \mathsf{ptuserproc} \\
\mid &\ \mathsf{ptdevproc}
\end{aligned}
$$

In this kernel, all device drivers are assigned a kind of ptdevproc.

Processes have various data areas associated with them. In the kernel modelled in this chapter, a process can have a stack, a data area and a code area; other kernels might allow a separate heap area. The store allocated to each of these areas is modelled as a storage descriptor.

The types *PSTACK*, *PCODE* and *PDATA* are required by the kernel modelled in this chapter. They are defined as synonym types as follows:

$PSTACK == MEMDESC$
$PCODE == MEMDESC$
$PDATA == MEMDESC$

The descriptor types are defined as synonyms for the *MEMDESC* type, a type describing regions of store (it is defined as an address-limit pair in Section 4.6 of this chapter). For the time being, these three types can be considered to have values that are atomic.

Next, there is the type modelling the process descriptor. In this chapter, the process descriptor is an extension of that defined in the previous chapter. As is standard, there is one process descriptor per actual process in the system and, possibly, as the idle process, as well.

The process descriptor, in this model, contains a representation of the hardware state (general registers, status word, instruction pointer, stack descriptor), as well as a descriptor for each of its code and data areas; the *memsize* slot is a descriptor holding the base address and size of the storage area allocated to this process. (It is assumed that the storage is allocated in one contiguous segment—this makes storage management much easier than allocating in discontinuous regions.) It also has a slot for the process status at the last transition and one for its kind (system, device or user process). In addition, the process descriptor has slots for the process' time quantum and scheduling level (in effect, its priority).

The operations defined for this type are basically concerned with setting and accessing the values stored in its slots. There is little need to comment on them. The two interesting operations deal with the *process context* and comments will be made after their definition.

ProcessDescr

$\upharpoonright(INIT, ProcessStatus, SetProcessStatusToNew,$
$SetProcessStatusToTerminated, SetProcessStatusToReady,$
$SetProcessStatusToRunning, SetProcessStatusToWaiting,$
$SetProcessStatusToSwappedOut, SetProcessStatusToZombie,$
$ProcessKind, SetProcessKindToDevice,$
$SetProcessKindToSystem, SetProcessKindToUserProc,$
$WaitingFor, SetWaitingType, SchedulingLevel,$
$BlocksProcesses, AddBlockedProcesses,$
$RemoveBlockedProcess, ClearBlockedProcesses,$

$TimeQuantum, SetTimeQuantum,$
$StoreSize, StoreDescr, SetStoreDescr, FullContext,$
$SetFullContext)$

$status : PROCSTATUS$
$kind : PROCESSKIND$
$schedlev : SCHDLVL$
$regs : GENREGSET$
$time\_quantum : TIME$
$statwd : STATUSWD$
$ip : \mathbb{N}$
$stack : PSTACK$
$data : PDATA$
$code : PCODE$
$mem : MEMDESC$
$memsize : \mathbb{N}$

$INIT$

$stat? : PROCSTATUS$
$knd? : PROCESSKIND$
$slev? : SCHDLVL$
$tq? : TIME$
$pstack? : PSTACK$
$pdata? : PDATA$
$pcode? : PCODE$
$mem? : MEMDESC$
$msz? : \mathbb{N}$

$status' = stat?$
$kind' = knd?$
$schedlev' = slev?$
$regs.INIT$
$time\_quantum' = tq?$
$statwd' = 0_S$
$ip' = 0$
$data' = pdata?$
$code' = pcode?$
$stack' = pstack?$
$mem' = mem?$
$memsize' = msz?$

$ProcessStatus \mathrel{\widehat{=}} \ldots$

$SetProcessStatusToNew \mathrel{\widehat{=}} \ldots$

$SetProcessStatusToTerminated \mathrel{\widehat{=}} \ldots$

$SetProcessStatusToReady \mathrel{\widehat=} \ldots$

$SetProcessStatusToRunning \mathrel{\widehat=} \ldots$

$SetProcessStatusToWaiting \mathrel{\widehat=} \ldots$

$SetProcessStatusToSwappedOut \mathrel{\widehat=} \ldots$

$SetProcessStatusToZombie \mathrel{\widehat=} \ldots$

$ProcessKind \mathrel{\widehat=} \ldots$

$SetProcessKindToDevice \mathrel{\widehat=} \ldots$

$SetProcessKindToSystem \mathrel{\widehat=} \ldots$

$SetProcessKindToUserProc \mathrel{\widehat=} \ldots$

$WaitingFor \mathrel{\widehat=} \ldots$

$SetWaitingType \mathrel{\widehat=} \ldots$

$SchedulingLevel \mathrel{\widehat=} \ldots$

$BlocksProcesses \mathrel{\widehat=} \ldots$

$AddBlockedProcesses \mathrel{\widehat=} \ldots$

$AddBlockedProcess \mathrel{\widehat=} \ldots$

$RemoveBlockedProcess \mathrel{\widehat=} \ldots$

$ClearBlockedProcesses \mathrel{\widehat=} \ldots$

$TimeQuantum \mathrel{\widehat=} \ldots$

$SetTimeQuantum \mathrel{\widehat=} \ldots$

$StoreSize \mathrel{\widehat=} \ldots$

$StoreDescr \mathrel{\widehat=} \ldots$

$SetStoreDescr \mathrel{\widehat=} \ldots$

$FullContext \mathrel{\widehat=} \ldots$

$SetFullContext \mathrel{\widehat=} \ldots$

ProcessStatus

$st! : PROCSTATUS$

$st! = status$

SetProcessStatusToNew

$\Delta(status)$

$status' = \mathsf{pstnew}$

SetProcessStatusToTerminated

$\Delta(status)$

$status' = \mathsf{pstterm}$

SetProcessStatusToReady

$\Delta(status)$

$status' = \mathsf{pstready}$

SetProcessStatusToRunning

$\Delta(status)$

$status' = \mathsf{pstrunning}$

SetProcessStatusToWaiting

$\Delta(status)$

$status' = \mathsf{pstwaiting}$

SetProcessStatusToSwappedOut

$\Delta(status)$

$status' = \mathsf{pstswappedout}$

SetProcessStatusToZombie

$\Delta(status)$

$status' = \mathsf{pstzombie}$

ProcessKind

$knd! : PROCESSKIND$

$knd! = kind$

SetProcessKindToDevProc

$\Delta(pkind)$

$kind' = \mathsf{ptdevproc}$

SetProcessKindToSysProc

$$\Delta(pkind)$$

$$kind' = \mathsf{ptsysproc}$$

SetProcessKindToUserProc

$$\Delta(pkind)$$

$$kind' = \mathsf{ptuserproc}$$

SchedulingLevel

$$sl! : SCHDLVL$$

$$sl! = schedlev$$

BlocksProcesses

$$bw! : \mathbb{F}\,APREF$$

$$bw! = blockswaiting$$

AddBlockedProcesses

$$\Delta(blockswaiting)$$
$$bs? : \mathbb{F}\,APREF$$

$$blockswaiting' = blockswaiting \cup bs?$$

AddBlockedProcess

$$\Delta(blockswaiting)$$
$$b? : APREF$$

$$blockswaiting' = blockswaiting \cup \{b?\}$$

RemoveBlockedProcess

$$\Delta(blockswaiting)$$
$$b? : APREF$$

$$blockswaiting' = blockswaiting \setminus \{b?\}$$

ClearBlockedProcesses

$$\Delta(blockswaiting)$$

$$blockswaiting' = \varnothing$$

TimeQuantum

$tq! : TIME$

$tq! = time\_quantum$

SetTimeQuantum

$tq? : TIME$

$time\_quantum' = tq?$

StoreSize

$memsz! : \mathbb{N}$

$memsize = memsz!$

StoreDescr

$memdescr! : MEMDESC$

$memdescr! = mem$

This is the descriptor containing the base address of the process' storage area, together with its length. It is set (by the next operation) when the process is first allocated and reset whenever it is swapped out and back in again.

SetStoreDescr

$\Delta(pmem, pmemsize)$
$newmem? : MEMDESC$

$mem' = newmem?$
$memsize' = hole\_size(newmem?)$

The next two operations are worthy of comment. The first is used to store the current hardware context (general registers, instruction pointer, status word and stack pointer—here modelled simply as the entire descriptor) when the process is suspended by an interrupt or by pre-emption. In addition to the hardware context, the operation also stores the value of the current time quantum from the scheduler if the process is at user priority.

FullContext

$pregs! : GENREGSET$
$pip! : \mathbb{N}$
$ptq! : TIME$
$pstatwd! : STATUSWD$
$pstack! : PSTACK$

$pregs! = regs$

$pip! = ip$
$ptq! = time\_quantum$
$pstatwd! = statwd$
$pstack! = stack$

The *SetFullContext* operation is used to restore the process' context to the hardware and also the time quantum value. It is called when a process is executed.

SetFullContext

$pregs? : GENREGSET$
$pip? : \mathbb{N}$
$ptq? : TIME$
$pstatwd? : STATUSWD$
$pstack? : PSTACK$

$regs' = pregs?$
$ip' = pip?$
$time\_quantum' = ptq?$
$statwd' = pstatwd?$
$stack' = pstack?$

The two context-manipulating operations are used to suspend and restore the process. Suspension can be performed by an ISR or by waiting on a semaphore. Resumption occurs when the scheduler selects the process as the one to run next.

It should be noted that these schemata operate *only* on the process descriptor. The actual context switch is performed by generic ISR code or by semaphores.

Next comes the process table. This is basically a table of process descriptors. When a process is created, a new entry in the process table is allocated if possible; if it is not possible, creation of the process fails. The *AddProcess* operation sets a new process' data in the table. When a process is to be removed (when it has terminated), it is removed from the table by *DelProcess*. The descriptor of a process is obtained from the table by the *DescrOfProcess* operation. (In a full model, if the designated process does not exist, an error would be signalled.)

The remaining operations included in the table fall into three categories:

1. The creation of the idle process. In this model, the idle process has an entry in the process table. Its descriptor can be used to store hardware context not otherwise catered for (e.g., the context of the kernel itself).
2. Handling of child processes.
3. Handling of zombie processes.

The last two classes of operation will be described in more detail when the process hierarchy is described and modelled and when termination is considered in detail.

The reason for including the last two classes of information in the process table is that they relate processes rather than describing individual processes; the process table collects all the necessary information in one place and this seemed rather better than scattering it in different places in the model. In any case, the process table deals with sets of processes, not single ones; child processes and zombies clearly deal with sets of processes, so there is a real semantic reason for the inclusion of this information here.

The organisation of the table is similar to the generic one presented in Chapter 2. There are slight differences between the structures; for example, the generic table is organised around a partial mapping, $\nrightarrow$, while the process table uses a partial injection ($\rightarrowtail\!\!\!\!\!\nrightarrow$). The appropriate results proved for the generic table transfer with only minor changes to the case of the process table.

In addition, it is worth pointing out that eac process descriptor in the process table is annotated with ©. This is an Object-Z symbol denoting the fact that the annotated entity (here the process descriptors in the table) is private to the class containing it. (This has the implication that there are no reference relations to be taken into account when manipulating or reasoning about process descriptors.)

The class exports the operations that are to be used by other components of the kernel. Of particular interest is the operation to retrieve a process descriptor for a given process (*DescrOfProcess*—this is an operation that will be particularly common. There are also operations that do not relate to single processes but to collections of related processes, for example those operating on descendant processes. The class exports a number of operations for the association (and disassociation) of child processes with their parents. For example, *AddChildOfProcess* adds a child process' identifier to a structure that relates it to its parent. The existence of child processes also requires the establishment of who owns the code of any particular process. If a process has no children, it owns its code. If, on the other hand, a process has created a child process, the child then shares its parent's code; this fact must be recorded.

There are also operations dealing with so-called zombies. These are processes that have almost terminated but cannot release their storage because they have children that have not yet terminated.

ProcessTable

$\upharpoonright(INIT, CreateIdleProcess, IsKnownProcess, AddProcess, DelProcess,$
$DescrOfProcess, AddCodeOwner, DelCodeOwner, ProcessHasChildren,$
$AddChildOfProcess, DelChildOfProcess, AllDescendants, IsCodeOwner,$
$AddProcessToZombies, MakeZombieProcess, ProcessIsZombie,$
$RemoveAllZombies, KillAllZombies, ProcessHasParent, ParentOfProcess,$
$AddProcessToTable, CanGenPId, NewPId, RemoveProcessFromParent)$

$$
\begin{array}{l}
procs : IPREF \rightarrowtail\!\!\!\!\!\!\!\rightarrow ProcessDescr_{\copyright} \\
known\_procs : \mathbb{F}\, IPREF \\
freeids, zombies, code\_owners : \mathbb{F}\, APREF \\
parent : APREF \twoheadrightarrow\!\!\!\!\!\!+ APREF \\
children, blockswaiting : APREF \twoheadrightarrow\!\!\!\!\!\!+ \mathbb{F}\, APREF \\
childof, parentof, share\_code : APREF \leftrightarrow APREF \\
\hline
(\forall p : APREF \bullet p \in freeids \Leftrightarrow p \notin known\_procs) \\
known\_procs = \mathrm{dom}\, procs \land zombies \subset known\_procs \\
\mathrm{dom}\, children \subseteq known\_procs \land \mathrm{dom}\, childof \subseteq known\_procs \\
\mathrm{ran}\, childof \subseteq known\_procs \land \mathrm{ran}\, childof = \mathrm{ran}\, parent \\
childof^{\sim} = parentof \land code\_owners \subseteq \mathrm{dom}\, parentof \\
(\forall p_1, p_2 : APREF \bullet \\
\quad p_1 \in \mathrm{dom}\, blockswaiting \land \\
\quad p_2 \in blockswaiting(p_1) \Rightarrow \\
\quad\quad (p_1 \in code\_owners \lor parentof^{+}(p_1, p_2)))
\end{array}
$$

$$
\begin{array}{l}
INIT \\
\hline
known\_procs' = \{IdleProcRef\} \\
freeids' = 1 \ldots maxprocs - 1 \\
code\_owners' = \{IdleProcRef\} \\
\mathrm{dom}\, shares\_code' = \varnothing \\
\mathrm{dom}\, childof' = \varnothing \\
\mathrm{dom}\, blockswaiting' = \varnothing \\
zombies' = \varnothing
\end{array}
$$

$CreateIdleProcess \mathrel{\widehat{=}} \ldots$

$IsKnownProcess \mathrel{\widehat{=}} \ldots$

$AddProcess \mathrel{\widehat{=}} \ldots$

$DelProcess \mathrel{\widehat{=}} \ldots$

$DescrOfProcess \mathrel{\widehat{=}} \ldots$

$AddCodeOwner \mathrel{\widehat{=}} \ldots$

$DelCodeOwner \mathrel{\widehat{=}} \ldots$

$ProcessHasChildren \mathrel{\widehat{=}} \ldots$

$AddChildOfProcess \mathrel{\widehat{=}} \ldots$

$DelChildOfProcess \mathrel{\widehat{=}} \ldots$

$IsCodeOwner \mathrel{\widehat{=}} \ldots$

$AddProcessToZombies \mathrel{\widehat{=}} \ldots$

$MakeZombieProcess \mathrel{\widehat{=}} \ldots$

$ProcessIsZombie \mathrel{\widehat{=}} \ldots$

$RemoveAllZombies \mathrel{\widehat=} \ldots$

$KillAllZombies \mathrel{\widehat=} \ldots$

$GotZombies \mathrel{\widehat=} \ldots$

$ProcessHasParent \mathrel{\widehat=} \ldots$

$RemoveProcessFromParent \mathrel{\widehat=} \ldots$

$ParentOfProcess \mathrel{\widehat=} \ldots$

$CanGenPId \mathrel{\widehat=} \ldots$

$NewPId \mathrel{\widehat=} \ldots$

$releasePId \mathrel{\widehat=} \ldots$

$AddProcessToTable \mathrel{\widehat=} \ldots$

$deleteProcessFromTable \mathrel{\widehat=} \ldots$

The following operation creates the idle process. The operation sets up basic data about the idle process, including the status (it will be a ready process, so can be immediately considered by the scheduler) and process kind (it is a system process); the operation assigns an arbitrary time quantum to the process (∞) and its status word is cleared to 0_s. Next, the storage areas are created; the idle process does not have any storage (since it does nothing other than loop), so anything can be assigned (here, empty storage descriptors are assigned). Then, the idle process' process descriptor is created by calling the *Init* method belonging to its type and the descriptor is stored in the process table.

CreateIdleProcess

$$
\begin{array}{l}
(\exists\, stat : PROCSTATUS;\ knd : PROCESSKIND;\ schdlv : SCHEDLVL; \\
\qquad tq : TIME;\ stwd : STATUSWD;\ emptymem : MEMDESC; \\
\qquad stkdesc : MEMDESC;\ memsz : \mathbb{N};\ ipd : ProcessDescr \bullet \\
\quad stat = \mathsf{pstready} \\
\quad \land\ knd = \mathsf{ptuserproc} \land schdlv = userq \\
\quad \land\ tq = \infty \land stwd = 0_s \land emptymem = (0, 0) \\
\quad \land\ stkdesc = (0, 0) \land memsz = 0 \\
\quad \land\ ipd.INIT[stat/stat?, knd/knd?, schdlv/slev?, tq/tq?, \\
\qquad\qquad stkdesc/pstack?, emptymem/pdata?, \\
\qquad\qquad emptymem/pcode?, emptymem/mem?, memsz/msz?] \\
\quad procs' = procs \oplus \{IdleProcRef \mapsto ipd\})
\end{array}
$$

It is necessary to generate new process identifiers. The following three operations are for this purpose. There is a maximum size associated with the process table: there can, at any time, be a maximum of *maxprocs* processes in the table. This is done in this model by manipulating a set of identifiers, as follows. When an identifier is in this set, it is considered available for use by a new process; when it is not in this set, it is considered to be the identifier of a process in the process table.

The following schema defines a predicate determining whether there are any process names that are free. The set *freeids* contains all those identifiers that have not been assigned to a process.

CanGenPId

$$freeids \neq \varnothing$$

NewPId

$$\Delta(freeids)$$
$$p! : APREF$$

$$(\exists\, p : APREF \bullet \\ \quad p \in freeids \\ \quad \wedge\ p! = p \\ \quad \wedge\ freeids' = freeids \setminus \{p\})$$

The *NewPId* operation returns a new process identifier. The predicate can be simplified, obtaining:

NewPId

$$\Delta(freeids) p! : APREF$$

$$p! \in freeids$$
$$freeids' = freeids \setminus \{p!\}$$

The operation selects an element of *freeids* at random, removes it from *freeids* and returns it as the next process identifier for use.

releasePId

$$\Delta(freeids)$$
$$p? : APREF$$

$$freeids' = freeids \cup \{p?\}$$

This operation returns an identifier to the free pool of identifiers. The identifier, denoted by *p*?, is added to *freeids* and, therefore, can no longer be used as the identifier of a process in the process table, as the class invariant states.

The *IsKnownProcess* operation is a predicate which tests whether a given identifier (*pid*?) is in the set *known_procs*, the set of known process identifiers. By the invariant of the class, an identifier is an element of *known_procs* if and only if it is not a member of *freeids*. Therefore, every member of *known_procs* is the identifier of a process in the process table.

$$\begin{array}{|l}
\multicolumn{1}{l}{__ \textit{IsKnownProcess} ____} \\
pid? : APREF \\
\hline
pid? \in known_procs \\
\hline
\end{array}$$

Process descriptors are added to the process table by the following operation. In a full model, it would be an error to attempt to add a descriptor that is already present in the table or to use an identifier that is in *freeids*. For present purposes, the following suffices:

$$\begin{array}{|l}
\multicolumn{1}{l}{__ \textit{AddProcessToTable} ____} \\
\Delta(procs) \\
pid? : APREF \\
pd? : ProcessDescr \\
\hline
procs' = procs \oplus \{pid? \mapsto pd?\} \\
\hline
\end{array}$$

This is an operation local to the *ProcessTable*. The public operation is the following:

$$\begin{array}{l}
AddProcess \mathrel{\widehat=} \\
\quad newPId \mathbin{;} addProcessToTable
\end{array}$$

The addition of the identifier generator implies that there is no need to check the validity of the new process' identifier (the fact that identifiers are unique and not in the table should be proved as a property of the model, as is done below, after the process table's operations have been defined).

Removal of a process descriptor from the process table is performed by the following local operation:

$$\begin{array}{|l}
\multicolumn{1}{l}{__ \textit{deleteProcessFromTable} ____} \\
\Delta(procs) \\
pid? : APREF \\
\hline
procs' = \{pid?\} \mathbin{⩤} procs \\
\hline
\end{array}$$

The deleted process' identifier is removed from the domain of the *procs* mapping using the domain subtraction operator $⩤$.

$$DelProcess \mathrel{\widehat=} deleteProcessFromTable \mathbin{;} releasePId$$

The public deletion operation also needs to ensure that the identifier of the deleted process is released (i.e., is added to *freeids*).

Throughout the model, access to each process' process descriptor is required. The following schema defines this operation. A fuller model, particularly one intended for refinement, would include an error schema to handle the case in which the process identifier, *pid?*, does not denote a value element of the *procs* domain (i.e., is not a member of *known_procs*).

$$
\begin{array}{l}
\textit{DescrOfProcess} \\
\hline
pid? : IPREF \\
pd! : ProcessDescr \\
\hline
pd! = procs(pid?)
\end{array}
$$

Methods for children and zombies now follow.

The owner of a code segment is recorded here. This allows the system to determine which process owns any segment of code when swapping occurs.

$$
\begin{array}{l}
\textit{AddCodeOwner} \\
\hline
\Delta(code\_owners) \\
p? : APREF \\
\hline
code\_owners' = code\_owners \cup \{p?\}
\end{array}
$$

$$
\begin{array}{l}
\textit{DelCodeOwner} \\
\hline
\Delta(code\_owners) \\
p? : APREF \\
\hline
code\_owners' = code\_owners \setminus \{p?\}
\end{array}
$$

Shared code is important when swapping is concerned. Since there can be many sharers of any particular process' code, it appears best to represent code sharing as a relation.

The following operation declares the process *owner?* as the owner of a code segment, while *sharer?* denotes a process that shares *owner?*'s code. For every such relation, there must be an instance in the *code_owners* relation in the process table.

$$
\begin{array}{l}
\textit{AddCodeSharer} \\
\hline
\Delta(code\_owners) \\
owner?, sharer? : APREF \\
\hline
code\_owners' = code\_owners \cup \{(owner?, sharer?)\}
\end{array}
$$

$$
\begin{array}{l}
\textit{DelCodeSharer} \\
\hline
\Delta(code\_owners) \\
owner?, sharer? : APREF \\
\hline
code\_owners' = code\_owners \setminus \{(owner?, sharer?)\}
\end{array}
$$

Zombies and sharing depend upon the process hierarchy. This is expressed in terms of parent and child processes. The hierarchy is most easily represented as a relation that associates a parent with its children. The following few operations handle the *childof* relation, which represents the process hierarchy in this model.

ProcessHasChildren

$$p? : APREF$$

$$\exists\, c : APREF \bullet childof(c, p?)$$

AddChildOfProcess

$$\Delta(childof)$$
$$parent?, child? : APREF$$

$$childof' = childof \cup (child?, parent?)$$

DelChildOfProcess

$$\Delta(childof)$$
$$parent?, child? : APREF$$

$$childof' = childof \setminus (child?, parent?)$$

AllDescendants

$$descs! : \mathbb{F}\, APREF$$
$$parent? : APREF$$

$$descs! = childof^{+}(\!|\, \{p?\} \,|\!)$$

When a process has children in this model, the children processes all share the parent's code. If the parent is swapped out, its code segment is transferred to backing store and, as a consequence, is no longer addressable by the child processes. Because of this, it is necessary to block (i.e., suspend) all child processes when their parent is swapped out.

The following schema is satisfied when the process, $p?$, owns the code it executes. Code-owning processes tend not to be descendants of other processes.

IsCodeOwner

$$p? : APREF$$

$$p? \in code\_owners$$

The operations for handling zombie processes now follow. Zombies are relatively easy to represent and manipulate in this model.

$$
\begin{array}{|l}
\multicolumn{1}{l}{\underline{\quad AddProcessToZombies \quad\quad\quad\quad\quad\quad\quad\quad\quad\quad\quad\quad}}\\
\Delta(zombies)\\
pid? : APREF\\
\hline
zombies' = zombies \cup \{pid?\}\\
\hline
\end{array}
$$

$$
\begin{array}{l}
MakeZombieProcess \mathrel{\widehat=}\\
\quad AddProcessToZombies \land\\
\quad\quad SetProcessStatusToZombie
\end{array}
$$

$$
\begin{array}{|l}
\multicolumn{1}{l}{\underline{\quad ProcessIsZombie \quad\quad\quad\quad\quad\quad\quad\quad\quad\quad\quad\quad}}\\
pid? : APREF\\
\hline
pid? \in zombies\\
\hline
\end{array}
$$

Operation *RemoveAllZombies* removes those processes from the *children* relation that are related to the zombie process *zmb*.

$$
\begin{array}{|l}
\multicolumn{1}{l}{\underline{\quad RemoveAllZombies \quad\quad\quad\quad\quad\quad\quad\quad\quad\quad\quad\quad}}\\
\Delta(parent, children, zombies)\\
deadzombs! : \mathbb{F}\, APREF\\
\hline
\exists\, zmbs : \mathbb{F}\, APREF \mid zmbs \subseteq zombies \land deadzombs! = zmbs \bullet\\
\quad zombies' = zombies \setminus zmbs\\
\quad \land (\forall\, zmb : APREF \mid zmb \in zombies \land children(zmb) = \varnothing \bullet\\
\quad\quad parent' = \{zmb\} \mathbin{⩤} parent\\
\quad\quad \land (\exists\, p : APREF;\ descs : \mathbb{F}\, APREF \mid\\
\quad\quad\quad\quad p = parent(zmb) \land descs = children(p) \bullet\\
\quad\quad\quad children' = children \oplus \{p \mapsto descs \setminus \{zmb\}\})\\
\hline
\end{array}
$$

When this operation is used, each *zmb* has no children. It must be removed from the *parent* table and it has to be removed as a child of its parent.

The *KillAllZombies* operation is defined as follows:

$$
\begin{array}{l}
KillAllZombies \mathrel{\widehat=}\\
\quad (RemoveAllZombies[dzombs/deadzombies!] \land\\
\quad\quad (\forall\, zmb : APREF \mid zmb \in dzombs \bullet\\
\quad\quad\quad DelProcess[zmb/p?])) \setminus \{dzombs\}
\end{array}
$$

This operation is performed on a periodic basic. It is called from the clock process.

The following schema defines a predicate that is true if the set of zombies contains at least one element. This operation is used in the clock driver.

$$
\begin{array}{|l}
\multicolumn{1}{l}{\underline{\quad GotZombies \quad\quad\quad\quad\quad\quad\quad\quad\quad\quad\quad\quad}}\\
zombies \neq \varnothing\\
\hline
\end{array}
$$

ProcessHasParent

$p? : APREF$

$(\exists p_1 : APREF \bullet$
$\quad parentof(p_1, p?))$

RemoveProcessFromParent

$\Delta(parentof)$
$parent?, child? : APREF$

$parentof' = parentof \setminus \{(parent?, child?)\}$

ParentOfProcess

$p? : APREF$
$parent! : APREF$

$(\exists p_1 : APREF \bullet$
$\quad parentof(p_1, p?) \land parent! = p_1)$

Note that the initialisation of the system should include a call to the operation that creates the idle process.

The definition of the *ProcessTable* class is now complete. It is now possible to state and prove some properties of this class.

Proposition 34. *The identifier of (reference to) the idle process, IdleProcRef, is unique.*

PROOF. *IdleProcRef* = *maxprocs*. The result follows by the uniqueness of natural numbers. □

Proposition 35. *The idle process is unique.*

PROOF. Each process is represented by:

(i) a unique identifier (its reference);
(ii) a single entry in the process table.

For (ii), $procs : IPREF \twoheadrightarrow PD$ since *procs* is a function:

$$procs(x) = procs(y) \Rightarrow x = y$$

Therefore, by Proposition 48, the idle process descriptor is unique. □

Proposition 36. *The identifier* NullProcRef *never appears in the process table.*

PROOF. The domain of $procs$ is $IPREF$ and $IPREF \subset PREF$. $NullProcRef \in PREF = 0 \mathinner{\ldotp\ldotp} maxprocs$, while $IPREF = 1 \mathinner{\ldotp\ldotp} maxprocs$. Since $NullProcRef = 0$, it is an element of $PREF$ but not of $IPREF$. □

Proposition 37. $\forall p : APREF \bullet p \in freeids \Leftrightarrow p \notin known\_procs$.

PROOF. This is a conjunct of the invariant. □

Proposition 38. $NewPId[p/p!] \Rightarrow p \in known\_procs'$.

PROOF.

NewPId

$\Delta(freeids)$
$p! : APREF$

$p! \in freeids$
$freeids' = freeids \setminus \{p!\}$

By the invariant:

$p \in freeids \Leftrightarrow p \notin known\_procs$

By propositional calculus:

$p \notin freeids \Leftrightarrow p \in known\_procs$

So, if $p \in freeids \Leftrightarrow p \notin known\_procs$,

$$
\begin{aligned}
& freeids' \\
& \quad = freeids \setminus \{p\} \\
& \quad = known\_procs \cup \{p\} \\
& \quad = known\_procs'
\end{aligned}
$$

Therefore, $p \in known\_procs'$. □

Proposition 39. $NewPId^n \Rightarrow freeids' = \varnothing$ *if* $n = maxprocs - 1$.

PROOF. The *NewPId* operation is:

NewPId

$\Delta(freeids)$
$p! : APREF$

$p! \in freeids$
$freeids' = freeids \setminus \{p!\}$

Initially, $freeids = 1 \mathrel{..} maxprocs - 1$, so $\#freeids = maxprocs - 1$.

Now, $NewPId^n = \overbrace{(NewPId \mathbin{;} \ldots NewPId)}^{n \text{ times}}$. From the definition of $NewPId$, it can be seen that $\#freeids' = \#freeids - 1$. Therefore, $newPId^n \Rightarrow \#freeids' = \#freeids - n$.

If $n = maxprocs - 1$, it is clear that $NewPId^n$ implies that:

$$\begin{aligned}&\#freeids'\\ &\quad = \#freeids - (maxprocs - 1)\\ &\quad = (maxprocs - 1) - (maxprocs - 1)\\ &\quad = 0\end{aligned}$$

So $freeids' = \varnothing$. □

Proposition 40. $NewPId \vdash \#freeids' = \#freeids - 1$.

PROOF. By the definition of $NewPId$, $p! \in freeids \wedge freeids' = freeids \setminus \{p!\}$. Therefore:

$$\begin{aligned}&\#freeids'\\ &\quad = \#(freeids \setminus \{p!\}\\ &\quad = \#freeids - \#\{p!\}\\ &\quad = \#freeids - 1\end{aligned}$$

□

Proposition 41. $DelProcess \vdash \#freeids' = \#freeids + 1$.

PROOF. The definition of $DelProcess$ is:

$$deleteProcessFromTable \mathbin{;} releasePId$$

The important conjunct is $releasePId$, whose predicate is:

$$freeids' = freeids \cup \{p?\}$$

The result is immediate:

$$\begin{aligned}&\#freeids'\\ &\quad = \#(freeids \cup \{p?\})\\ &\quad = \#freeids + \#\{p?\}\\ &\quad = \#freeids + 1\end{aligned}$$

□

Corollary 6. $deleteProcessFromTable \vdash p \notin known\_procs'$.

PROOF. By the definition of *deleteProcessFromTable*, $procs' = \{pid?\} \ndres procs$ and $known\_procs = \mathrm{dom}\, procs$. The predicate implies that $pid? \notin \mathrm{dom}\, procs'$, which, in turn, implies that $pid? \notin known\_procs$. □

Proposition 42. *If $p \in known\_procs$ and $p_1 \neq p$, the substitution instance of schema deleteProcessFromTable$[p_1/pid?]$ implies that $p \in known\_procs'$.*

PROOF. By the definition of *deleteProcessFromTable*:

$$procs' = \{p_1\} \ndres procs$$

The invariant states that $\mathrm{dom}\, procs = known\_procs$. Therefore:

$$\begin{aligned} &\mathrm{dom}\, procs' \\ &\quad = \mathrm{dom}(\{p_1\} \ndres procs) \\ &\quad = (\mathrm{dom}\, procs) \setminus \{p_1\} \\ &\quad = known\_procs \setminus \{p_1\} \\ &\quad = known\_procs' \end{aligned}$$

However, by assumption, $p \neq p_1$, so $p \in known\_procs$. □

Proposition 43. *Using the definition of $NewPId^n$ above, the composition $NewPId^n \mathbin{⨾} DelProcess^m$ implies $\#freeids = \#freeids'$ iff $n = m$.*

PROOF. The proof of this proposition requires the following (obvious) lemmata.

Lemma 13. *If $\#freeids = n$, $NewPId^n$ implies $\#freeids = 0$.*

PROOF. Since $NewPId \Rightarrow \#freeids' = \#freeids - 1$, then, by induction, for all n, $n \geq \#freeids$, $NewPId^n$ implies that $\#freeids' = \#freeids - n$. □

Lemma 14. *If $\#freeids = n$, DelProcess implies that $\#freeids' = n + 1$.*

PROOF. Immediate from the fact that *releasePId* implies that $\#freeids' = \#freeids + 1$. □

Lemma 15. *If $\#freeids = 0$, then $DelProcess^n$ implies that $\#freeids' = n$.*

PROOF. By induction, using Proposition 14. □

The proof of Proposition 43 follows immediately from the three lemmata. □

Proposition 44. $\neg$ *CanGenPId* $\Rightarrow \neg$ *NewPId*.

PROOF. The operations are defined by the following schemata:

$$
\begin{array}{|l}
\hline
\textit{CanGenPId} \\
\textit{freeids} \neq \varnothing \\
\hline
\end{array}
$$

and:

$$
\begin{array}{|l}
\hline
\textit{NewPId} \\
\Delta(\textit{freeids}) \\
p! : \textit{APREF} \\
\hline
p! \in \textit{freeids} \\
\textit{freeids}' = \textit{freeids} \setminus \{p!\} \\
\hline
\end{array}
$$

Given

$$
\begin{array}{|l}
\hline
\neg\ \textit{canGenPId} \\
\textit{freeids} = \varnothing \\
\hline
\end{array}
$$

it is clear that there can be no p s.t. $p! \in \textit{freeds}$. □

Proposition 45. *$\textit{NewPId}^n \mathbin{⨾} \textit{releasePId}^m \Rightarrow \textit{freeids}' = \varnothing$ iff $m = n$.*

PROOF. Immediate consequence of Proposition 43. □

Proposition 46. *There can be no $p \in \textit{APREF}$ such that $p \in \textit{freeids}$ and $p \in \textit{known\_procs}$.*

PROOF. By the invariant, $p \in \textit{freeids} \Leftrightarrow p \notin \textit{known\_procs}$, for all p. Using propositional calculus, the following can be derived:

1. If $p \in \textit{freeids}$, then $p \notin \textit{known\_procs}$.
2. If $p \in \textit{known\_procs}$, then $p \notin \textit{freeids}$.

Therefore, $\neg\ \exists\, p : \textit{PREF} \bullet p \in \textit{freeids} \land p \in \textit{known\_procs}$. □

Proposition 47. *$\textit{NewPId}[p_1/p!] \mathbin{⨾} \textit{NewPId}[p_2/p!] \Rightarrow p_1 \neq p_2$.*

PROOF. The schema for *NewPId* is:

$$
\begin{array}{|l}
\hline
\textit{NewPId} \\
\Delta(\textit{freeids}) \\
p! : \textit{APREF} \\
\hline
p! \in \textit{freeids} \\
\textit{freeids}' = \textit{freeids} \setminus \{p!\} \\
\hline
\end{array}
$$

By the definition of $\fatsemi$, $NewPId[p_1/p!] \fatsemi NewPId[p_2/p!]$ is:

$$\exists freeids'' : \mathbb{F}\, APREF \bullet \\ \quad p_1 \in freeids \land freeids'' = freeids \setminus \{p_1\} \\ \quad \land\, p_2 \in freeids'' \land freeids' = freeids'' \setminus \{p_2\}$$

This simplifies to:

$$p_1 \in freeids \land p_2 \in freeids \setminus \{p_1\} \\ \quad \land\, freeids' = (freeids \setminus \{p_1\}) \setminus \{p_2\}$$

This is clearly equivalent to:

$$p_1 \in freeids \land p_2 \in freeids \setminus \{p_1\} \\ \quad \land\, freeids' = freeids \setminus \{p_1, p_2\}$$

For $p_2 \in freeids \setminus \{p_1\}$ to be the case, $p_1 \neq p_2$, for the reason that $p \notin freeids\{p\}$ for any p. □

Proposition 48.

$$deleteProcessFromTable \Rightarrow known\_procs' = known\_procs \setminus \{pid?\}$$

PROOF.

deleteProcessFromTable
$\Delta(procs)$
$pid? : AREF$
$procs' = \{pid?\} \ntriangleleft procs$

By the invariant, $known\_procs = \operatorname{dom} procs$, so:

$$\operatorname{dom} procs' \\ \quad = \operatorname{dom}(\{pid?\} \ntriangleleft procs) \\ \quad = (\operatorname{dom} procs) \setminus \{pid?\} \\ \quad = known\_procs \setminus \{pid?\} \\ \quad = known\_procs'$$

□

4.5 The Scheduler

The scheduler is based on a three-level priority scheme. The basic type is:

$$SCHDLVL == 1 \,..\, 3$$

This type is used to identify queues of waiting processes. The queues are identified by the following constants:

$$userqueue, sysprocqueue, devprocqueue : SCHDLVL$$

$$userqueue = 3$$
$$sysprocqueue = 2$$
$$devprocqueue = 1$$

The first constant, *userqueue*, denotes the queue of user processes; the second constant, *sysprocqueue*, denotes the queue of system processes; and the third, *devprocqueue*, denotes the queue of device processes.

Device processes must always be preferred to other processes, so the constant *devprocqueue* denotes the queue of highest-priority processes. System processes must be preferred by the scheduler to user processes but should be pre-empted by device processes, so *sysprocqueue* denotes the next highest priority. The constant *userqueue* denotes the queue of user processes; they have lowest priority.

In addition, there is the idle process (denoted by the constant *IdleProcRef*) which runs when there is nothing else to do. Strictly speaking, the idle process has the lowest priority but is considered a special case by the scheduler (see the schema for *ScheduleNext*), so it appears in none of the scheduler's queues. (In the process table, the idle process is assigned to the user-process priority—this is just a value that is assigned to avoid an unassigned attribute in its process descriptor.)

The ordering on the priorities is:

$$devprocqueue < sysprocqueue < userqueue$$

This property will be exploited in the definition of the scheduler.

$$queuelevel : PROCESSKIND \to SCHDLVL$$

$$\begin{array}{l}\forall pt : PROCESSKIND \bullet \\ \quad (\exists l : SCHDLVL \bullet \\ \qquad queuelevel(pt) = \\ \qquad\qquad (pt = \mathsf{ptdevdrvr} \land l = devprocqueue) \\ \qquad\qquad \lor (pt = \mathsf{ptsysproc} \land l = sysprocqueue) \\ \qquad\qquad \lor (pt = \mathsf{ptuserproc} \land l = userqueue))\end{array}$$

As noted above, type *ProcessQueue* is not defined in terms of $QUEUE[X]$ but defined separately. This is because all elements of a process queue are unique (i.e., there are no duplicates), so the basic type iseq is used rather than *seq*.

The class *Context* is defined so that process context manipulation can be made simpler. The class accesses the process table, the scheduler (to be defined shortly) and the hardware registers. The class defines five operations.

Context

$\upharpoonright(INIT, SaveState, RestoreState, SwapIn, SwapOut, SwitchContext)$

$ptab : ProcessTable$
$sched : LowLevelScheduler$
$hw : HardwareRegisters$

INIT

$ptb? : ProcessTable$
$shd? : LowLevelScheduler$
$hwregs? : HardwareRegisters$

$ptab' = ProcessTable$
$sched' = LowLevelScheduler$
$hw' = hwregs?$

$SaveState \mathrel{\widehat=} \ldots$

$RestoreState \mathrel{\widehat=} \ldots$

$SwapIn \mathrel{\widehat=} \ldots$

$SwapOut \mathrel{\widehat=} \ldots$

$SwitchContext \mathrel{\widehat=} \ldots$

When an interrupt occurs, *SaveState* is called to save the state. Then, the device-specific stuff is executed (this might involve calling *SendInterruptMsg*). Finally, the *RestoreState* method is called to perform a context switch.

SaveState

$(\exists\, cp : IPREF \bullet$
$\quad sched.CurrentProcess[cp/cp!]$
$\quad (\exists\, pd : ProcessDescr \bullet$
$\qquad ptab.DescrOfProcess[cp/pid?, pd/pd!]$
$\qquad \land (\exists\, regs : GENREGSET;\ stk : PSTACK;\ ip : \mathbb{N};$
$\qquad\qquad stat : STATUSWD;\ tq : TIME \bullet$
$\qquad\quad hw.GetGPRegs[regs/regs!]$
$\qquad\quad \land\ hw.GetStackReg[stk/stk!]$
$\qquad\quad \land\ hw.GetIP[ip/ip!]$
$\qquad\quad \land\ hw.GetStatWd[stat/stwd!]$
$\qquad\quad \land\ sched.GetTimeQuantum[tq/tquant!]$
$\qquad\quad \land\ pd.SetFullContext[regs/pregs?, ip/pip?,$
$\qquad\qquad stat/pstatwd?, stk/pstack?, tq/ptq?])))$

The current process referred to here is *not necessarily* the same as the one referred to above. Basically, whatever is in (bound to) *currentp* runs next.

RestoreState

$$
\begin{array}{l}
(\exists\, cp : IPREF \bullet \\
\quad sched.CurrentProcess[cp/cp!] \\
\quad \land (\exists\, pd : ProcessDescr \bullet \\
\quad\quad ptab.DescrOfProcess[cp/pid?, pd/pd!] \\
\quad\quad \land (\exists\, regs : GENREGSET;\ stk : PSTACK;\ ip : \mathbb{N}; \\
\quad\quad\quad\quad stat : STATUSWD;\ tq : TIME \bullet \\
\quad\quad\quad pd.FullContext[regs/pregs!, ip/pip!, stat/pstatwd!, \\
\quad\quad\quad\quad stk/pstack!, tq/ptq!] \\
\quad\quad\quad \land hw.SetGPRegs[regs/regs?] \\
\quad\quad\quad \land hw.SetStackReg[stk/stk?] \\
\quad\quad\quad \land hw.SetStatWd[stat/stwd?] \\
\quad\quad\quad \land sched.SetTimeQuantum[tq/tquant?] \\
\quad\quad\quad \land hw.SetIP[ip/ip?])))
\end{array}
$$

For completeness, we define the *SwapOut* and *SwapIn* operations (although they are not used in this book): they are intended to be mutually inverse.

$$
\begin{array}{l}
SwapOut \mathrel{\widehat=} \\
\quad (\exists\, cp : IPREF;\ pd : ProcessDescr \bullet \\
\quad\quad sched.CurrentProcess[cp/cp!] \\
\quad\quad \land ptab.DescrOfProcess[pd/pd!] \land pd.SetProcessStatusToWaiting) \\
\quad \land SaveState \fatsemi \\
\quad (sched.MakeUnready[currentp/pid?] \land sched.ScheduleNext)
\end{array}
$$

The *SwapOut* operation uses *SaveState* and alters the status of the process concerned. The process is removed from the ready queue and a reschedule is performed, altering the current process.

$$
\begin{array}{l}
SwapIn \mathrel{\widehat=} \\
\quad (\exists\, cp : IPREF;\ pd : ProcessDescr \bullet \\
\quad\quad sched.CurrentProcess[cp/cp!] \land pd.SetProcessStatusToRunning \\
\quad\quad RestoreState)
\end{array}
$$

The *SwapIn* operation just performs simple operations on the process' descriptor and then switches the process' registers onto the hardware and makes it the current process (i.e., the currently running process).

$$SwitchContext \mathrel{\widehat=} SwapOut \fatsemi SwapIn$$

This is a combination operation that swaps out the current process, schedules and executes the next one.

The organisation of the scheduler implies that the idle process *must* be represented by a descriptor in the process table. This is for a number of reasons, including the need for somewhere to store the kernel's context.

The kernel's scheduler is called the *LowLevelScheduler*. It is defined as follows.

The scheduler has three queues (*readyqs*), one each for device, system and user process (in that order). It is worth noting that the process queues are all contained in the class. This scheme is a *multi-level* priority queue.

The currently executing process is represented by *currentp*. The time quantum of the current process is represented by *currentquant* (if it is a user-level process). For all processes, the priority is represented by *currentplev*.

The component *prevp* refers to the process that executed immediately before the one currently denoted by *currentp*.

As already observed, *readyqs* is an array of queues (represented by a bijection) and *nqueues* is the number of queues in *readyqs* (is the cardinality of *readyqs*' domain).

The scheduler is defined at a level in the kernel that is *below* that at which semaphores are defined. For this reason, it is important to find another way to obtain exclusive access to various data structures (e.g., the process table, a particular process descriptor, the hardware registers or the scheduler's own queues). At the level at which the scheduler is defined, the only way to do this in the present kernel is to employ locking. For this reason, the class initialises itself with an instance of *Lock*.

LowLevelScheduler

$\upharpoonright(INIT, GetTimeQuantum,$
$SetTimeQuantum, RunIdleProcess,$
$CurrentProcess, MakeReady,$
$UpdateProcessQuantum, MakeUnready,$
$ContinueCurrent, \quad ScheduleNext)$

$currentp : IPREF$
$currentquant : TIME$
$currentplev : SCHDLVL$
$prevp : IPREF$
$nqueues : \mathbb{N}_1$
$readyqs : SCHDLVL \rightarrowtail\!\!\!\!\twoheadrightarrow ProcessQueue_{©}$
$lck : Lock$
$ctxt : Context$
$proctab : ProcessTable$
$hw : HardwareRegisters$

$nqueues = 3$
$\#readyqs = nqueues$

INIT

$lk? : Lock$
$ptb? : ProcessTable$
$hwrs? : HardwareRegisters$

$lck' = lk?$
$proctab' = ptb?$
$currentp' = NullProcRef$
$currentquant' = 0$
$currentplev' = userqueue$
$prevp' = NullProcRef$
$prevplev' = 1 \land hw' = hwregs?$

$GetTimeQuantum \mathrel{\widehat=} \ldots$

$SetTimeQuantum \mathrel{\widehat=} \ldots$

$RunIdleProcess \mathrel{\widehat=} \ldots$

$CurrentProcess \mathrel{\widehat=} \ldots$

$MakeReady \mathrel{\widehat=} \ldots$

$UpdateProcessQuantum \mathrel{\widehat=} \ldots$

$MakeUnready \mathrel{\widehat=} \ldots$

$reloadCurrent \mathrel{\widehat=} \ldots$

$ContinueCurrent \mathrel{\widehat=} \ldots$

$ScheduleNext \mathrel{\widehat=} \ldots$

$runTooLong \mathrel{\widehat=} \ldots$

$allEmptyQueues \mathrel{\widehat=} \ldots$

$selectNext \mathrel{\widehat=} \ldots$

The operations defined for the scheduler can now be described.

User processes are associated with a time quantum. This is used to determine when a user process should be removed from the processor if it has not been blocked by other means. The current time quantum must be copied to and from the current process' descriptor in the process table. It must be possible to assign *currentquant* to a value. The following pair of operations specify these operations.

The first returns the value stored in the time quantum variable. This value represents the time quantum that remains for the current process.

GetTimeQuantum

$tquant! : TIME$

$currentquant = tquant!$

The second operation sets the value of the current process' time quantum; this operation is only used when the current process is a user-defined one. When a user process is not executing, its time quantum for a process is stored in its process descriptor.

$$
\begin{array}{|l}
\hline
\textit{SetTimeQuantum} \\
\textit{tquant?} : \textit{TIME} \\
\hline
(\exists \textit{pd} : \textit{ProcessDescr}; \ \textit{lv} : \textit{SCHDLVL} \bullet \\
\quad \textit{proctab.DescrOfProcess}[\textit{currentp}/\textit{pid?}, \textit{pd}/\textit{pd!}] \\
\quad \land \textit{pd.ProcessLevel}[\textit{lv}/\textit{lev!}] \\
\quad \land ((\textit{lv} = \textit{userqueue} \land \textit{currentquant}' = \textit{currentquant} - \textit{tquant?}) \\
\qquad \lor \textit{currentquant}' = 0)) \\
\hline
\end{array}
$$

When there are no other processes to execute, the idle process is run.

$$
\begin{array}{|l}
\hline
\textit{RunIdleProcess} \\
\Delta(\textit{currentp}) \\
\hline
\textit{currentp}' = \textit{IdleProcRef} \\
\textit{billp}' = \textit{IdleProcRef} \\
\hline
\end{array}
$$

This schema defines the operation to select the idle process as the next process to run. Note that it does not switch to the idle process' context because the code that calls this will perform that operation by default.

When a process is swapped off the processor, its identifier must be retrieved from *currentp*. The following schema defines that operation:

$$
\begin{array}{|l}
\hline
\textit{CurrentProcess} \\
\textit{cp!} : \textit{IPREF} \\
\hline
\textit{cp!} = \textit{currentp} \\
\hline
\end{array}
$$

With the *MakeReady*, we have come to the core set of scheduler operations. This operation adds the process named by *pid?* to the ready queue at the appropriate priority level.

The *MakeReady* operation first locks everything else out of the processor. It then obtains the process descriptor for the process referred to by *pid?*. The priority of this process is extracted from its process descriptor and the process is enqueued on the appropriate queue. (Actually, the *pid?*—a reference to the process—is enqueued.) Finally, the operation unlocks the processor.

$$
\begin{array}{|l}
\hline
\textit{MakeReady} \\
\Delta(readyqueues) \\
pid? : IPREF \\
\hline
lck.Lock \fatsemi \\
(\exists\, pd : ProcessDescr;\ lv : SCHDLVL \bullet \\
\quad proctab.DescrOfProcess[pd/pd!] \\
\quad \land\ pd.ProcessLevel[lv/lev!] \\
\quad \land\ pd.SetProcessStatusToReady \\
\quad \land\ (readyqueues(lv).Enqueue[pid?/x?] \\
\quad \land\ lck.Unlock) \\
\hline
\end{array}
$$

Proposition 49. *For any process, p, at priority level, l, MakeReady[p/pid?] implies that:*

$$\#readyqueues'(l) = \#readyqueues(l) + 1$$

Proof. The critical line in the predicate is:

$$readyqueue(l).Enqueue[p/x?]$$

The definition of *Enqueue*, after substituting p for $x?$, is:

$$elts' = elts \frown \langle p \rangle$$

So:

$$
\begin{array}{l}
\#elts' = \\
\quad \#(elts \frown \langle p \rangle) = \\
\quad \#elts + \#\langle p \rangle = \\
\quad \#elts + 1
\end{array}
$$

□

Operation *UpdateProcessQuantum* updates the time quantum in a process descriptor, provided that the process is a user-level one. The quantum (stored in *currentquant*) is decremented by one to denote the fact that the process has just executed.

$$
\begin{array}{|l}
\hline
\textit{UpdateProcessQuantum} \\
(\exists\, pd : ProcessDescr;\ lv : SCHDLVL \bullet \\
\quad proctab.DescrOfProcess[currentp/pid?, pd/pd!] \\
\quad \land\ pd.ProcessLevel[lv/lev!] \\
\quad \land\ ((lv = userqueue \\
\qquad \land\ currentquant' = currentquant - 1 \\
\qquad \land\ ((currentquant' \leq minpquantum \land runTooLong) \\
\qquad \lor\ (currentquant' > minpquantum \land ContinueCurrent))) \\
\qquad \lor\ Skip)) \\
\hline
\end{array}
$$

This operation deals with the case in which a user process has run for too long a period. Its operation is very much as one might expect.

$$
\begin{array}{l}
\textit{runTooLong} \\
\hline
(\exists p : APREF \bullet \\
\quad readyqueues(userqueue).RemoveFirst[p/x!] ⨾ \\
\quad (readyqueues(userqueue).Enqueue[p/x?] ⨾ \\
\quad\quad (prevp' = currentp \land ScheduleNext)))
\end{array}
$$

This is called by the clock driver (Section 4.6.4) to cause pre-emption of the current user process.

Proposition 50. *UpdateProcessQuantum implies that if* currentp*'s priority is userqueue and currentquant* $>$ *minpquantum, then* $currentp' = currentp$.

Proof. By the predicate, if $currentquant' > minpquantum$, *ContinueCurrent* is executed. The predicate of *ContinueCurrent* contains $currentp' = currentp$ as a conjunct. □

Proposition 51. *If there are no processes of higher priority in the scheduler's ready queue, UpdateProcessQuantum implies that* $currentp' = currentp$ *if the user-level queue only contains process* p.

Proof. Let $readyqueues(userqueue) = \langle p \rangle$. The predicate of *runTooLong* is (in shortened form): *RemoveFirst* ⨾ *Enqueue*. This implies that $elts = elts'$ if $elts = \langle p \rangle$.

The sequential composition expands into:

$$
\begin{array}{l}
\exists\, elts'' : \mathrm{iseq} APREF;\ x! : APREF \bullet \\
\quad x! = head\ elts \\
\quad \land\ elts'' = tail\ elts \\
\quad \land\ elts' = elts'' \frown \langle x! \rangle
\end{array}
$$

This simplifies to $elts' = (tail\ elts) \frown \langle head\ elts \rangle$. If $elts = \langle p \rangle$, $tail\ elts = \langle\,\rangle$, and:

$$
\begin{array}{l}
elts' \\
\quad = (tail\ elts) \frown \langle head\ elts \rangle \\
\quad = \langle\,\rangle \frown \langle head\ elts \rangle \\
\quad = \langle\,\rangle \frown (head \langle x \rangle) \\
\quad = \langle\,\rangle \frown \langle x \rangle \\
\quad = \langle x \rangle \\
\quad = elts
\end{array}
$$

By *selectNext*, $currentp' = p$ since *selectNext* contains *QueueFront* as a conjunct and *QueueFront* is defined in terms of *head elts*. □

Proposition 52. *Assuming there are no processes of higher priority in the ready queue, if the level of the executing process is userqueue and*

$$currentquant \leq minpquantum$$

then

$$currentp' \neq head\ readyqueues(userqueue)$$

if $\#readyqueues(userqueue) > 1$.

PROOF. Assume that the queue $readyqueues(userqueue)$ is of length greater than 1. Then $head\ elts = currentp$ and $head\ tail\ elts \neq currentp$. By the composition $RemoveFirst \mathbin{;} Enqueue$, $elts' = (tail\ elts) \frown \langle currentp \rangle$. Since there are no duplicates in $elts$, this implies that $currentp' = head\ tail\ elts' \neq currentp$. □

The *MakeUnready* operation is another key operation. Its intent is to remove the process denoted by *pid?* from the ready queue. What happens to the process thereafter is a matter for the caller of *MakeUnready*. The *MakeUnready* operation does not manipulate the context of the victim process because it is not clear what that state is; instead, it just removes the process from the queue. If the process is the head of its queue, it is removed and a rescheduling operation occurs; otherwise, the process is just removed.

The operation removes any process reference that is in the queue. The reference can be the head of the queue or somewhere inside the queue. What happens to the reference that is removed is a matter for the user of this operation. Typically, the process reference is enqueued on another queue (e.g., a device queue or the clock). Because what is to happen to the removed process cannot be determined, the *MakeUnready* schema contains no reference to operations manipulating the representation of the process' state in the process descriptor. (The reader should note that *RemoveElement* will also remove the head of the queue—the schema is written to be as clear as possible.)

MakeUnready

$pid? : APREF$

$$
\begin{array}{l}
(\exists q : ProcessQueue;\ pd : ProcessDescr;\ lv : SCHDLVL \bullet \\
\quad proctab.DescrOfProcess[pd/pd!] \\
\quad \land pd.ProcessLevel[lv/lev!] \land q = readyqueues(lv) \\
\quad \land (\neg\ q.IsEmpty \\
\quad\quad \land (\exists f : APREF \bullet \\
\quad\quad\quad q.QueueFront[f/x!] \\
\quad\quad\quad \land (f = pid? \land (q.RemoveFirst \mathbin{;} \\
\quad\quad\quad\quad\quad q.selectNext[q/q?, lv/lev?])) \\
\quad\quad\quad \lor q.RemoveElement[pid?/x?])))
\end{array}
$$

Proposition 53. *If a process, p, has priority l, and is in the ready queue, then MakeUnready[p/pid?] implies that:*

$$\#readyqueues'(l) = \#readyqueues(l) - 1$$

PROOF. There are two cases to consider: in the first, p is at the head of the queue and in the other, p is not at the head.

In both cases, let $readyqueues(l) = elts$.

Case 1. In the predicate, $f = p$ or $p = head\ elts$, so $elts = \langle p \rangle \frown elts'$. Then:

$$\begin{array}{l} \#elts = \\ \quad \#(\langle p \rangle \frown elts') = \\ \quad \#\langle p \rangle + \#elts' = \\ \quad 1 + \#elts' \end{array}$$

So $\#elts' = \#elts - 1$.

Case 2. $p \neq head\ elts$. Therefore:

$$\begin{array}{l} \exists\, s_1, s_2 : \mathrm{iseq}\ APREF \bullet \\ \quad elts = s_1 \frown \langle p \rangle \frown s_2 \wedge \\ \quad elts' = s_1 \frown s_2 \end{array}$$

Therefore:

$$\begin{array}{l} \#elts = \\ \quad \#(s_1 \frown \langle p \rangle \frown s_2) = \\ \quad \#s_1 + \#\langle p \rangle + \#s_2 = \\ \quad \#s_1 + 1 + \#s_2 = \\ \quad 1 + \#s_1 + \#s_2 = \\ \quad 1 + \#elts' \end{array}$$

□

Proposition 54. *Let q be readyqueues(p_r), where p_r is the priority of process, p. If p is an element of q, then MakeUnready[p/pid?] implies that p is not an element of q'.*

PROOF. There are two cases to consider:

Case 1. Process p is the head of q. The appropriate conjunct of *MakeUnready* is:

$$\begin{array}{l} q.QueueFront[f/x!] \\ \wedge\ (f = pid? \wedge q.RemoveFirst) \end{array}$$

The predicate of *RemoveFirst* is:

$$elts' = tail\ elts$$

which is equivalent to:

$$elts = \langle p \rangle \frown elts'$$

from which it is clear that p is not an element of $elts'$; hence, p cannot be an element of q'.
Case 2. Process p is not the head element of q. Therefore, in *MakeUnready*, $f \neq pid?$, so $q.RemoveElement[pid?/x?]$ is required. The *RemoveElement*'s predicate is:

$$\exists\, s_1, s_2 : \text{iseq}\ APREF \bullet \\ \quad s_1 \frown \langle pid? \rangle \frown s_2 = elts \\ \quad \land\ s_1 \frown s_2 = elts'$$

It is again immediate that p cannot be an element of $elts'$ and, hence, not of q'. □

reloadCurrent

$$\Delta(currentp, prevp)$$

$$currentp' = currentp \\ prevp' = prevp$$

$$ContinueCurrent \mathrel{\widehat=} \\ \quad reloadCurrent \land ctxt.RestoreState$$

This operation is really just the identity on the scheduler's state. The intent is that the current process' execution is continued after a possible rescheduling operation. If the rescheduling operation determines that the same process be continued, the operation specified by this schema is performed.

The scheduler needs to determine whether all of its queues are empty. The following schema defines this test:

allEmptyQueues

$$\forall\, i : SCHDLVL \bullet \\ \quad readyqueues(i).IsEmpty$$

If all the queues are empty, the idle process must be run.

The *selectNext* operation is a central part of the scheduler.

$$\begin{array}{l}
\textit{selectNext} \\
\hline
(\exists q : ProcessQueue \bullet \\
\quad q = readyqueues(devprocqueue) \\
\qquad\quad \otimes readyqueues(sysprocqueue) \\
\qquad\qquad\quad \otimes readyqueues(userqueue) \\
\quad \land (\exists p : APREF;\ l : SCHDLVL;\ pd : ProcessDescr \bullet \\
\qquad q.QueueFront[p/x!] \\
\qquad \land\ proctab.DescrOfProcess[p/pid?, pd/pd!] \\
\qquad \land\ pd.SchedulingLevel[l/sl!] \\
\qquad \land\ prevp' = currentp \\
\qquad \land\ currentp' = p \\
\qquad \land\ currentplev' = l))
\end{array}$$

The schema first concatenates the three queues so that the head can be determined. This is licensed by the fact that, for any sequence, injective or not, $\langle\rangle \frown q = q$. The process at the head of the queue is determined and its priority is obtained from the process descriptor. The current and previous process variables are updated, as is the record of the current process' priority. The priority, *currentplev*, is only of significance when it is *userqueue*: in this case, pre-emption using time quanta is employed.

Proposition 55. *If:*

$$\begin{array}{l}
q = readyqueues(devprocqueue) \\
\quad \otimes readyqueues(sysprocqueue) \\
\qquad \otimes readyqueues(userqueue)
\end{array}$$

then $0 \leq \#q \leq maxprocs - 1$.

PROOF. The element type of $readyqueue(i)$, $i : SCHDLVL$, is $APREF$. There can be a maximum of $maxprocs - 1$ elements in $APREF$. If *all* processes in the system are readied, they will be in exactly one of the queues, depending upon priority. Since q is the concatenation of all priority queues, its maximum length is therefore $maxprocs - 1$.

If, on the other hand, there are no ready processes (and the idle process is the next to run), *readyqueues* is empty for all $i : SCHDLVL$, so $\#q = 0$. □

Finally, we reach *ScheduleNext*. This is the scheduling operation.

$$\begin{array}{l}
ScheduleNext \mathrel{\widehat=} \\
\quad (allEmptyQueues \land RunIdleProcess) \\
\quad \lor selectNext
\end{array}$$

If all scheduler queues are empty, the idle process is executed; otherwise, another process is selected for execution. It should be noted that this operation *must* always be executed in a locked context: it is called either in an ISR or in the body of a semaphore.

Time quantum manipulation and pre-emption are performed by the clock process. The appropriate operations are defined there (see Section 4.6.3).

The scheduler's data structures and operations have now been defined. It is now time to prove some of the more important properties of the scheduler as defined above.

Proposition 56. *If a process, p, is of priority l, operation MakeReady[p/pid?] enqueues p on the queue for level l.*

PROOF. Expanding *MakeReady*[*p*/*pid*?], we obtain:

$$
\begin{array}{l}
pd.procs(p) \\
\quad \wedge\ pd.SchedulingLevel[l/sl!] \\
\quad \wedge\ q = readyqueues(l) \\
\quad \wedge\ q.elts' = q.elts \frown \langle p \rangle
\end{array}
$$

This is equivalent to:

$$readyqueues(procs(p).lev).elts' = elts \frown \langle p \rangle$$

The mapping *readyqueues* is a bijection, so its range elements are uniquely determined by those of its domain. □

Proposition 57. *If a process, p, is of priority l, all other queues are unaffected by the execution of MakeReady[p/pid?].*

PROOF. In the schema for *MakeReady*, the important conjunct is:

$$readyqueues(lv).Enqueue[pid?/x?]$$

where *lv* is the priority of the process and *pid*? its identifier. This expands into:

$$readyqueues(lv).elts' = readyqueues(lv).elts \frown \langle pid \rangle$$

Since *readyqueues* is a bijection, *readyqueues*(*lv*) uniquely determines *elts*. Therefore, only one queue is affected by *MakeReady*. □

Proposition 58. *If a process, p, is of priority l, all other queues are unaffected by the execution of MakeUnready[p/pid?].*

PROOF. This follows from the fact that *readyqueues* is a bijection. The core is:

$$
\begin{array}{l}
q = readyqueues(lv) \\
\wedge\ q.QueueFront[f/x!] \\
\wedge\ ((f = pid? \wedge q.RemoveFirst[f/x!]) \\
\quad \vee\ (f \neq pid? \wedge q.RemoveElement[f/x?]))
\end{array}
$$

Since *readyqueues*(*lv*) is uniquely determined, only one queue can be the result. □

Proposition 59. *runTooLong implements a round-robin régime on the user queue.*

PROOF. The core is:

$$\exists p : AREF \bullet \\ \quad readyqueues(userqueue).RemoveFirst[p/x?] \fatsemi \\ \quad\quad readyqueues(userqueue).Enqueue[p/x?]$$

This expands into:

$$\exists elts'' : \text{iseq } APREF \bullet \\ \quad elts'' = tail\ elts \\ \quad \land p = head\ elts \\ \quad \land elts' = elts'' \frown \langle p \rangle$$

which simplifies to:

$$elts' = (tail\ elts) \frown \langle head\ elts \rangle$$

That is, the first element becomes the last. This is exactly the round-robin scheme.

It has already been established that the queue is uniquely determined by *readyqueues*(*userqueue*). □

Proposition 60. *ScheduleNext* $\land$ *allEmptyQueues implies that the idle process is the current process; that is,* $currentp' = IdleProcRef$.

PROOF. This is an or-elimination proof, so there are two cases.
Case 1: $allEmptyQueues \land runIdleProcess$. By propositional calculus:

$$p \land q \Rightarrow q$$

so $allEmptyQueues \Rightarrow runIdleProcess$, or (just taking the most relevant components):

$$(\forall q : ProcessQueue.q.IsEmpty) \Rightarrow currentp' = IdleProcRef$$

by MP, $currentp' = IdleProcRef$ follows.
Case 2: *selectNext*.

First, assume that $currentp' \neq IdleProcRef$.

Now, the essential part of *ScheduleNext* is:

$$(\exists q : ProcessQueue \bullet \\ \quad q = readyqueues(devprocqueue) \\ \quad\quad\quad \otimes readyqueues(sysprocqueue) \\ \quad\quad\quad \otimes readyqueues(userqueue) \\ \quad \land (\exists p : APREF;\ l : SCHDLVL;\ pd : ProcessDescr \bullet \\ \quad\quad q.QueueFront[p/x!] \\ \quad\quad \land currentp' = p))$$

This simplifies to:

$$\begin{array}{l} q = readyqueues(devprocqueue) \\ \quad \otimes readyqueues(sysprocqueue) \\ \quad\quad \otimes readyqueues(userqueue) \\ \wedge\ currentp' = head\ q.elts \end{array}$$

The fact about queues (and sequences more generally) that:

$$\forall q \bullet q.elts \neq \langle\rangle \Leftrightarrow \exists h \bullet h = head\ q.elts$$

can be used to show that $q.elts \neq \langle\rangle$ because $currentp' = head\ q.elts$. This contradicts the assumption that $allEmptyQueues$. □

Proposition 61. *If* $\neg\ allEmptyQueues$*, then the predicate of* $ScheduleNext$ *implies that* $currentp' \neq IdleProcRef$.

PROOF. Again, this is an or-elimination proof.
Case 1: The property of queue heads:

$$\begin{array}{l} \forall q \bullet q \neq \langle\rangle \Leftrightarrow \exists h \bullet h = head\ q.elts \\ \quad \Leftrightarrow \forall q \bullet q = \langle\rangle \Leftrightarrow \neg\ \exists h \bullet h = head\ q.elts \end{array}$$

Assuming that $currentp' \neq IdleProcRef$, $q.elts = \langle\rangle$. This contradicts the assumption that $\neg\ allEmptyQueues$. Therefore $currentp' \neq IdleProcRef$.
Case 2: $\neg\ allEmptyQueues \wedge selectNext \Rightarrow currentp' \neq IdleProcRef$. Using the property of queues and sequences:

$$\forall q \bullet q \neq \langle\rangle \Leftrightarrow \exists h \bullet h = head\ q.elts$$

and applying MP, $\neg\ allEmptyQueues$ implies that there is a head of $q.elts$. By the definition of $selectNext$, $currentp' = head\ q.elts$ and $currentp' \neq IdleProcRef$. □

Proposition 62. *If the device-level queue is not empty, then* $ScheduleNext$ *implies that the priority of* $currentp'$ *is the device level (i.e., the highest priority).*

PROOF. Let dq denote the device process queue, $readyqueues(devprocqueue)$, let wq denote the other system process queue, $readyqueues(sysprocqueue)$, and, finally, let uq denote the queue of user processes, $readyqueues(userqueue)$. Then, by the definition of $selectNext$, the queue from which the next process is chosen is:

$$q.elts = dq.elts \frown sq.elts \frown uq.elts$$

By assumption, $dq.elts \neq \langle\rangle$. By the existential quantification,

$$\begin{aligned} p &= head\ q.elts \\ &= head(dq.elts \frown sq.elts \frown uq.elts) \\ &= head\ dq.elts \quad \text{since} \quad head\ q = q(1) \\ &= currentp' \end{aligned}$$

The element that is selected from the queue is an element of the device queue, and so must have the highest (device) priority. □

Proposition 63. *If the device queue is empty and the queue of system processes is not empty, ScheduleNext will select a system process as the next value of currentp.*

PROOF. This proof is similar to the immediately preceding one. The details, though, are given.

Using the same abbreviations as in the last proof and expanding the definition of q, we have:

$$q.elts = dq.elts \frown sq.elts \frown uq.elts$$

By assumption, $sq.elts \neq \langle\rangle$, while $dq.elts = \langle\rangle$, so:

$$\begin{aligned} & q.elts \\ &= \langle\rangle \frown sq.elts \frown uq.elts \\ &= sq.elts \frown uq.elts \quad (\text{since } \langle\rangle \frown q = q, \text{ for all } q) \end{aligned}$$

Therefore:

$$\begin{aligned} & head\ q.elts \\ &= head\langle\rangle \frown sq.elts \frown uq.elts \\ &= head\ sq.elts \frown uq.elts \quad (\text{since } \langle\rangle \frown q = q) \\ &= head\ sq.elts \quad \text{since} \quad head\ q = q(1) \\ &= currentp' \end{aligned}$$

Therefore, $currentp' = head\ sq.elts$ and $currentp'$ is bound to a reference to a system process; that process must have system-process priority (middle priority). □

Proposition 64. *If both the device queue and the system-process queue are empty and the user-process queue is not empty, ScheduleNext selects a user process as the next value of currentp.*

PROOF. Again, using the same abbreviations as above, the queue from which the next value of *currentp* is taken is:

$$\begin{aligned} & q.elts \\ &= dq.elts \frown sq.elts \frown uq.elts \\ &= \langle\rangle \frown \langle\rangle \frown uq.elts \\ &= \langle\rangle \frown uq.elts \\ &= uq.elts \end{aligned}$$

Following the usual reasoning:

$$
\begin{aligned}
& \mathit{head}\ q.\mathit{elts} \\
& \quad = \mathit{head}(dq.\mathit{elts} \frown sq.\mathit{elts} \frown uq.\mathit{elts}) \\
& \quad = \mathit{head}(\langle\rangle \frown \langle\rangle \frown uq.\mathit{elts}) \\
& \quad = \mathit{head}(\langle\rangle \frown uq.\mathit{elts}) \\
& \quad = \mathit{head}\ uq.\mathit{elts} \\
& \quad = \mathit{currentp}'
\end{aligned}
$$

Therefore, a user process is selected. □

Proposition 65. *If $\neg$ allEmptyQueues, then ScheduleNext implies that the highest-priority process is referred to by currentp′.*

PROOF. By the three immediately preceding propositions, *SelectNext* assigns to *currentp* processes in the order:

1. device processes;
2. system processes;
3. user processes.

This corresponds to the organisation of priorities defined immediately prior to the definition of the scheduler. □

Proposition 66. *After ScheduleNext, the current process, currentp, is bound either to the identifier of the idle process or to the identifier of a process that resides in one of the three scheduling queues. It is not possible for any other process identifier to be bound by ScheduleNext to currentp.*

PROOF. By examination of the predicate of *ScheduleNext*, there are two cases to consider.
Case 1: All the process queues are empty (i.e., *allEmptyQueues*). In this case, $\mathit{currentp}' = \mathit{IdleProcRef}$.
Case 2: At least one of the scheduler's three queues is not empty. By the preceding results, *currentp′* can only be the head of one of these queues. Furthermore, there is no operation defined by the scheduler for setting the value of *currentp* from an external source. □

Proposition 67. *It is always the case that currentp′ after ScheduleNext is not identical to NullProcRef.*

PROOF. The type of *currentp* is *IPREF*. The constant *NullProcRef* is of type *PREF*. However, $\mathit{IPREF} \subset \mathit{IPREF}$, so *NullProcRef* is not an element of *IPREF* and cannot be bound to *currentp* without a type-assignment error.

□

The basic policy is that the scheduler is called quite frequently, so decisions can change. The current process is, therefore, left on the head of its queue. When a user-level process has run out of time, it is pre-empted and the *runTooLong* method is executed, removing the current head and placing it at the end (provided, that is, that the user-level queue is not empty). When a user-level process terminates, it is removed from the queue, in any case. Otherwise, *currentp* always points to the head of the system or device queue. It is occasionally necessary to remove these processes. Device processes must end by waiting on the corresponding device semaphore. System processes either suspend themselves (by making a call to the self-suspend primitive) or they wait on a semaphore.

4.6 Storage Management

This kernel performs a certain amount of storage management. The management scheme is relatively simple but still more complex than the one employed in the last chapter (which, the reader will recall, employed a totally static allocation method).

This section begins with a relatively long series of definitions. Most of the definitions are of axiomatically defined functions. A number of proofs appear among the definitions. The reader will note that not all of the functions defined below are used in the model that follows; some are introduced because they can be used in an alternative version of this model. Furthermore, some functions have interesting consequences or properties that are included just for their interest value.

It is assumed that the store is not infinite in size. The limit is:

$$memlim : \mathbb{N}_1$$

This is assumed to be the maximum address for the storage configuration.

In defining the *ADDRESS* type, address 0 is omitted (it makes little difference).

$$ADDRESS == 1 \mathrel{..} memlim$$
$$MEMDESC == ADDRESS \times \mathbb{N}$$

The second type is the storage descriptor encountered towards the start of this chapter. The intention is that an element of *MEMDESC* describes a region of store whose address is given by the first component and whose size is given by the second. These memory descriptors are constructed by the following function:

$$mkrmemspec : ADDRESS \times \mathbb{N} \rightarrow MEMDESC$$
$$\forall a : ADDRESS; \; s : \mathbb{N} \bullet$$
$$\quad mkrmemspec(a, s) = (a, s)$$

As a pair, there are naturally two selector (projection) functions:

$$
\begin{array}{l}
memstart : MEMDESC \rightarrow ADDRESS \\
memsize : MEMDESC \rightarrow ADDRESS \\
\hline
\forall r : MEMDESC \bullet \\
\quad memstart(r) = fst\ r \\
\quad memsize(r) = snd\ r
\end{array}
$$

Of utility are the following:

$$
\begin{array}{l}
memend : MEMDESC \rightarrow ADDRESS \\
nextblock : MEMDESC \rightarrow ADDRESS \\
\hline
\forall s : MEMDESC \bullet \\
\quad memend(s) = (memstart(s) + memsize(s)) \\
\quad nextblock(s) = memstart(s) + memsize(s) + 1
\end{array}
$$

Next, it is necessary to define a type for whatever is stored. The type *PSU* is the type of the contents of each cell in the store

$[PSU]$

This is the type of the *Primary Storage Unit* and it can be a word or byte. We assume a byte.

We can also assume:

$$
NullVal : PSU
$$

The entire store is a sequence of content type, i.e.:

$MEM == \mathrm{seq}\, PSU$

Below, it will be assumed that each process has one storage area.

As far as processes and the storage manager are concerned, the store is represented by collections of objects of type *MEMDESC*.

It is necessary to determine a few properties about storage. More specifically, we need to know the properties of storage descriptors.

There is the case of overlap.

$$
\begin{array}{l}
memsegoverlap : MEMDESC \leftrightarrow MEMDESC \\
\hline
\forall ms_1, ms_2 : MEMDESC \bullet \\
\quad memsegoverlap(ms_1, ms_2) \Leftrightarrow \\
\qquad (memstart(ms_1) \leq memstart(ms_2) \land \\
\qquad\quad nextblock(ms_1) \leq nextblock(ms_2)) \lor \\
\qquad (memstart(ms_1) > memstart(ms_2) \land \\
\qquad\quad nextblock(ms_1) \geq nextblock(ms_2))
\end{array}
$$

A symmetric version of this predicate can be defined as follows:

$$
\begin{array}{|l}
memsegsymoverlap : MEMDESC \leftrightarrow MEMDESC \\
\hline
\forall ms_1, ms_2 : MEMDESC \bullet \\
\quad memsegsymoverlap(ms_1, ms_2) \Leftrightarrow \\
\quad\quad memsegoverlap(ms_1, ms_2) \lor memsegoverlap(ms_2, ms_1)
\end{array}
$$

Proposition 68. *memsegsymoverlap is symmetric.*

PROOF. Obvious from the definition and the fact that $p \lor q \Leftrightarrow q \lor p$. □

It is necessary that all holes be disjoint. This is expressed in the third conjunct of the invariant.

The following function returns a subsequence of a sequence m, starting at element *offset* and running to the end of m (the definition being taken from [15]):

$$
\begin{array}{|l}
\_after\_ : \mathrm{seq}\, X \times \mathbb{N} \rightarrow \mathrm{seq}\, X \\
\hline
\forall m : \mathrm{seq}\, X;\ offset : \mathbb{N} \bullet \\
\quad \mathrm{dom}(m\ after\ offset) = (1 \mathinner{..} \#m - offset) \land \\
\quad (\forall n : \mathbb{N} \bullet \\
\quad\quad (n + offset) \in \mathrm{dom}\, m \Rightarrow (m\ after\ offset)(n) = m(n + offset))
\end{array}
$$

The store is considered to be two sets of segments. The segments used by processes are, in effect, invisible to the operating system. The segments that are not used by any processes constitute the free store; free store is represented by a sequence of zero or more segments described by *MEMDESC*s. Initially, the free store consists of exactly one segment: it is the entire store.

For fairly obvious reasons, regions in free store are called "holes".

$$
\begin{array}{|l}
lower\_hole\_addr : MEMDESC \times MEMDESC \rightarrow ADDRESS \\
upper\_hole\_addr : MEMDESC \times MEMDESC \rightarrow ADDRESS \\
\hline
\forall h_1, h_2 : MEMDESC \bullet \\
\quad lower\_hole\_addr(h_1, h_2) = \begin{cases} memstart(h_1), & \text{if } memstart(h_1) < memstart(h_2) \\ memstart(h_2), & \text{otherwise} \end{cases} \\
\quad upper\_hole\_addr(h_1, h_2) = \begin{cases} memstart(h_1), & \text{if } memstart(h_1) > memstart(h_2) \\ memstart(h_2), & \text{otherwise} \end{cases}
\end{array}
$$

These two functions return the lower of the start addresses of the arguments.

The holes in the free store need to be merged to form larger blocks when a compaction is performed. The merge function can be defined as:

$$
\begin{array}{|l}
mergememholes : MEMDESC \times MEMDESC \rightarrow MEMDESC \\
\hline
\forall h_1, h_2 : MEMDESC \bullet \\
\quad (lower\_hole\_addr(h_1, h_2), upper\_hole\_addr(h_1, h_2))
\end{array}
$$

$$
\begin{array}{|l}
hole\_size : MEMDESC \rightarrow \mathbb{N} \\
\hline
\forall h : MEMDESC \bullet \\
\quad hole\_size(h) = memsize(h)
\end{array}
$$

This function merely returns the size of a hole.

$$
\begin{array}{|l}
room\_in\_hole : MEMDESC \rightarrow \mathbb{N} \\
room\_left\_in\_hole : \mathbb{N} \times MEMDESC \rightarrow \mathbb{N}_1 \\
\hline
\forall n : \mathbb{N}, h : MEMDESC \bullet \\
\quad room\_in\_hole(n, h) \Leftrightarrow n \leq hole\_size(h) \\
\quad room\_left\_in\_hole(n, h) = hole\_size(h) - n
\end{array}
$$

The function *room_in_hole* returns the size of the hole supplied as its argument. The second function returns the amount of space left in the hole after the first argument has been removed.

Finally, it is assumed that the store in which allocations are made starts at some address that is sufficiently far away from the kernel to avoid problems. This address is:

$$
| \; startaddr : ADDRESS
$$

The main store is modelled as a description. The class is as follows:

$$
\begin{array}{|l}
\hline
REALMAINSTORE \\
\upharpoonright (INIT, RSCanAllocateInStore, RSAllocateFromHole, \\
\quad MergeAdjacentHoles, FreeMainstoreBlock, \\
\quad RSFreeMainstore, RSAllocateFromUsed, \\
\quad RSCopyMainStoreSegment, RSWriteMainStoreSegment, \\
\quad CreateProcessImage) \\
\\
\quad \begin{array}{|l}
\hline
mem : \operatorname{seq} PSU \\
holes : \operatorname{seq} MEMDESC \\
usermem : \operatorname{seq} MEMDESC \\
\hline
\#mem = memlim \\
\#holes \leq memlim \\
((hole\_size(holes(1)) = \#mem) \\
\vee (\sum_{i=1}^{i=\#holes} hole\_size(holes(i)) + \sum_{j=1}^{j=\#usermem} hole\_size(usermem(j)) \\
\qquad = \#mem)) \\
(\forall h : MEMDESC \mid h \in \operatorname{ran} holes \bullet \\
\quad \neg (\exists h_1 : MEMDESC \mid h_1 \in \operatorname{ran} holes \bullet \\
\qquad h \neq h_1 \wedge memsegsymoverlap(h, h_1))) \\
(\forall h : MEMDESC \mid h \in \operatorname{ran} holes \bullet \\
\neg (\exists m : MEMDESC \mid m \in usermem \bullet \\
\quad memsegsymoverlap(m, h))) \\
\hline
\end{array}
\end{array}
$$

$$\begin{array}{l}
\underline{\ \ INIT\ \ } \\
holes' = \langle (startmem, memlim) \rangle \\
usermem' = \langle\,\rangle
\end{array}$$

$RSCanAllocateInStore \mathrel{\widehat=} \ldots$

$RSAllocateFromHole \mathrel{\widehat=} \ldots$

$MergeAdjacentHoles \mathrel{\widehat=} \ldots$

$FreeMainstoreBlock \mathrel{\widehat=} \ldots$

$RSFreeMainstore \mathrel{\widehat=} \ldots$

$RSAllocateFromUsed \mathrel{\widehat=} \ldots$

$RSCopyMainStoreSegment \mathrel{\widehat=} \ldots$

$RSWriteMainStoreSegment \mathrel{\widehat=} \ldots$

$CreateProcessImage \mathrel{\widehat=} \ldots$

The class divides the store into two main areas. The first is composed of all the store currently allocated to user processes. This is called *usermem* in the class. The second area is the *free space*, or all the store that is not currently allocated to processes. The free space is called *holes* for the reason that there can be free areas within allocated ones—such areas of unallocated store are often called "holes".

The *usermem* sequence is required only because the store of processes that are currently swapped out must be able to be recycled for use by other processes. This (perhaps extreme) requirement forces the recording of allocated store. In other kernels, such as those that only allocate once or do not recycle storage in the same way as this one, it is only necessary to record the descriptors to unallocated store. The descriptors for allocated store are never passed to user processes: instead, the base address and size of the storage block are passed instead. This makes descriptor management somewhat easier.

The rather convoluted allocation and recycling approach has been chosen because it introduces a way of handling store that is implied by much of the literature but not explicitly described. In a swapping system without virtual store, *how* does the storage manager handle store? One simple way is to swap a process back to the storage area it originally occupied (MINIX [30] does this); this means that one or more processes might need to then be relocated and/or swapped out. There are clearly other strategies that could be adopted; the one adopted here was chosen because it does recycle storage and it relocates processes when they are swapped back into store. It also enables the question of the integration of heap storage with main storage design. (The strategy modelled here is distantly related to heap storage methods.) This remains an open problem, one that is worth some consideration in our opinion.

The *mem* variable could also be replaced by two variables: one denoting the start of the available store, the other denoting its size. This would provide all the information required to check allocations. The *mem* variable is included just for those readers who wonder "where" the store to be allocated is.

It should be noted that various operations over holes are also used on allocated storage chunks. The reason for this is that holes and chunks of user store are defined in terms of the same mathematical structures. (In any case, there is a duality: a "hole" is free space inside a region of allocated store and it can also be a piece of allocated store inside a region of free store.)

The invariant of this class is somewhat complex. It first states that *memlim* specifies the size of the store and that the limit to the number of holes is the size of the store itself (in the worst case, all holes will be of unit size). The next conjunct states that the store is as large as the sum of the sizes of all holes plus the size of all allocated store. (This is a way of stating that all storage can be accounted for in this model—memory leaks are not permitted.) Finally, the two quantified formulæ state that all holes are disjoint, as are all allocated regions.

The following schema is used as a predicate. It is true iff there is a hole of sufficient size to satisfy the request for storage. The request requires $rqsz?$ units for satisfaction.

RSCanAllocateInStore

$rqsz? : \mathbb{N}$

$(\exists h : MEMDESC \mid h \in \operatorname{ran} holes \bullet$
$\quad hole\_size(h) > 0 \land$
$\quad room\_in\_hole(rqsz?, h))$

Clearly, store can be allocated iff there is a hole of at least the requested size. The operation *RSAllocateFromHole* performs storage allocation from free store. It is expected that allocation from the *holes* will be the norm. However, if there is insufficient free store, the storage manager can also reallocate user storage (using the swapping mechanisms).

The *RSAllocateFromHole* operation's definition naturally falls into two cases:

1. The chosen hole is exactly of $rqsz?$ units (bytes).
2. The chosen hole is larger than $rqsz?$ units (bytes).

In the first case, the hole is removed from the free list (*holes*) and added to allocated store (*usermem*). In the second case, the hole is split into two parts with the one of $rqsz?$ bytes being transferred to *usermem* and the remainder allocated in a new hole in *holes*.

RSAllocateFromHole

$\Delta(holes, usermem)$
$rqsz? : \mathbb{N}$
$mspec! : MEMDESC$

$(\exists h : MEMDESC; \; n : \mathbb{N} \mid n \in \operatorname{dom} holes \land h = holes(n) \bullet$

$$
\begin{array}{l}
room\_in\_hole(rqsz?, h) \land \\
((room\_left\_in\_hole(rqsz?, h) = 0 \land \\
\quad mspec! = h \land usermem' = usermem \frown \langle mspec! \rangle \land \\
\quad holes' = holes \rhd \{h\}) \\
\lor (room\_left\_in\_hole(rqsz?, h) > 0 \land \\
\quad (\exists la : ADDRESS; hsz : \mathbb{N} \bullet \\
\quad\quad la = memstart(h) \land hsz = memsize(h) - rqsz? \land \\
\quad\quad mspec! = (la, rqsz?) \land \\
\quad\quad usermem' = usermem \frown \langle mspec! \rangle \land \\
\quad\quad holes' = (holes \rhd \{h\})^\frown \\
\quad\quad\quad \langle mkrmemspec(nextblock(mspec!), hsz) \rangle)))))
\end{array}
$$

Every so often (actually when a used block is freed by a process), the free store in *holes* is scanned and adjacent holes merged to form larger ones. This is defined by the following operation:

$$
\begin{array}{l}
\textit{MergeAdjacentHoles} \\
\hline
\Delta(holes) \\
\hline
(\forall h_1, h_2 : MEMDESC \mid h_1 \in \operatorname{ran} holes \land h_2 \in \operatorname{ran} holes \land \\
\quad (memstart(h_1) + memsize(h_1) + 1) = memstart(h_2) \bullet \\
\quad holes' = ((holes \rhd \{h_1\}) \rhd \{h_2\}) \frown \langle mergememholes(h_1, h_2) \rangle)
\end{array}
$$

The freeing of an allocated block is achieved by the next operation:

$$
\begin{array}{l}
\textit{FreeMainstoreBlock} \\
\hline
\Delta(holes, usermem) \\
start? : ADDRESS \\
sz? : \mathbb{N} \\
\hline
holes' = holes \frown \langle mkrmemspec(start?, sz?) \rangle \\
usermem' = usermem \rhd \{mkrmemspec(start?, sz?)\}
\end{array}
$$

The operation just adds the block (region) to *holes* (rendering it a free block) and removes the block from user storage in *usermem*; this operation is modelled by a range subtraction ($\rhd$).

Proposition 69. *FreeMainstoreBlock increases the length of the free list by 1.*

PROOF. The free list is called *holes* in this class. The predicate of schema *FreeMainstoreBlock* contains the following identity:

$$holes' = holes \frown \langle mkrmemspec(start?, sz?) \rangle$$

From this, the following calculation establishes the result.

$$\begin{aligned}&\#holes'\\&\quad = \#(holes \frown \langle mkrmemspec(start?, sz?)\rangle)\\&\quad = \#holes + \#\langle mkrmemspec(start?, sz?)\rangle\\&\quad = \#holes + 1\end{aligned}$$

□

Proposition 70. *FreeMainstoreBlock removes one block from user store.*

Proof. The predicate of the schema states that: $usermem' = usermem \mathbin{⩥} \{mkrmemspec(start?, sz?)\}$.

There are two ways (at least) to prove this proposition.
(1) By taking ranges:

$$\operatorname{ran} usermem' = (\operatorname{ran} usermem) \setminus \{mkrmemspec(start?, sz?)\}$$

so:

$$\begin{aligned}&\# \operatorname{ran} usermem' = \#((\operatorname{ran} usermem) \setminus \{mkrmemspec(start?, sz?)\})\\&\quad = \# \operatorname{ran} usermem - \#\{mkrmemspec(start?, sz?)\}\\&\quad = \# \operatorname{ran} usermem - 1\end{aligned}$$

(2) By writing the deletion in the equivalent form,

$$\begin{aligned}&\exists s_1, s_2 : \operatorname{seq} MEMDESC \bullet\\&\quad usermem = s_1 \frown \langle mkrmemspec(start?, sz?)\rangle \frown s_2 \land\\&\quad usermem' = s_1 \frown s_2\end{aligned}$$

it is clear that:

$$\begin{aligned}&\#usermem\\&\quad = \#s_1 + \#\langle mkrmemspec(start?, sz?)\rangle + \#s_2\\&\quad = \#s_1 + \#s_2\\&\quad = \#usermem'\end{aligned}$$

Therefore $\#usermem' = \#usermem - 1$. □

The combined operation that frees an allocated block and merges all adjacent blocks in *holes* is the following:

$$RSFreeMainStore \mathrel{\widehat{=}} FreeMainstoreBlock \mathbin{⨾} MergeAdjacentHoles$$

Sometimes, a process will require storage that is already allocated. This happens, in particular, when *holes* is empty and a new process is created. In this case, the resident process is swapped out to disk and its storage reallocated to the new process. The operation to perform the basic reallocation is as follows:

$$
\begin{array}{|l}
\hline
\textit{RSAllocateFromUsed} \\
\Delta(holes, usermem) \\
rqsz? : \mathbb{N} \\
n? : \mathbb{N}_1 \\
start! : ADDRESS \\
\hline
\exists h : MEMDESC \mid h = usermem(n?) \bullet \\
\quad (room\_left\_in\_hole(rqsz?, h) = 0 \land start! = memstart(h)) \\
\quad \lor (room\_left\_in\_hole(rqsz?, h) \geq 0 \land \\
\qquad start! = memstart(h) \land \\
\qquad holes' = \\
\qquad\quad holes \frown \langle mkrmemspec((start! + rqsz? + 1), \\
\qquad\qquad\qquad memsize(h) - rqsz?)\rangle \land \\
\qquad usermem'(n?) = mkrmemspec(start!, rqsz?) \\
\hline
\end{array}
$$

Again, this operation is defined in terms of two cases: where the hole is of the exact size and where the hole is of greater size.

The swapping process requires store segments to be written to and read from disk. The first of the next two operations returns a segment of store that is a copy of the one designated by the pair (*start*?, *end*?).

$$
\begin{array}{|l}
\hline
\textit{RSCopyMainStoreSegment} \\
start?, end? : ADDRESS \\
mseg! : MEM \\
\hline
mseg! = (\lambda\, i : start? \,..\, end? \bullet mem(i)) \\
\hline
\end{array}
$$

The second is an operation that overwrites a segment of store. The overwriting starts at the location specified by *loadpoint*. The input *mseg*? contains the piece of store that is to be written to main store.

$$
\begin{array}{|l}
\hline
\textit{RSWriteMainStoreSegment} \\
\Delta(mem) loadpoint? : \mathbb{N} \\
mseg? : MEM \\
\hline
\exists size : \mathbb{N} \mid size = \#mseg? \bullet \\
\quad mem' = (\lambda\, i : 1 \,..\, (loadpoint - 1) \bullet mem(i)) \\
\qquad\qquad \frown mseg? \frown (mem \; after \; ((loadpoint + size) - 1)) \\
\hline
\end{array}
$$

If there is just no space in store, write a new process to disk, setting store to zero as required.

The following is used to convert an object of type *PCODE* to a segment of store. It is a loose definition (even though, given the equivalences at the start of this chapter, it could be completed). The function is included so that the specification remains well-typed. (The constructs in the next two chapters are under-specified and will not typecheck correctly—more on this in the next chapter.)

| $codeToPSUs : PCODE \rightarrow MEM$

The next operation creates the sequence of bytes that will actually be copied to disk on a swap. It uses *codeToPSUs* as well as two λ expressions that operate more as one would find in a complete model. (When this schema is used, that use will be a little incorrect because the extraction of start and size from data and stack segments is ignored.)

— *CreateProcessImage* —

$code? : PCODE$
$stkstrt?, datastrt? : ADDRESS$
$stksz?, datasz? : \mathbb{N}_1; \; image! : MEM$

$image! = codeToPSUs(code?) ^\frown (\lambda\, i : datastrt? \,..\, datasz? \bullet 0) ^\frown$
$\quad (\lambda\, i : stkstrt? \,..\, stksz? \bullet 0)$

It is now possible to prove a few propositions about the main store and its operations.

Proposition 71. *RSCanAllocateStore is false iff there are no holes of positive size.*

PROOF. By the predicate, $hole\_size(h) > 0$ for some hole, h, in $\operatorname{ran} holes$. □

Proposition 72. *Each use of RSAllocateFromHole monotonically decreases available free storage.*

PROOF. Assume there have already been allocations. Then, by the invariant:

$$\sum_{i=1}^{i=\#holes} hole\_size(holes(i)) + \sum_{j=1}^{j=\#usermem} hole\_size(usermem(j)) = \#mem$$

There are two cases.
Case 1. $rqsz?$ = hole size. Then the number of holes decreases by one. The sum decreases by the corresponding amount.
Case 2. $rqsz?$ < hole size. The hole is split into two blocks, one of size $rqsz?$ and the other of size $memsize(h) - rqsz?$. The size of this new hole is necessarily less than $memsize(h)$. Therefore, the available storage decreases. □

The following two propositions establish the fact that free store decreases by the action of *RSAllocateFromHole* (when it is applicable) and the action of *RSFreeMainstore* increases the amount of free store.

Proposition 73. *The action of $RSAllocateFromHole[k/rqsz?]$ decreases the available free store by k units.*

PROOF. Again, without loss of generality, assume there have already been allocations. Then, by the invariant:

$$\sum_{i=1}^{i=\#holes} hole\_size(holes(i)) + \sum_{j=1}^{j=\#usermem} hole\_size(usermem(j)) = \#mem$$

If k units are allocated from free store, it follows that $\#mem'$ is given by:

$$\sum_{i=1}^{i=\#holes} hole\_size(holes(i)) - k + \sum_{j=1}^{j=\#usermem} hole\_size(usermem(j)) + k =$$
$$\sum_{i=1}^{i=\#holes'} hole\_size(holes'(i)) + \sum_{j=1}^{j=\#usermem'} hole\_size(usermem'(j))$$

□

Proposition 74. *The action of RSFreeMainstore*$[k/sz?]$ *increases the available free store by k units.*

PROOF. This is the converse of the last proposition.
Again, we use the same conjunct of the invariant:

$$\sum_{i=1}^{i=\#holes} hole\_size(holes(i)) + \sum_{j=1}^{j=\#usermem} hole\_size(usermem(j)) = \#mem$$

If k units are returned to free store, it follows that $\#mem'$ is given by:

$$\sum_{i=1}^{i=\#holes} hole\_size(holes(i)) + k + \sum_{j=1}^{j=\#usermem} hole\_size(usermem(j)) - k =$$
$$\sum_{i=1}^{i=\#holes'} hole\_size(holes'(i)) + \sum_{j=1}^{j=\#usermem'} hole\_size(usermem'(j))$$

□

Proposition 75. *If a hole is exactly the size of a request, it disappears from the free list.*

PROOF. The predicate of *RSAllocateFromHole* states that

$$room\_left\_in\_hole(rqsz?, h) = 0 \land$$
$$\quad \operatorname{ran} usermem' = \operatorname{ran} usermem \setminus \{mspec!\}$$

Since $h = mspec!$, $\operatorname{ran} usermem' = \operatorname{ran} usermem \setminus \{h\}$, so $h \notin \operatorname{ran} usermem'$. □

Proposition 76. *If a hole is larger than that requested, it is split into two and the smaller block is returned to the free list.*

PROOF. The predicate of *RSAllocateFromHole* states that:

$$mspec! = (la, rqsz!) \wedge$$
$$holes' = (holes \rhd \{h\}) \frown \langle mkrmemspec(nextblock(mspec!), hsz)\rangle$$

where $hsz = memsize(h) - rqsz?$ and *nextblock* yields the index of the start of the next block: $nextblock(mkrmemspec(strt, sz)) = strt + sz$.

Since hsz is the size of the block added to holes and $hsz = memsize(h) - rqsz?$ and $hsz > 0$ (by the predicate), it follows that:

$$memsize(mkrmemspec(nextblock(mspec!), hsz)) < memsize(h)$$

□

Proposition 77. *If all holes have size* $< rqsz?$, *RSAllocateFromHole cannot allocate any store.*

PROOF. Let $rqsz? = n$ and let n be larger than the greatest block size. Then $room\_left\_in\_hole(rqsz?, h) < 0$ for all h. This falsifies the predicate of the schema. □

Proposition 78. *If the allocating hole is* $\geq rqsz?$, *the hole is split into two parts: one of size* $= rqsz?$, *the other of size,* s, $s \geq 0$.

PROOF. There are two cases to consider, given *RSAllocateFromHole*'s predicate:

1. $memsize(h) = rqsz?$, and $mspec$ is of size $rqsz?$, so $s = 0$ (the smaller part is of zero length);
2. $memsize(h) = rqsz?$ and $mspec$ is of size $rqsz?$, so $memsize(h) - rqsz?$ is the size of one part and $s > 0$ is the size of the other.

□

The next proposition establishes the fact that merging adjacent free blocks (holes) decreases the number of blocks in free store.

Proposition 79. $MergeAdjacentBlocks \Rightarrow \#\,ran\,holes' < \#\,ran\,holes$.

PROOF. For the purposes of this proposition, the critical line is:

$$holes' = [((holes \rhd \{h_1\}) \rhd \{h_2\}) \frown \langle mergememholes(h_1, h_2)\rangle]$$

So:

$\# \operatorname{ran} holes'$

$$
\begin{aligned}
&= \# \operatorname{ran}[((holes \rhd \{h_1\}) \rhd \{h_2\}) \frown \langle mergememholes(h_1, h_2)\rangle] \\
&= \# \operatorname{ran}((holes \rhd \{h_1\}) \rhd \{h_2\}) + \# \operatorname{ran}\langle mergememholes(h_1, h_2)\rangle \\
&= \# \operatorname{ran}((holes \setminus \{h_1\}) \setminus \{h_2\}) + \# \operatorname{ran}\langle mergememholes(h_1, h_2)\rangle \\
&= \# \operatorname{ran}((holes \setminus \{h_1\}) \setminus \{h_2\}) + 1 \\
&= (\#(\operatorname{ran} holes \setminus \{h_1\}) - 1) + 1 \\
&= (\#(\operatorname{ran} holes) - 2) + 1 \\
&= \# \operatorname{ran} holes - 1 \\
&\leq \# \operatorname{ran} holes
\end{aligned}
$$

$\square$

If the free blocks are reduced in number, what happens to their size? The following proposition establishes the fact that the merging of adjacent free blocks creates a single new block whose size is the sum of all of the merged blocks.

Proposition 80. *If h_1 and h_2 are adjacent holes in the store of size n_1 and n_2, respectively, then MergeAdjacentHoles implies that there exists a hole of size $n_1 + n_2$.*

PROOF. Since h_1 and h_2 are adjacent, they can be merged. The definition of *mergememholes* is:

$\forall h_1, h_2 : MEMDESC \bullet$
$\quad (lower\_hole\_addr(h_1, h_2), memsize(h_1) + memsize(h_2))$

The size of the merged hole is therefore $memsize(h_1) + memsize(h_2)$. Letting $memsize(h_1) = n_1$ and $memsize(h_2) = n_2$, it is clear, by the definition of *mergememholes*, that:

$memsize(h_1) + memsize(h_2) = n_1 + n_2$

$\square$

It is clear that we do not want operations on the free store to affect the store allocated to processes. The following proposition assures us that nothing happens to user store when adjacent blocks of free store are merged.

Proposition 81. *MergeAdjacentHoles leaves user store invariant.*

PROOF. The predicate does not alter *usermem*. $\square$

Proposition 82. *If h_1 and h_2 are adjacent holes and MergeAdjacentHoles is applied to merge them, then $\# \operatorname{ran} holes' = \# \operatorname{ran} holes - 1$.*

PROOF. By Proposition 79. $\square$

Proposition 83. *The predicate of schema FreeMainstoreBlock implies that* $\# \operatorname{ran} holes' > \# \operatorname{ran} holes$ *and that* $\# \operatorname{ran} usermem' < \# \operatorname{ran} usermem$.

PROOF. By the definition of *FreeMainstoreBlock*:

$$holes' = holes \frown \langle mkrmemspec(start, sz?)\rangle$$

so:

$$\begin{aligned}&\# \operatorname{ran} holes\\ &\quad = \# \operatorname{ran}(holes \frown \langle mkrmemspec(start, sz?)\rangle)\\ &\quad = \# \operatorname{ran} holes + \# \operatorname{ran}(\langle mkrmemspec(start, sz?)\rangle)\\ &\quad = \# \operatorname{ran} holes + 1\end{aligned}$$

and so, $\# \operatorname{ran} holes' > \operatorname{ran} holes$.

Now,

$$\begin{aligned}&\# \operatorname{ran} usermem'\\ &\quad = \# \operatorname{ran}(usermem \rhd \{\langle mkrmemspec(start, sz?)\rangle\})\\ &\quad = \#(\operatorname{ran} usermem \setminus \{\langle mkrmemspec(start, sz?)\rangle\})\\ &\quad = \# \operatorname{ran} usermem - 1\end{aligned}$$

Therefore $\# \operatorname{ran} usermem' < \# \operatorname{ran} usermem$. □

Proposition 84. *If n calls to the allocator request k units of store, followed immediately by n calls to RSFreeMainStore, each returning k units of store, return the store to its original state.*

PROOF. We need to show that the sizes of *usermem* and *holes* are unchanged.

By Proposition 73, the size of the store after the n allocations is:

$$\begin{aligned}&\textstyle\sum_{i=1}^{i=\#holes} hole\_size(holes(i)) - nk +\\ &\quad \textstyle\sum_{j=1}^{j=\#usermem} hole\_size(usermem(j)) + nk =\\ &\textstyle\sum_{i=1}^{i=\#holes''} hole\_size(holes''(i)) +\\ &\quad \textstyle\sum_{j=1}^{j=\#usermem''} hole\_size(usermem''(j))\end{aligned}$$

while that after the n deallocations is, by Proposition 74:

$$\begin{aligned}&\textstyle\sum_{i=1}^{i=\#holes''} hole\_size(holes''(i)) + nk +\\ &\quad \textstyle\sum_{j=1}^{j=\#usermem''} hole\_size(usermem''(j)) - nk =\\ &\textstyle\sum_{i=1}^{i=\#holes'} hole\_size(holes'(i)) +\\ &\quad \textstyle\sum_{j=1}^{j=\#usermem'} hole\_size(usermem'(j)) =\\ &\textstyle\sum_{i=1}^{i=\#holes} hole\_size(holes(i)) +\\ &\quad \textstyle\sum_{j=1}^{j=\#usermem} hole\_size(usermem(j))\end{aligned}$$

□

Proposition 85. $\#image! = \#code + stksz? + datasz?$.

PROOF. Note that *codeToPSUs* is of type $PCODE \rightarrow MEM$, so $\#code = \#codeToPSUs$ since $MEM = \mathrm{seq}\, PSU$.

Now

$$
\begin{aligned}
&\#image! = \\
&\quad \#(codeToPSU(code?) \frown (\lambda\, i : 1 \mathrel{..} datasz? \bullet 0) \frown (\lambda\, i : 1 \mathrel{..} stksz? \bullet 0)) \\
&\quad = \#(codeToPSU(code?) + \#(\lambda\, i : 1 \mathrel{..} datasz? \bullet 0) + \#(\lambda\, i : 1 \mathrel{..} stksz? \bullet 0)) \\
&\quad = \#code + \#datasz? + \#stksz?
\end{aligned}
$$

□

The real store on the hardware is represented by a unique instance of *SharedMainStore*. This is a store that refers to the real store but whose operations are protected by locks. All that is required is that the operations be indivisible. The class is defined as follows:

SharedMainStore

$\upharpoonright(INIT, CanAllocateInStore, AllocateFromHole,$
$\quad AllocateFromUsed, FreeMainStore, CopyMainStore, WriteMainStore)$

$lms : LINERAMAINSTORE$
$lck : Lock$

INIT

$lms.INIT$

$CanAllocateInStore \mathrel{\widehat=}$
$\quad lck.Lock \mathbin{;} lms.RSCanAllocateInStore \mathbin{;} lms.Unlock$

$AllocateFromHole \mathrel{\widehat=}$
$\quad lck.Lock \mathbin{;} RSAllocateFromHole \mathbin{;} lck.Unlock$

$AllocateFromUsed \mathrel{\widehat=}$
$\quad lck.Lock \mathbin{;} RMAllocateFromUsed \mathbin{;} lck.Unlock$

$FreeMainStore \mathrel{\widehat=}$
$\quad lck.Lock \mathbin{;} RSFreeMainStore \mathbin{;} lck.Unlock$

$CopyMainStore \mathrel{\widehat=}$
$\quad lck.Lock \mathbin{;} RSCopyMainStoreSegment \mathbin{;} lck.Unlock$

$WriteMainStore \mathrel{\widehat=}$
$\quad lck.Lock \mathbin{;} RSWriteMainStoreSegment \mathbin{;} lck.Unlock$

4.6.1 Swap Disk

This section contains a high-level model of the swap disk. The swap disk is where swapped process images are stored. It is assumed to be more or less infinite in size.

Communication with the swap disk is in terms of a buffer containing an operation code. The codes are defined as:

$$
\begin{aligned}
SWAPRQMSG ::= {} & NULLSWAP \\
\mid {} & SWAPOUT\langle\langle PREF \times ADDRESS \times ADDRESS\rangle\rangle \\
\mid {} & SWAPIN\langle\langle PREF \times ADDRESS\rangle\rangle \\
\mid {} & NEWSPROC\langle\langle PREF \times MEM\rangle\rangle \\
\mid {} & DELSPROC\langle\langle PREF\rangle\rangle
\end{aligned}
$$

The *NULLSWAP* operation is a no-operation: if the opcode is this value, the swap disk should do nothing. A *SWAPOUT* code specifies the identifier of the process whose store is to be swapped out and the start and end addresses of the segment to be written to disk. A *SWAPIN* code requests the disk to read a segment and transfer it to main store. A *NEWSPROC* specifies that the store represented by *MEM* is to be stored on disk and that *PREF* denotes a newly created process that cannot be allocated in store at present. Finally, the *DELSPROC* code indicates that the named process is to be removed completely from the disk (it should be removed from the swap disk's index).

The buffer that supplies information to the swap disk is *SwapRQBuffer*. A semaphore is used to provide synchronisation between the swapper process and the swap disk process.

The buffer is modelled by a class and is defined as follows:

SwapRQBuffer

$\upharpoonright(INIT, Write, Read)$

$$
\begin{aligned}
& mutex, msgsema : Semaphore \\
& buff : SWAPRQMSG
\end{aligned}
$$

INIT

$$
\begin{aligned}
& mt? : Semaphore \\
& ptab? : ProcessTable \\
& sched? : LowLevelScheduler \\
& lck? : Lock \\
\hline
& mutex' = mt? \\
& (\exists iv : \mathbb{Z} \mid iv = 1 \bullet \\
& \quad msgsema' = Semaphore.Init[iv/iv?, ptab?/pt?, \\
& \qquad\qquad sched?/sch?, lck?/lk?]) \\
& buff' = NULLSWAP
\end{aligned}
$$

$Write \mathrel{\widehat=} \ldots$

$Read \mathrel{\widehat=} \ldots$

This class has two main operations, one for reading a request buffer and one for writing a reply buffer. The buffers are protected by semaphores. Semaphores are correct at this level because the code that calls *Read* and *Write* is executed by system processes, not by kernel primitives.

The *Write* operation is simple and defined as:

Write

$\Delta(buff)$
$rq? : SWAPRQMSG$

$msgsema.Wait \land buff' = rq? \land msgsema.Signal$

The *Read* operation is also simple:

Read

$rq! : SWAPRQMSG$

$mutex.Wait$
$msgsema.Wait$
$mutex.Signal$
$rq! = buff$
$buff' = NULLSWAP$
$mutex.Wait$
$msgsema.Signal$
$mutex.Signal$

Readers should note that the above buffer protocol is asymmetric. If a reader is already reading and a writer is waiting to write, the code will permit other readers to perform reads before the writer is permitted to write new data. In this particular case, this is permissible because there is exactly one reader, the swap-disk driver, and two writers, the swapper and store manager processes.

The driver process for the swap disk is relatively simple. Its basic tasks are to store process images and to retrieve them again when required. The images are indexed by process reference or identifier ($APREF$). Only "genuine" processes can have their images swapped out, and thus only processes whose reference is an element of $APREF$. The image stored on the swap disk is a copy of a contiguous segment of main store, so the objects stored on the swap disk are elements of type MEM (sequences of PSU).

The swap-disk model uses a finite partial map to represent the disk storage and index. Two semaphores are used, one to synchronise with the device driver that passes requests to the disk-controller process and a semaphore to synchronise with the storage management module. The second semaphore is used to signal the fact that the transfer has been completed; if this semaphore were not included, there is the risk that the storage management module would assume that a transaction had been completed, while, in fact, it had not.

Requests to the swap disk process are placed in the *SwapRQBuffer*. This is a piece of shared storage and is guarded by its own semaphore.

The read and write operations are to main store. Main store is, of course, shared, so locking is used to prevent interrupts from occurring while read and write operations are under way. It would be natural to assume that, since this is the only process running at the time reads and writes are performed, main store would, in effect, belong to this process. However, an interrupt could cause another process to be resumed and that process might interact with this one. This is, it must be admitted, a bit unlikely, but it is safer to use the scheme employed here. The alternative is to guard main store with a semaphore. This is not an option here because the storage-management software is implemented as a module, not a process.

The driver uses a semaphore to synchronise with the swapper process for reading the *SwapRQBuffer*. This is the semaphore called *devsema* in the definition of the class. It also uses a second semaphore, called *donesema*, which is used to indicate the fact that the disk read has been completed (the reason for this will become clear below).

The class that follows is, in fact, a combination of the process that performs the copy to and from disk and the disk itself. The reason for this is that the disk image is as important a part of the model as the operations to read and write the byte sequences and process references.

The swap disk's driver process is defined as:

$SWAPDISKDriverProcess$

$\upharpoonright(INIT, RunProcess)$

$devsema : Semaphore$
$donesema : Semaphore$
$dmem : APREF \twoheadrightarrow\!\!\!\!\!\shortmid\;\; MEM$
$sms : SharedMainStore$
$rqs : SwapRqBuffer$

$INIT$

$dsma? : Semaphore$
$devsemaphore? : Semaphore$
$rqbuff? : SwapRqBuffer$
$store? : SharedMainStore$

$donesema' = dsma?$
$devsema' = devsemaphore?$
$\mathrm{dom}\ dmem' = \varnothing$
$rqs' = rqbuff?$
$sms' = store?$

$writeProcessStoreToDisk \mathrel{\widehat=} \ldots$

$readProcessStoreFromDisk \mathrel{\widehat=} \ldots$

$deleteProcessFromDisk \mathrel{\widehat=} \ldots$

$sleepDriver \mathrel{\widehat=} \ldots$

$handleRequest \mathrel{\widehat=} \ldots$

$RunProcess \mathrel{\widehat=} \ldots$

Even though this is a system process, the main store is locked when read and write operations are performed. This is because arbitrary interrupts might occur when these operations are performed; even though it is controlled by a semaphore (so processes cannot interfere with any operation inside it), the body of critical regions is still open to interrupts. The lock is used as an additional safety measure, even though it is not particularly likely that an interrupt would interfere with the store in question.

writeProcessStoreToDisk

$\Delta(dmem)$
$p? : APREF$
$ms? : MEM$

$dmem' = dmem \oplus \{p? \mapsto ms?\}$

readProcessStoreFromDisk

$p? : APREF$
$ms! : MEM$

$ms! = dmem(p?)$

deleteProcessFromDisk

$\Delta(dmem)$
$p? : APREF$

$dmem' = \{p?\} \lhd dmem$

When the driver is not performing any operations, it waits on its *devsema*. The driver is awakened up by a *Signal* on *devsema*. When the request has been handled, the *Wait* operation is performed to block the driver. This is a safe and somewhat standard way to suspend a device process.

$sleepDriver \mathrel{\widehat=} devsema.Wait$

The remaining operation is the one that handles requests. When the device process has the semaphore, it reads the data in the request block; in particular, it examines the operation. The operation requested is used to perform the appropriate operation. The schema modelling this is:

$$
\begin{array}{l}
\textit{handleRequest} \\
\hline
rq? : SWAPRQMSG \\
\hline
(\exists p : APREF;\ start, end : ADDRESS;\ mem : MEM \bullet \\
\qquad rq? = SWAPOUT\langle\langle p, start, end\rangle\rangle \\
\qquad \land\ sms.CopyMainStore[start/start?, end/end?, mem/mseg!] \\
\qquad \land\ writeProcessStoreToDisk[p/p?, mem/ms?]) \\
\quad \lor (\exists p : APREF;\ ldpt : ADDRESS;\ mem : MEM \bullet \\
\qquad rq? = SWAPIN\langle\langle p, ldpt\rangle\rangle \\
\qquad \land\ readProcessStoreFromDisk[p/p?, mem/ms!] \\
\qquad \land\ sms.WriteMainStore[ldpt/loadpoint?, mem/mseg?] \\
\qquad \land\ donesema.Signal) \\
\quad \lor (\exists p : APREF \bullet \\
\qquad rq? = DELSPROC\langle\langle p\rangle\rangle \land deleteProcessFromDisk[p/p?]) \\
\quad \lor (\exists p : APREF;\ img : MEM \bullet \\
\qquad rq? = NEWSPROC\langle\langle p, img\rangle\rangle \\
\qquad \land\ writeProcessStoreToDisk[p/p?, img/img?]) \\
\hline
\end{array}
$$

The semaphore, *donesema*, is used to synchronise with the swapper process directly. It is used to ensure that the write request has completed before the swapper process updates the storage tables associated with the process that is being swapped. This is to ensure consistency.

The main loop for the swap disk process is as follows. The reader should note the *ad hoc* use of a universal quantifier to model an infinite loop:

$$
\begin{array}{l}
RunProcess \mathrel{\widehat=} \\
\quad \forall i : 1 \mathinner{\ldotp\ldotp} \infty \bullet \\
\qquad sleepDriver \mathbin{;} \\
\qquad (\exists rq : SWAPRQMSG \bullet \\
\qquad\quad rqs.ReadRequest[rq/rq!] \\
\qquad\quad \land (rq = NULLSWAP \land sleepDriver) \\
\qquad\quad \lor (handleRequest \land sleepDriver))
\end{array}
$$

Proposition 86. *$p? \notin \operatorname{dom} dmem$ and $dmem' = dmem = \oplus\{p? \mapsto ms?\}$ implies that $p? \in \operatorname{dom} dmem'$. In addition, if $p? \in \operatorname{dom} dmem$ and $dmem' = dmem \oplus \{p? \mapsto ms?\}$, this implies that $p? \in \operatorname{dom} dmem'$.*

PROOF. Both parts are a consequence of the definition of $\oplus$: $f \oplus g(x) = g(x)$ if $x \in \operatorname{dom} g$ and $f(x)$ otherwise. □

4.6.2 Swapper

This subsection is about the process swapper. In fact, the swapper is better described as a storage-management module. The software is a module because it implements a set of tables describing the state of each user process' storage. In particular, the module contains tables recording the identifiers of those processes that are currently swapped out to disk (*swapped_out*) and the time that each process has spent out of main store on the swap disk (*swappedout_time*). Swapping, in this kernel, is based on the time processes have spent swapped out, so these two tables are of particular importance. However, the time a process has been resident in main store is significant and is used to determine which process to swap out when its store is required to hold a process that is being swapped in from disk. The time each process resides in main store is recorded in the *residency_time* table.

The operations on the class *ProcessStorageDescr* are composed of structures that record the time each (user) process has resided in main store and the time it has resided on disk. Marking operations are also provided so that the system can keep track of which processes are in store and which are not. The remaining operations are concerned with housekeeping and which determining which processes to swap in and out of main store.

It was decided (somewhat unfairly) that main-store residency time would include the time processes spend in queues of various sorts. This has the unfortunate consequence that a process could be swapped in, immediately make a device request and block; as soon as the request is serviced and the process is readied, it is swapped out again. However, other schemes are very much more complicated to model and therefore to implement.

The class is defined as follows:

ProcessStorageDescrs

$\upharpoonright$(*INIT*, *MakeInStoreProcessSwappable*, *MakeProcessOnDiskSwappable*,
UpdateAllStorageTimes, *MarkAsSwappedOut*, *MarkAsInStore*,
ClearProcessResidencyTime,
ClearSwappedOutTime, *IsSwappedOut*, *SetProcessStartResidencyTime*,
SetProcessStartSwappedOutTime, *UpdateProcessStoreInfo*,
RemoveProcessStoreInfo,
AddProcessStoreInfo,
ProcessStoreSize, *ReadyProcessChildren*,
CodeOwnerSwappedIn, *ReadyProcessChildren*, *NextProcessToSwapIn*,
BlockProcessChildren, *HaveSwapoutCandidate*, *FindSwapoutCandidate*)

$proctab : ProcessTable$
$sched : LowLevelScheduler$
$swapped\_out : \mathbb{F}\, APREF$
$residencytime : APREF \twoheadrightarrow\!\!\!\!\!\!\mapsto TIME$
$swappedout\_time : APREF \twoheadrightarrow\!\!\!\!\!\!\mapsto TIME$

$swapped\_out \subseteq \mathrm{dom}\, pmem \land swapped\_out \subseteq \mathrm{dom}\, pmemsize$
$\mathrm{dom}\, swappedout\_time = swapped\_out$
$\mathrm{dom}\, residencytime \cap swapped\_out = \varnothing$

INIT

$pt? : ProcessTable$
$sch? : LowLevelScheduler$

$proctab' = pt? \land sched' = sch?$
$swapped\_out' = \varnothing \land \mathrm{dom}\ residencytime' = \varnothing$
$\mathrm{dom}\ swappedout\_time' = \varnothing$

$MakeInStoreProcessSwappable \mathrel{\widehat{=}} \ldots$

$MakeProcessOnDiskSwappable \mathrel{\widehat{=}} \ldots$

$UpdateAllStorageTimes \mathrel{\widehat{=}} \ldots$

$MarkAsSwappedOut \mathrel{\widehat{=}} \ldots$

$MarkAsInStore \mathrel{\widehat{=}} \ldots$

$ClearProcessResidencyTime \mathrel{\widehat{=}} \ldots$

$ClearSwappedOutTime \mathrel{\widehat{=}} \ldots$

$IsSwappedOut \mathrel{\widehat{=}} \ldots$

$SetProcessStartResidencyTime \mathrel{\widehat{=}} \ldots$

$SetProcessStartSwappedOutTime \mathrel{\widehat{=}} \ldots$

$AddProcessStoreInfo \mathrel{\widehat{=}} \ldots$

$UpdateProcessStoreInfo \mathrel{\widehat{=}} \ldots$

$RemoveProcessStoreInfo \mathrel{\widehat{=}} \ldots$

$ProcessStoreSize \mathrel{\widehat{=}} \ldots$

$CodeOwnerSwappedIn \mathrel{\widehat{=}} \ldots$

$BlockProcessChildren \mathrel{\widehat{=}} \ldots$

$ReadyProcessChildren \mathrel{\widehat{=}} \ldots$

$NextProcessToSwapIn \mathrel{\widehat{=}} \ldots$

$HaveSwapoutCandidate \mathrel{\widehat{=}} \ldots$

$FindSwapoutCandidate \mathrel{\widehat{=}} \ldots$

As can be seen, the class has a rather large number of operations.

The following schema defines the operation that makes a process swappable. It does this by setting its main-store residency time to 0.

MakeInStoreProcessSwappable

$pid? : APREF$

$residencytime' = residencytime \oplus \{pid? \mapsto 0\}$

Processes can be created on disk when there is insufficient main store available. As user processes, they can be made swappable. The following operation does this. It just sets the swapped-out time to 0 and adds the process reference to the set of swapped-out processes.

$$
\begin{array}{|l}
\multicolumn{1}{l}{\_\_ \mathit{MakeProcessOnDiskSwappable} \_\_\_\_} \\
pid? : AREF \\
\hline
swappedout\_time' = swappedout\_time \oplus \{pid? \mapsto 0\} \\
swapped\_out' = swapped\_out \cup \{pid?\} \\
\hline
\end{array}
$$

The management module interacts with the clock. On every clock tick, the time that each process has been main-store and swap-disk resident is incremented by one tick (actually by the amount of time represented by a single tick). The following schema defines this operation:

$$
\begin{array}{|l}
\multicolumn{1}{l}{\_\_ \mathit{UpdateAllStorageTimes} \_\_\_\_} \\
\Delta(swappedout\_time, residencytime) \\
\hline
(\forall p : APREF \mid p \in \mathrm{dom}\, residencytime \bullet \\
\quad residencytime' = residencytime \oplus \{p \mapsto residencytime(p) + 1\}) \\
(\forall p : APREF \mid p \in \mathrm{dom}\, swappedout\_time \bullet \\
\quad swappedout\_time' = swappedout\_time \oplus \\
\qquad \{p \mapsto swappedout\_time(p) + 1\}) \\
\hline
\end{array}
$$

When a process is swapped out to disk, it must be marked as being no longer in main store. The following schema defines this operation:

$$
\begin{array}{|l}
\multicolumn{1}{l}{\_\_ \mathit{MarkAsSwappedOut} \_\_\_\_} \\
\Delta(swapped\_out) \\
p? : APREF \\
\hline
swapped\_out' = swapped\_out \cup \{p?\} \\
\hline
\end{array}
$$

Conversely, when a process is copied into main store, the management software needs to make a record of this fact. The operation *MarkAsInStore* performs this marking and is defined as:

$$
\begin{array}{|l}
\multicolumn{1}{l}{\_\_ \mathit{MarkAsInStore} \_\_\_\_} \\
\Delta(swapped\_out) \\
p? : APREF \\
\hline
swapped\_out' = swapped\_out \setminus \{p?\} \\
\hline
\end{array}
$$

Note that the marking is modelled as a simple set operation. The assumption is that a process that is not marked as swapped out is resident in main store.

When a process enters main store, or terminates, its residency time has to be cleared:

$$
\begin{array}{|l}
\hline
\textit{ClearProcessResidencyTime} \\
\Delta(residencytime) \\
p? : APREF \\
\hline
residencytime' = residencytime \oplus \{p? \mapsto 0\} \\
\hline
\end{array}
$$

Similarly, when a process is swapped out, or terminates, the time that it has spent on disk has to be set to zero:

$$
\begin{array}{|l}
\hline
\textit{ClearSwappedOutTime} \\
\Delta(swappedout\_time) \\
p? : APREF \\
\hline
swappedout\_time' = swappedout\_time \oplus \{p? \mapsto 0\} \\
\hline
\end{array}
$$

The following pair of schemata define operations to set the start times for main-store and swap-disk residency. The idea is that the actual time is set, rather than some number of clock ticks.

$$
\begin{array}{|l}
\hline
\textit{SetProcessStartResidencyTime} \\
\Delta(residencytime) \\
p? : APREF \\
t? : TIME \\
\hline
residencytime' = residencytime \oplus \{p? \mapsto t?\} \\
\hline
\end{array}
$$

$$
\begin{array}{|l}
\hline
\textit{SetProcessStartSwappedOutTime} \\
\Delta(swappedout\_time) \\
p? : APREF \\
t? : TIME \\
\hline
swappedout\_time' = swappedout\_time \oplus \{p? \mapsto t?\} \\
\hline
\end{array}
$$

The following predicate is used to determine whether a process is on disk.

$$
\begin{array}{|l}
\hline
\textit{IsSwappedOut} \\
p? : APREF \\
\hline
p? \in swapped\_out \\
\hline
\end{array}
$$

When a process is created, entries in the storage-management tables must be created. The storage descriptor describing the process' main-store region is set in the process' descriptor.

$$
\begin{array}{l}
AddProcessStoreInfo \mathrel{\widehat=} \\
\quad (\exists\, pd : ProcessDescr \bullet \\
\qquad proctab.DescrOfProcess[p?/pid?, pd/pd!] \\
\qquad \land\ pd.SetStoreDescr[mdesc?/newmem?])
\end{array}
$$

The following operation updates the storage descriptor should a process be relocated when swapped into main store. The storage descriptor input to this operation ($mdesc?$) need not be the same as the one already stored. This is because the swap-in operation stores the process image in the first available hole in main store that is of sufficient size.

$$
\begin{array}{l}
UpdateProcessStoreInfo \mathrel{\widehat=} \\
\quad (\exists\, pd : ProcessDescr \bullet \\
\qquad proctab.DescrOfProcess[p?/pid?, pd/pd!] \\
\qquad pd.SetStoreDescr[mdesc?/newmem?])
\end{array}
$$

The following operation removes a process from the storage-management module's tables. It also removes the storage descriptor from the process' descriptor in the process table.

$$
\begin{array}{|l}
\hline
\textit{RemoveProcessStoreInfo} \\
\Delta(residencytime, swappedout\_time) \\
p? : APREF \\
\hline
residencytime' = \{p?\} \mathbin{\lhd\!\!\!-} residencytime \\
swappedout\_time' = \{p?\} \mathbin{\lhd\!\!\!-} swappedout\_time \\
(\exists\, md : MEMDESC;\ pd : ProcessDescr \bullet \\
\quad md = (0, 0) \\
\quad \land\ proctab.DescrOfProcess[p?/pid?, pd/pd!] \\
\quad \land\ pd.SetStoreDescr[md/newmem?]) \\
\hline
\end{array}
$$

The next schema defines an operation that computes the size of the storage occupied by a process:

$$
\begin{array}{l}
ProcessStoreSize \mathrel{\widehat=} \\
\quad (\exists\, pd : ProcessDescr \bullet \\
\qquad proctab.DescrOfProcess[p?/pid?, pd/pd!] \\
\qquad pd.StoreSize)
\end{array}
$$

The next few schemata operate on the children of a process. When a process blocks, its children, according to this process model, must also be blocked. The reason for this is that the children of a process share its code. Child processes do not copy their parent's code and become a totally independent unit. The reason for this is clear: if child processes were to copy their parent's code, the demand upon store would increase and this would decrease the number of processes that could be maintained in main store at any one time. The advantage to independent storage of code is that processes can be swapped out more easily. However, the consumption of main store is considered, in this design at least, to be more important than the ease of swapping. Therefore, the swapping rules for this kernel are somewhat more complex than for some other possible designs.

The process model (such as it is) for this kernel is somewhat similar to that used by Unix: processes can create child processes (and child processes can create child processes up to some limit on depth[3]. Child processes share their parent's code but have their own private stack and data storage. When a parent is swapped out, its code is also swapped out (which makes an already complex swapping mechanism a little simpler). Because the parent process' code is swapped out, child processes have no code to execute. It is, therefore, necessary to unready the children of a swapped-out parent. The following schema defines this operation.

The schema named *BlockProcessChildren* blocks the descendant processes of a given parent. The complete set of descendants is represented by the transitive closure of the *childof* relation; the complete set of descendants of a given process are represented by $childof^+ (\!| \{p?\} |\!)$ for any process identifier $p?$. In *BlockProcessChildren*, *ps* is the set of descendants of $p?$ (should they exist). The operation then adds the processes in *ps* to the *blockswaiting* set (which is used to denote those processes that are blocked because the code they execute has been swapped out); and it sets their status to pstwaiting.

$$
\begin{array}{l}
\textit{BlockProcessChildren} \\
\hline
p? : APREF \\
\hline
\exists\, ps, \mathit{offspring} : \mathbb{F}\, APREF;\ pd : ProcessDescr \bullet \\
\quad proctab.DescrOfProcess[p?/pid?, pd/pd!] \\
\quad \land\ proctab.AllDescendants[p?/parent, \mathit{offspring}/descs!] \\
\quad \land\ pd.BlocksProcesses[ps/bw!] \\
\quad \land\ (\forall\, p : APREF \mid p \in ps \cup \mathit{offspring} \bullet \\
\qquad (\exists\, pd_1 : ProcessDescr \bullet \\
\qquad\quad proctab.DescrOfProcess[p/pid?, pd_1/pd!] \\
\qquad\quad \land\ pd_1.SetProcessStatusToWaiting) \\
\qquad \land\ sched.MakeUnready[p/pid?])
\end{array}
$$

The schema could be simplified.

When a parent is returned to main store, its children can be readied (i.e., added to the ready queue). The following schema defines this operation in a fairly obvious fashion.

First, the identifiers of all processes that become blocked when the process denoted by *parent?* is blocked are determined by *pd.BlocksProcesses*. Next, identifiers of all the descendants of the process are determined by *AlDescendants*. Next, each of the identifiers in the union of these two sets is marked as being present in store and then added to the ready queue so that it can be scheduled.

[3] The actual limit is imposed by the maximum number of entries in the process table. This fact is a clear problem for a kernel's security: a malicious process could deliberately create child processes.

$$
\begin{array}{l}
\textit{ReadyProcessChildren} \\
\hline
\Delta(\textit{swapped\_out}) \\
\textit{parent}? : \textit{APREF} \\
\hline
\exists\, \textit{pd} : \textit{ProcessDescr};\ \textit{bw}, \textit{offspring} : \mathbb{F}\, \textit{APREF} \bullet \\
\quad \textit{proctab}.\textit{DescrOfProcess}[\textit{parent}?/\textit{pid}?, \textit{pd}/\textit{pd}!] \\
\quad \wedge\ \textit{pd}.\textit{BlocksProcesses}[\textit{bw}/\textit{bw}!] \\
\quad \wedge\ \textit{proctab}.\textit{AllDescendants}[\textit{offspring}/\textit{descs}!] \\
\quad \wedge\ (\forall\, c : \textit{APREF} \mid c \in \textit{bw} \cup \textit{offspring} \bullet \\
\qquad (\exists\, \textit{cdesc} : \textit{ProcessDescr} \bullet \\
\qquad\quad \textit{proctab}.\textit{DescrOfProcess}[c/\textit{pid}?, \textit{cdesc}/\textit{pd}!] \\
\qquad\quad \wedge\ \textit{MarkAsInStore}[c/p?] \wedge \textit{sched}.\textit{MakeReady}[c/\textit{pid}?]))
\end{array}
$$

What if a child is waiting for a device request completion? It cannot suddenly be stopped. A quick and totally horrid solution is to require that all children be in the ready queue when the swap occurs.

The reader is invited to find better alternatives and to specify them in an appropriate notation.

Proposition 87. *For any parent process, p,*

$$\textit{BlockProcessChildren} \Rightarrow (\forall\, p_1 : \textit{APREF} \mid \textit{childof}(p_1, p) \bullet p \notin \operatorname{ran} \textit{userqueue})$$

PROOF. The predicate of *BlockProcessChildren* contains an instance of *MakeUnready* inside the scope of the universal quantifier. The universal quantifier ranges over all possible descendants of input process $p?$. Since *MakeUnready* removes its argument from the ready queue, the result is proved. □

Proposition 88. *If there are n processes in the ready queue at the user level and process φ has p descendants, then after BlockProcessChildren, the length of the user-level queue will be $n - 1$.*

PROOF. Without loss of generality, it can be assumed that all descendants of process φ, and all processes that it blocks, have user-level priority. Let $\textit{blocks} = \textit{ps} \cup \textit{offspring}$ and $\#\textit{blocks} = p$. By the predicate of the schema, it follows that $\forall\, p \in \textit{blocks} \bullet \textit{MakeUnready}[p/\textit{pid}?]$, so there must be p applications of *MakeUnready* to *blocks*. By Proposition 53:

$$\#\textit{readyqueues}'(\textit{userqueue}) = \#\textit{readyqueues}(\textit{userqueue}) - p$$

□

Proposition 89. *If there are n processes in the ready queue at the user level and process φ has p descendants, then after ReadyProcessChildren, the length of the user-level queue will be $n + p$.*

PROOF. Again, without loss of generality, it can be assumed that all descendants of process φ, and all processes that it blocks, have user-level priority. Again, let $blocks = bw \cup offspring$ and let $\#blocks = p$. By reasoning similar to that in the last proposition:

$$\#readyqueues'(userqueue) = \#readyqueues(userqueue) + p$$

□

The following is an immediate consequence of the last proposition.

Corollary 7. *Operation ReadyProcessChildren changes the state of all processes affected by it to* pstready.

Proposition 90. *BlockProcessChildren* $\fatsemi$ *ScheduleNext implies that currentp′ is not a descendant of the ancestor of the process just blocked.*

PROOF. This requires the proof of the following lemma.

Lemma 16. *For any process, p, BlockProcessChildren implies that there are no children of p in the ready queue after the operation completes.*

PROOF. In the predicate of schema *BlockProcessChildren*, *ps* represents the descendants of process *p*. From this, using the predicate, it can be seen that *MakeUnready*[*p*/*pid*?] for all $p \in ps$ implies $p \notin \operatorname{ran} userqueue$. In other words, the process, *p*, is removed from the ready queue by the operation *MakeUnready*. Therefore, there are no children of *p* in the ready queue. □

By Lemma 16, no child of *p* can be in the ready queue. More specifically, that *head*(*tail userqueue*) cannot be a child of *p*. This establishes the desired result. □

The following schema defines a predicate that is true when a process that owns its code is swapped into main store. Code owners are either independent processes or are parents.

```
── CodeOwnerSwappedIn ──────────────────────────
│ p? : APREF
├──────────────────
│ (∃ p1 : APREF; pd : ProcessDescr •
│      proctab.DescrOfProcess[p1/pid?, pd/pd!]
│      ∧ (pd.SharesCodeWith[p1/pid?]
│           ∨ pd.HasChild[p1/ch?])
│      ∧ pd.IsCodeOwner
│      ∧ p1 ∉ swapped_out)
└────────────────────────────────────────────────
```

The next schema is another predicate. This time, it is one that determines which process next to swap into main store. The candidate is the process that has been swapped out for the longest time. The identifier of the process ($pid!$), together with the amount of store it requires ($sz!$), is returned.

$$
\begin{array}{l}
\textit{NextProcessToSwapIn} \\
\hline
pid! : APREF \\
sz! : \mathbb{N} \\
\hline
(\exists p : APREF \mid p \in swapped\_out \bullet \\
\quad swappedout\_time(p) = max(\mathrm{ran}\ swappedout\_time) \land \\
\quad pid! = p \land \\
\quad sz! = pmemsize(p))
\end{array}
$$

Only user processes in the ready state can be swapped out. It is essential that this condition be recorded in the model. Instead of stating it directly, a less direct way is preferred. It is expressed by the following constant definition:

$$
\begin{array}{l}
\hline
illegalswapstatus : PROCSTATUS \\
\hline
illegalswapstatus = \{\mathsf{pstrunning}, \mathsf{pstwaiting}, \mathsf{pstswappedout}, \\
\qquad \mathsf{pstnew}, \mathsf{pstterm}, \mathsf{pstzombie}\}
\end{array}
$$

Again, the following schema defines a predicate. This predicate is true when the storage-management module has a candidate process to swap out to disk.

$$
\begin{array}{l}
\textit{HaveSwapoutCandidate} \\
\hline
rqsz? : \mathbb{N} \\
\hline
(\exists pd : ProcessDescr;\ st : PROCSTATUS;\ k : PROCESSKIND;\ sz : \mathbb{N}; \\
\qquad tm : TIME;\ tms : \mathbb{F}\ TIME \bullet \\
\quad proctab.DescrOfProcess[p_1/pid?, pd/pd!] \\
\quad \land pd.ProcessKind[k/knd!] \land k \neq \mathsf{ptsysproc} \land k \neq \mathsf{ptdevproc} \\
\quad \land pd.ProcessStatus[st/st!] \land st \notin illegalswapstatus \\
\quad \land pd.StoreSize[sz/memsz!] \land sz \geq rqsz? \\
\quad \land pdescrs.ResidencyTime[tm/tm!] \land pdescrs.AllResidencyTimes[tms/tms!] \\
\quad \land tm = max\ tms
\end{array}
$$

The swapout candidate is a user process that is in the ready queue ($illegalstatus$ is a set of state names that excludes pstready). The amount of store the candidate occupies must be at least the same as that requested for the incoming process (this is represented by $rqsz?$). The candidate must also have the greatest main-store residency time. System and device processes do not appear in any of the storage-management tables, so it is not possible for an attempt to be made to swap one of them out.

The candidate process to be swapped out is located by the following operation. It is, again, fairly straightforward. It locates a process that is not in one of the "banned" states defined by *illegalswapstatus*. The process must not be a device or system process; that is, it must be a user process. The victim process must also occupy a storage region whose size is at least that required ($rqsz?$) to fit the incoming process.

$$
\begin{array}{l}
\textit{FindSwapoutCandidate} \\
\hline
p? : APREF \\
cand! : APREF \\
rqsz? : \mathbb{N} \\
slot! : MEMDESC \\
\hline
(\exists p : APREF;\ t : TIME \mid residencytime(p) = t \bullet \\
\quad p \neq p? \land \\
\quad pstatus(p) \notin illegalswapstatus \land \\
\quad pkind(p) \neq \mathsf{ptsysproc} \land \\
\quad pkind(p) \neq \mathsf{ptdevproc} \land \\
\quad pmemsize(p) \geq rqsz? \land \\
\quad (\forall p_1 : APREF \bullet \\
\qquad (p_1 \neq p \land p_1 \neq p? \\
\qquad\qquad \land\ p_1 \in \operatorname{dom} residencytime \land t \geq residencytime(p_1)) \\
\qquad\quad \Rightarrow p = cand! \land pmem(p) = slot!)) \\
\hline
\end{array}
$$

A proposition can now be proved about the priority of swap-out candidates.

Proposition 91. *Only user-level processes that are ready to run can be swapped out.*

PROOF. By the predicate of *HaveSwapoutCandidate*, the kind of process is k, and $k \neq \mathsf{sysproc} \land k \neq \mathsf{devproc}$ implies that $k = \mathsf{userproc}$, so the process is at the user level. By the condition that $st \notin illegalswapstatus$, and

$$illegalswapstatus = \{\mathsf{pstrunning}, \mathsf{pstwaiting}, \mathsf{pstswappedout}, \mathsf{pstnew}, \mathsf{pstterm}\}$$

it follows that the process can only be in the ready state (pstready). □

4.6.3 Clock Process

The clock process is an extremely important component of the system. As has been seen, the clock is used to pre-empt user-level processes. In addition, processes of all kinds can use the clock to suspend themselves for a specified period of time; when that time has expired, the processes receive an "alarm" call which wakes them up and places them on the ready queue. This first organisation for the clock process is shown in Figure 4.2.

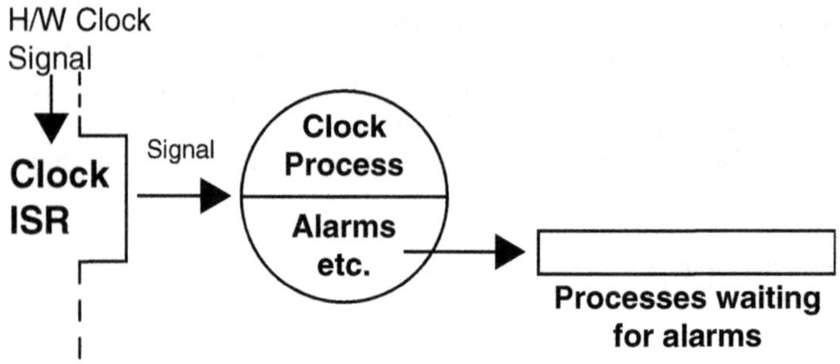

Fig. 4.2. *The clock process in relation to its interrupt and alarm requests.*

$$ticklength : TIME$$

GenericISR

$\upharpoonright(INIT,$
$\quad OnInterrupt,$
$\quad AfterProcessingInterrupt,$
$\quad WakeDriver)$

$hw : HardwareRegisters$
$ptab : ProcessTable$
$driversema : Semaphore$
$sched : LowLevelScheduler$

INIT

$sema? : Semaphore$
$schd? : LowLevelScheduler$
$hwregs? : HardwareRegisters$
$proctb? : ProcessTable$

$ptab' = proctb? driversema' = sema?$
$hw' = hwregs?$
$sched' = schd?$

$OnInterrupt \mathrel{\widehat=} \ldots$

$AfterProcessingInterrupt \mathrel{\widehat=} \ldots$

$WakeDriver \mathrel{\widehat=} \ldots$

$saveState \mathrel{\widehat=} \ldots$

$restoreState \mathrel{\widehat=} \ldots$

$$\begin{array}{|l}
\hline
\textit{WakeDriver} \\
\hline
\textit{driversema.Signal} \\
\hline
\end{array}$$

When an interrupt occurs, *SaveState* is called to save the state. The schema defines an operation that retrieves the current process' descriptor from the process table. Then, the contents of the hardware's general registers are copied from the hardware, as are the contents of the stack register, the instruction pointer and the status word. The time quantum value is also copied and the values set in the appropriate slots in the process descriptor.

The reader should note that there is a slight fiction in the *saveState* operation. It concerns the instruction pointer. Clearly, as *saveState* executes, the *IP* register will point to instructions in *saveState*, not in the code of the current process (the process pointed to by *currentp*). The *saveState* operation is called from ISRs. This implies that an interrupt has occurred and that the hardware state has already been stored somewhere (certainly, the instruction pointer *must* have been stored somewhere so that the ISR could execute). Because this model is at a relatively high level and because we are not assuming any specific hardware, we can only assume that operations such as *GetGPRegs* and *GetIP* can retrieve the general-purpose and instruction registers' contents from *somewhere*.

What has been done in the model is to abstract from all hardware. The necessary operations have been provided, even though we are unable to define anything other than the name and signature of the operations at this stage. (In a refinement, these issues would, of necessity, be confronted and resolved.)

Once *saveState* has terminated, device-specific code is executed. Finally, the operation to restore the hardware state is called to perform a context switch.

The first part of the context switch is performed by *saveState*. This operation copies the hardware state, as represented by the programmable registers, the instruction pointer and the status word, as well as the variable containing the process' time quantum. (Non-user processes just have an arbitrary value stored.) The state information is then copied into the outgoing process' process descriptor.

$$\begin{array}{|l}
\hline
\textit{saveState} \\
\hline
(\exists\, cp : IPREF \bullet \\
\quad sched.CurrentProcess[cp/cp!] \\
\quad (\exists\, pd : ProcessDescr \bullet \\
\quad\quad ptab.DescrOfProcess[cp/pid?, pd/pd!] \\
\quad\quad \land (\exists\, regs : GENREGSET;\ stk : PSTACK;\ ip : \mathbb{N}; \\
\quad\quad\quad\quad\quad\quad stat : STATUSWD;\ tq : TIME \bullet \\
\quad\quad\quad hw.GetGPRegs[regs/regs!] \\
\quad\quad\quad \land\ hw.GetStackReg[stk/stk!] \\
\quad\quad\quad \land\ hw.GetIP[ip/ip!]
\end{array}$$

$$
\begin{array}{l}
\quad\quad \land\ hw.GetStatWd[stat/stwd!] \\
\quad\quad \land\ sched.GetTimeQuantum[tq/tquant!] \\
\quad\quad \land\ pd.SetFullContext \\
\quad\quad\quad\quad [regs/pregs?, ip/pip?, stat/pstatwd?, \\
\quad\quad\quad\quad\quad stk/pstack?, tq/ptq?])))
\end{array}
$$

The current process referred to here is *not necessarily* the same as the one referred to above. Basically, whatever is in *currentp* runs next. The reason for this is that the scheduler might be called by the device-specific code that is not defined here.

The code supplied for each specific device should be as short as possible. It is a general principle that ISRs should be as fast as possible, preferably just handing data to the associated driver process.

Once the device-specific code has been run, the state is restored. As noted above, the actual state might be that of a process different from the one bound to *currentp* when *saveState* executed. This is because the low-level scheduler might have been called and *currentp*'s contents replaced by another value. The operation for restoring state (of whatever process) is defined by the following schema:

restoreState

$$
\begin{array}{l}
(\exists\, cp : IPREF \bullet \\
\quad sched.CurrentProcess[cp/cp!] \\
\quad \land\ (\exists\, pd : ProcessDescr \bullet \\
\quad\quad ptab.DescrOfProcess[cp/pid?, pd/pd!] \\
\quad\quad \land\ (\exists\, regs : GENREGSET;\ stk : PSTACK; \\
\quad\quad\quad\quad\quad ip : \mathbb{N};\ stat : STATUSWD;\ tq : TIME \bullet \\
\quad\quad\quad pd.FullContext[regs/pregs!, ip/pip!, stat/pstatwd!, \\
\quad\quad\quad\quad\quad stk/pstack!, tq/ptq!] \\
\quad\quad\quad \land\ hw.SetGPRegs[regs/regs?] \\
\quad\quad\quad \land\ hw.SetStackReg[stk/stk?] \\
\quad\quad\quad \land\ hw.SetStatWd[stat/stwd?] \\
\quad\quad\quad \land\ sched.SetTimeQuantum[tq/tquant?] \\
\quad\quad\quad \land\ hw.SetIP[ip/ip?])))
\end{array}
$$

In this case, the various registers are all stored in known locations inside the kernel (in the descriptor of the process that is to run next). The transfers are moves to the hardware's registers. The instruction pointer is the last to be set (for obvious reasons).

These are the generic interrupt service routines. The first is called before performing the interrupt-specific operations:

$$
\begin{array}{l}
OnInterrupt \mathrel{\widehat=} \\
\quad (saveState \fatsemi \\
\qquad (\exists\, p : IPREF \bullet \\
\qquad\qquad sched.CurrentProcess[p/cp!] \land sched.MakeReady[p/pid?])) \\
\quad \fatsemi WakeDriver
\end{array}
$$

The second operation is called when the ISR is about to terminate:

$$
\begin{array}{l}
AfterProcessingInterrupt \mathrel{\widehat=} \\
\quad (sched.ScheduleNext \fatsemi restoreState)
\end{array}
$$

It is assumed that the clock interrupt does just that—raise an interrupt. A shared variable, encapsulated in *TimeNow*, stores the current time. The actual value passed to *TimeNow* is the length of one tick (expressed in arbitrary units here). The shared variable is only updated by *CLOCKISR*, so there is no contention problem because all other accesses are reads that are protected by locking. The update of the clock is atomic because it is performed within an ISR; the reads are also atomic because they are performed inside locks. This mechanism is quite sufficient.

The clock's ISR now follows, presented as a class. Note that it notionally inherits methods from a *GenericISR*.

$$
\begin{array}{|l}
\textbf{CLOCKISR} \\
\hline
\upharpoonright(INIT, ServiceISR) \\
\\
GenericISR \\
\\
\quad zsema : Semaphore \\
\quad tmnow : TimeNow \\
\\
\quad \underline{INIT} \\
\quad tn? : TimeNow \\
\quad zs? : Semaphore \\
\quad \hline
\quad tmnow' = tn? \\
\quad zsema' = zs? \\
\\
setTime \mathrel{\widehat=} \\
\qquad (\exists\, tn : TIME \mid tn = ticklength \bullet \\
\qquad\qquad tmnow.SetTime[tn/t?]) \\
ServiceISR \mathrel{\widehat=} \\
\qquad OnInterrupt \fatsemi \\
\qquad\qquad setTime \\
\qquad\qquad \land\ zsema.Signal \\
\qquad\qquad \fatsemi AfterProcessingInterrupt \\
\hline
\end{array}
$$

The ISR uses a semaphore to wake the driver when an interrupt occurs. It is assumed that the semaphore is initialised by some kernel start-up operation before it is passed to the ISR. The main operation of the ISR is *ServiceISR*.

This is the *TimeNow* shared variable. It has two operations: *SetTime* and *CurrentTime*. The *CurrentTime* operation retrieves the current value of the *time_now* variable.

TimeNow

$\upharpoonright(INIT, SetTime, CurrentTime)$

$time\_now : TIME$

INIT

$sttm? : TIME$

$time\_now' = sttm?$

SetTime

$\Delta(time\_now)$
$t? : TIME$

$time\_now' = t? + time\_now$

CurrentTime

$t! : TIME$

$t! = time\_now$

The process that removes zombies is now specified so that it is out of the way. It is encapsulated as a class, as follows:

DeZombifier

$\upharpoonright(INIT, RunProcess)$

$zsema : Semaphore$
$lck : Lock$
$proctab : ProcessTable$

INIT

$zs? : Semaphore$
$lk? : Lock$
$pt? : ProcessTable$

$zsema' = zs?$
$lck' = lk?$
$proctab' = pt?$

$$
\begin{array}{l}
RunProcess \mathrel{\widehat=} \\
\quad \forall i : 1 \mathrel{..} \infty \bullet \\
\qquad zsema.Wait \mathbin{;} \\
\qquad lck.Lock \mathbin{;} \\
\qquad ((proctab.GotZombies \land proctab.KillAllZombies \land lck.Unlock) \\
\qquad\quad \lor lck.Unlock)
\end{array}
$$

The main entry point, *RunProcess*, is readied by the *zsema.Wait* operation, as is standard for this kind of process (it counts as a driver process). The main routine then disables interrupts with a *lck.Lock*. Next, it determines whether there are any zombies (*proctab.GotZombies*); if there are, it kills them and unlocks. Otherwise, there are no zombies, so interrupts are re-enabled (*lck.Unlock* on the last line).

Immediately, a couple of results can be proved about the de-zombifier.

Lemma 17. *Operation KillAllZombies removes all zombies from the system.*

PROOF. Zombies are only stored in the *zombies* list. The *KillAllZombies* operation is defined as the conjunction of two operations. The crucial parts of the definition (after simplification and substitution) are:

$$
\begin{array}{l}
deadzombs! \subseteq zombies \\
\land\ zombies' = zombies \setminus deadzombs! \\
\land\ procs' = deadzombies! \lhd procs
\end{array}
$$

where *deadzombs!* is a set composed of the identifiers of those zombies whose children have all been deleted from the process table. □

Proposition 92. *DeZombifier.RunProcess removes all childless zombies from the system if any exist.*

PROOF. If there are any such zombies, the *KillAllZombies* operation is executed. The result follows from Lemma 17 that *KillAllZombies* removes zombies from everything except the scheduling queues and the process table, and from Proposition 14 (*DelProcess* removes a process from the process table). □

The clock and alarms raise an interesting question: how do user programs communicate with the kernel. A way of performing system calls must be defined. Immediately, there are two alternatives:

- a semaphore to ensure mutual exclusion between user processes, ;
- an interrupt.

The first alternative requires that all user processes signal on a semaphore when they are required to perform an SVC (system call). When inside the critical region, the user process can call system-interface functions.

The second alternative is to use an interrupt. In this case, an interrupt not expected to be used by hardware is reserved. When a user process needs to perform an SVC, it calls an interface routine that raises that interrupt. The user process passes parameters to the SVC either on its stack or in predefined locations (both pose problems of crossing address-space boundaries but the stack option generally appears the better). The parameters must include an operation code denoting the operation to be performed. ISR picks up the SVC's parameters and opcode, places them in appropriate locations within the kernel and then wakes a driver process. At this point, there are choices to be made.

The first option is for the SVC ISR to wake up a special driver and pass all the parameters to it. The driver then passes the opcode and associated data to the necessary processes (e.g., allocate store, add a request to the clock's alarm queue, or, in bulkier kernels, perform an I/O request); should the operation not involve a kernel process, the driver performs the operation directly. The ISR must unready the calling process and pass its identifier to the driver process which readies it again after the request has been serviced. The ISR wakes up the driver using a *Signal* operation on a semaphore they share. It also unreadies the calling process by a call to *MakeUnready* (the driver will ready it at a later time).

The second option is for the SVC ISR to perform as many of the requests itself as it possibly can. This means that the ISR has to inspect the opcode to determine what to do. For example, if the SVC is to request a period of sleep, it will add the identifier of the calling process (always *currentp*), together with the requested sleep period, to the alarm queue in the clock driver.

The second alternative appears attractive but suffers from some problems. First, it might easily violate the principle that ISRs do only as much as they absolutely must—ISRs should be as fast as possible. Second, it could entail significant periods during which interrupts are disabled—this is clearly not a good idea. Third, the operation of the SVC ISR might interfere with other interrupts (e.g., the system clock).

For these reasons, the first alternative is adopted here. The ISR has a structure roughly as follows:

$SVCISR$

$\lceil(INIT, HandleSVC)$

$GenericISR$

$$
\begin{array}{l}
proctab : ProcessTable \\
sched : LowLevelScheduler \\
ctxt : Context \\
alarmrqs : AlarmRQBuffer \\
\ldots
\end{array}
$$

INIT

$$\ldots$$

$$HandleSVC \mathrel{\widehat{=}} \ldots\ldots$$

The SVC ISR must place clock-related requests in a buffer. To this end, it needs to inspect the opcode associated with each SVC. The following is a fragment of the specification of the ISR:

HandleSVC

$$
\begin{array}{l}
opcode? : \mathbb{N} \\
params? :
\end{array}
$$

$$
\begin{array}{l}
\exists\, cp : IPREF \bullet \\
\quad sched.CurrentProcess[cp/cp!] \land ctxt.SaveState \mathbin{;} \\
\quad (sched.MakeUnready[currentp/pid?] \\
\qquad \land (\exists\, pd : ProcessDescr \bullet \\
\qquad\quad proctab.DescrOfProcess[cp/pid?, pd/pd!] \\
\qquad\quad \land pd.SetProcessStatusToWaiting) \land \ldots) \\
\qquad \lor (\exists\, tm : TIME \bullet \\
\qquad\quad opcode? = SLEEP\langle\!\langle tm \rangle\!\rangle \\
\qquad\quad \land alarmrqs.AddAlarm[cp/p?, tm/t?] \land WakeDriver) \\
\quad sched.ScheduleNext \mathbin{;} ctxt.RestoreState
\end{array}
$$

In this system, the user can request to sleep for a period, denoted $SLEEP\langle\!\langle tm \rangle\!\rangle$ above, where tm denotes the period of sleep. The handler sets the sleep period and the process identifier of the caller in the request buffer using *AddAlarm*. The current process is the caller (it is the process that raised the interrupt) and the period of sleep is specified as a parameter to the SVC (passed on the stack or in another known location).

It should be noted that the type used to represent requests (of which *SLEEP* is one component) is omitted. The reason for this is that the remainder of the range of SVCs is only partially defined in this chapter.

The request buffer is just a shared variable. There is no need for the request buffer to be protected by a semaphore because, as noted above, the only writer of requests is the SVC ISR. The alarm buffer also contains a variable denoting the current time. This variable is also updated only within an ISR, this time the one associated with the hardware clock. These two updates cannot occur at the same time because they are performed by ISRs. All reads to the variables inside the request buffer are protected by locks, so there can be no contention there, either.

The request buffer has three visible operations. The first (*AddAlarm*) adds a sleep request to the internal queue held in the buffer (the queue is represented by the finite partial function *alarms*); here, sleeping is interpreted as the time before the process is to be resumed (called an "alarm"). The second, *CancelAlarm*, is used to remove an alarm request from the buffer; its use, in this kernel, is restricted to tidying up when processes are killed (not covered here). The third, *HaveAlarms*, is a predicate used to determine whether there are any alarm requests in the queue; it only reads *alarms*. The final operation is *CallAlarms*, which runs over the queue determining which processes are ready to wake.

It should be noted that the use of the finite partial function from process identifiers (*APREF*) to time values (*TIME*) to represent the queue of sleeping processes (or, alternatively, those processes waiting for an alarm call), *alarms*, ensures that a process can only make one sleep request at any time. This does not appear to be a restriction. It should also be noted that, because of *alarm*'s domain type, the idle process cannot sleep:

Proposition 93. *The idle process cannot sleep.*

PROOF. Immediate from the definition of *APREF*, the domain type of *alarms*. □

The buffer is called *AlarmRQBuffer* and is defined as follows (the identifier "*AlarmRQQueue*" was resisted on the grounds of euphony):

AlarmRQBuffer

$\upharpoonright(INIT, AddAlarm, CancelAlarm, HaveAlarms, CallAlarms)$

$sched : LowLevelScheduler$
$alarms : APREF \nrightarrow\!\!\!\!\nrightarrow TIME$
$timenow : TimeNow$

INIT

$tn? : TimeNow$
$sch? : LowLevelScheduler$

$timenow' = tn?$
$sched' = sch?$
$\mathrm{dom}\ alarms' = \varnothing$

$$
\begin{array}{l}
\textit{AddAlarm} \\
\hline
\Delta(alarms) \\
p? : APREF \\
t? : TIME \\
\hline
\exists\, tm : TIME \bullet \\
\quad timenow.CurrentTime[tm/t!] \land alarms' = alarms \oplus \{p? \mapsto t? + tm\} \\
\hline
\end{array}
$$

$$
\begin{array}{l}
\textit{CancelAlarm} \\
\hline
\Delta(alarms) \\
p? : APREF \\
\hline
alarms' = \{p?\} \lhd alarms \\
\hline
\end{array}
$$

$$
\begin{array}{l}
\textit{HaveAlarms} \\
\hline
\exists\, tm : TIME \bullet \\
\quad timenow.CurrentTime[tm/t!] \\
\quad \land \{p : APREF \mid p \in \operatorname{dom} alarms \land alarms(p) \leq tm \bullet \\
\qquad\qquad (p, alarms(p))\} \neq \varnothing \\
\hline
\end{array}
$$

$$
\begin{array}{l}
\textit{CallAlarms} \\
\hline
\exists\, tm : TIME \bullet \\
\quad timenow.CurrentTime[tm/t1] \\
\quad \land (\exists\, pairs : \mathbb{F}\, APREF \times Time;\ pids : \mathbb{F}\, APREF \bullet \\
\qquad pairs = \{p : APREF \mid p \in \operatorname{dom} alarms \land alarms(p) \leq tm \bullet \\
\qquad\qquad (p, alarms(p))\} \\
\qquad \land alarms' = alarms \setminus pairs \\
\qquad \land pids = \{p : APREF;\ tm : TIME \mid (p, tm) \in pairs \bullet p\} \\
\qquad \land (\forall\, p : APREF \mid p \in pids \bullet \\
\qquad\quad \land sched.MakeReady[p/p?])) \\
\hline
\end{array}
$$

The model of the driver process now follows. The process is represented by a class that exports two operations: its initialisation operation and the *RunProcess* operation. The *RunProcess* operation is an infinite loop that merely updates the swap times in the swapper process and determines whether there are any alarms to be called. All alarm operations are performed by the *CallAlarms* operation inside the request buffer, so the driver does not see the structure of alarm requests. The driver also does not see the structures inside the swapper process. These encapsulations ensure that the clock driver's operations are simple and easy to understand; they also localise any problems with timing.

The definition of the *ClockDriver* process now follows.

$$
\begin{array}{l}
\textit{ClockDriver} \\
\hline
\upharpoonright(\textit{INIT}, \textit{RunProcess}) \\
\\
\textit{lck} : \textit{Lock};\ \textit{devsema} : \textit{Semaphore};\ \textit{swaptabs} : \textit{ProcessStorageDescrs} \\
\textit{swappersema} : \textit{Semaphore};\ \textit{timenow} : \textit{TimeNopw};\ \textit{alarms} : \textit{AlarmRQBuffer} \\
\\
\textit{INIT} \\
\hline
\textit{lk}? : \textit{Lock};\ \textit{alarms}? : \textit{AlarmRQBuffer} \\
\textit{swaptb}? : \textit{ProcessStorageDescrs};\ \textit{swapsema}? : \textit{Semaphore} \\
\textit{tn}? : \textit{TimeNow} \\
\hline
\textit{lck}' = \textit{lk}? \wedge \textit{alarms}' = \textit{alarms}? \\
\textit{tiemnow}' = \textit{tn}? \wedge \textit{swaptabs}' = \textit{swaptb}? \\
\textit{swappersema}' = \textit{swapsema}? \\
\hline
\\
\textit{putDriverToSleep} \mathrel{\widehat=} \ldots \\
\textit{updateSwapperTimes} \mathrel{\widehat=} \ldots \\
\textit{RunProcess} \mathrel{\widehat=} \ldots
\end{array}
$$

The operation of the driver is relatively simple, as will be seen from the description of its component routines.

The driver is made to wait for the next interrupt by the following operation. It waits on the *devsema*, the device semaphore:

$$
\begin{array}{l}
\textit{putDriverToSleep} \\
\hline
\textit{devsema.Wait} \\
\hline
\end{array}
$$

The swapper uses time to determine which is to be swapped out. This requires updating swapper tables at every clock tick. The operation called by the clock driver is the following.

$$
\begin{array}{l}
\textit{updateSwapperTimes} \mathrel{\widehat=} \\
\quad \textit{swaptabs.UpdateAllStorageTimes} \mathbin{;} \textit{swappersema.Signal}
\end{array}
$$

The main clock-driver routine is as follows. Its basic operation is to update the swapper's timers and call alarms:

$$
\begin{array}{l}
\textit{RunProcess} \mathrel{\widehat=} \\
\quad \textit{putDriverToSleep} \mathbin{;} \\
\quad \wedge\ (\forall i : 1 \mathinner{\ldotp\ldotp} \infty \bullet \\
\qquad \textit{lck.Lock} \mathbin{;} \\
\qquad (\exists\, \textit{tm} : \textit{TIME} \bullet \\
\qquad\quad \textit{timenow.CurrentTime}[\textit{tm}/t!] \\
\qquad\quad \wedge\ \textit{updateSwapperTimes}[\textit{tm}/\textit{tm}?] \\
\qquad\quad \wedge\ \textit{sched.UpdateProcessQuantum}) \\
\quad \wedge\ ((\textit{alarms.HaveAlarms} \wedge \textit{alarms.CallAlarms}) \\
\qquad \vee\ \textit{Skip})
\end{array}
$$

$\wedge$ *lck.Unlock*
$\wedge$ *putDriverToSleep*)

Proposition 94. *The operation alarmsToCall implies that, if $alarms \neq \varnothing$, $\forall p : APREF \mid p \in domalarms' \bullet alarms'(p) > now$.*

PROOF. By the predicate, $alarms' = alarms \setminus pairs$, where:

$$pairs = \{p : APREF \mid p \in \mathrm{dom}\, alarms \wedge alarms(p) \leq now \bullet (p, alarms(p))\}$$

Therefore, on each call to *alarmsToCall*, it is true that:

$$\forall p : APREF \mid p \in domalarms' \bullet alarms'(p) > now$$

(since this is just *alarms'*). □

Proposition 95. *All swapped-out processes age by one tick when the clock driver is executed.*

PROOF. The critical schema is *UpdateAllStorageTimes*. This is a component of *updateSwapperTimes*.

The schema *UpdateAllStorageTimes* contains the identity

$$swappedout\_time' = swappedout\_time \oplus \{p \mapsto swappedout\_time(p) - 1\}$$

□

Proposition 96. *All resident processes age by one tick when the clock driver is executed.*

PROOF. Similar to the above but replacing *swappedout_time'* by *residencytime'*. □

Proposition 97. *The current process' time quantum is reduced by one unit (if the current process is at the user level) each time the clock driver is executed.*

PROOF. The body of the *RunProcess* operation in the clock driver contains, as a conjunct, a reference to the schema *sched.UpdateProcessQuantum*. The predicate of this last schema contains the identity $currentquant' = currentquant - 1$. □

Proposition 98. *If, in alarmsToCall, $\#pairs > 0$, the ready queue grows by $\#pairs$.*

PROOF. By induction using Proposition 49. □

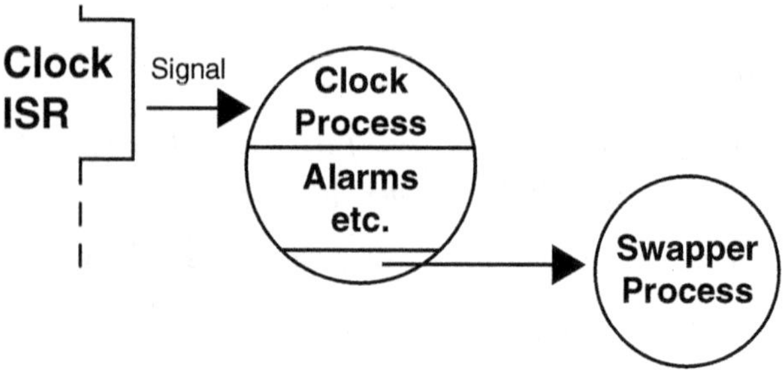

Fig. 4.3. *Interaction between clock and swapper processes.*

4.6.4 Process Swapping

In this kernel, user processes are swapped in and out of main store. This mechanism is introduced so that there can be more processes in the system than main store could support. It is a simple storage-management principle that pre-dates virtual store and requires less hardware support. In our scheme, processes are swapped after they have been resident in main store for a given amount of time. When a process is swapped, its entire image is copied to disk, thus freeing a region of main store for another user process to be swapped in.

The relationship between this process, the *Swapper* process, and the clock process is depicted in Figure 4.3

Swapping is performed by two main processes: one to select victims and another to copy process images to and from a swapping disk.

The swapper process is modelled by the following class. The main routine is *RunProcess*.

SwapperProcess

↾(*INIT*, *swapProcessOut*, *swapCandidateOut*, *swapProcessIn*,
swapProcessIntoStore, *DoDiskSwap*, *RunProcess*)

donesema : *Semaphore*;
swapsema : *Semaphore*;
pdescrs : *ProcessStorageDescrs*;
proctab : *ProcessTable*;
sched : *LowLevelScheduler*;
sms : *SharedMainStore*;
hw : *HardwareRegisters*;
diskrqbuff : *SwapRQBuffer*;
realmem : *SharedMainStore*

$$
\begin{array}{l}
\textit{INIT} \\
\hline
dsma? : Semaphore \\
pdescs? : ProcessStorageDescrs \\
sched? : LowLevelScheduler \\
pt? : ProcessTable \\
store? : SharedMainStore \\
hwr? : HardwareRegisters \\
dskrq? : SwapRQBuffer \\
swpsema? : Semaphore \\
ms? : SharedMainStore \\
\hline
donesema' = dsma? \\
swapsema' = swpsema? \\
pdescrs' = pdescr? \\
sched' = sched? \\
proctab' = pt? \\
sms' = store? \\
hw' = hwr? \\
diskrqbuff' = dskrq? \\
realmem' = ms?
\end{array}
$$

$requestWriteoutSegment \mathrel{\widehat=} \ldots$

$requestReadinSegment \mathrel{\widehat=} \ldots$

$swapProcessOut \mathrel{\widehat=} \ldots$

$swapCandidateOut \mathrel{\widehat=} \ldots$

$swapProcessIn \mathrel{\widehat=} \ldots$

$swapProcessIntoStore \mathrel{\widehat=} \ldots$

$doDiskSwap \mathrel{\widehat=} \ldots$

$waitForNextSwap \mathrel{\widehat=} \ldots$

$RunProcess \mathrel{\widehat=} \ldots$

The following operation requests that a segment of main store be written to disk. It supplies the start and end addresses of the segment to be copied:

$$
\begin{array}{l}
\textit{requestWriteoutSegment} \\
\hline
p? : APREF \\
start?, end? : ADDRESS \\
\hline
(\exists rq : SWAPRQMSG \bullet \\
\quad rq = SWAPOUT\langle\!\langle p?, start?, end? \rangle\!\rangle \\
\quad \land\ diskrqbuff.SetRequest[rq/rq?])
\end{array}
$$

The next operation models the operation to read a segment into main store. The name of the process to which the image belongs, as well as the address at which to start copying, are supplied as parameters. The disk image contains the length of the segment.

$$
\begin{array}{|l}
\hline
\textit{requestReadinSegment} \\
p? : APREF \\
loadpoint? : ADDRESS \\
\hline
(\exists\, rq : SWAPRQMSG \bullet \\
\quad rq = SWAPIN \langle\langle p?, loadpoint? \rangle\rangle \\
\quad \land\ diskrqbuff.SetRequest[rq/rq?]) \\
\hline
\end{array}
$$

The operation that actually swaps process images out is given by the following schema. The operation, like many of those that follow, is deceptively simple when written in this form. It should be noted that it is disk residency time that determines when swapping occurs; the basic principle on which the swapper operates is that processes compete for main store, not disk residency.

$$
\begin{array}{l}
swapProcessOut \mathrel{\widehat=} \\
\quad (\exists\, pd : ProcessDescr \bullet \\
\qquad proctab.DescrOfProc[p?/pid?, pd/pd!] \\
\qquad \land\ requestWriteoutSegment \\
\qquad \land\ sched.MakeUnready[p?/pid?] \\
\qquad \land\ realmem.FreeMainStore \\
\qquad \land\ pdescr.ClearProcessResidencyTime \\
\qquad \land\ pdescr.SetProcessStartSwappedOutTime \\
\qquad \land\ pdescr.BlockProcessChildren \\
\qquad \land\ pd.SetStatusToSwappedOut \\
\qquad \land\ pdescr.MarkAsSwappedOut)
\end{array}
$$

A high-level description is relatively easy. The process descriptor of the process to be swapped out is retrieved from the process table. The segment corresponding to the selected process is determined (and copied) and the process is unreadied. The residency and start of swapout times for the process are then cleared and the children of the selected process are then blocked. The status of the selected process is set to swappedout.

Processes have to be swapped into store. This operation is defined as:

$$
\begin{array}{l}
swapCandidateOut \mathrel{\widehat=} \\
\quad (\exists\, pd : ProcessDescr \bullet \\
\qquad proctab.DescrOfProcess[p?/pid?, pd/pd!] \\
\qquad \land\ (proctab.ProcessHasChildren \\
\qquad\quad \land\ ((proctab.IsCodeOwner \\
\qquad\qquad \land\ swapProcessOut \\
\qquad\qquad \land\ proctab.BlockProcessChildren) \\
\qquad\quad \lor\ swapProcessOut))
\end{array}
$$

$$\begin{array}{l}\quad\lor\ ((proctab.IsCodeOwner \land swapProcessOut)\\ \qquad\quad \lor\ swapProcessOut))\end{array}$$

When a process is to be swapped into main store, the following operation is employed. It determines whether the process has any child processes. If it has, it swaps the process into store and readies its children. If the newly swapped-in process owns the code it executes, it marks its code as in store and then performs the swap-in operation. If the process has no children, there is no need to ready them; the rest of the operation is the same as just described.

$$\begin{array}{l}swapProcessIn \mathrel{\widehat=}\\ \quad(proctab.ProcessHasChildren\\ \qquad\land\ ((proctab.IsCodeOwner\\ \qquad\qquad\land\ swapProcessIntoStore\\ \qquad\qquad\qquad\land\ pdescr.ReadyProcessChildren)\\ \qquad\lor\ (pdescr.CodeOwnerSwappedIn \land swapProcessIntoStore)))\\ \quad\lor\ ((proctab.IsCodeOwner \land swapProcessIntoStore)\\ \qquad\lor\ pdescr.CodeOwnerSwappedIn)\\ \qquad\qquad\qquad\qquad\land\ swapProcessIntoStore\end{array}$$

The following operation performs the swap-in operation. It allocates store and reads in the process image. It then updates the storage descriptors associated with the newly swapped-in process and then updates the relocation registers so that the image can be accessed correctly at its new address. The process is marked as in store and its status set to pstready; the swap parameters are then updated. Finally, the newly swapped-in process is readied and a reschedule occurs.

$$\begin{array}{l}swapProcessIntoStore \mathrel{\widehat=}\\ \quad(sms.AllocateFromHole[mspec/mspec!]\\ \qquad\land\ ([mspec : MEMDESC;\ \ ldpt : \mathbb{N};\ \ sz : \mathbb{N} \mid\\ \qquad\qquad\qquad\land\ ldpt = memstart(mspec)\\ \qquad\qquad\qquad\qquad\land\ sz = memsize(mspec)]\\ \qquad\land\ requestReadinSegment[ldpt/loadpoint?]\\ \qquad\land\ donesema.Wait\\ \qquad\land\ pdescrs.UpdateProcessStoreInfo[sz/sz?, mspec/mdesc?])\\ \qquad\qquad\qquad\qquad\setminus\{ldpt, sz, mspec\}\\ \quad\land\ hw.UpdateRelocationRegisters\\ \quad\land\ pdescrs.MarkAsInStore\\ \quad\land\ pd.SetStatusToReady[p?/pid?]\\ \quad\land\ pdescrs.SetProcessStartResidencyTime\\ \quad\land\ pdescrs.ClearSwappedOutTime\\ \quad\land\ (sched.MakeReady[p?/pid?] \mathbin{;} sched.ScheduleNext)\end{array}$$

The *doDiskSwap* operation is the main swapper routine. It determines the process to swap in and then finds out whether it can allocate its image in free store. If it can, it just performs the swap. If not, it determines whether it can swap some process out of store—that process should have an image size that is at least as big as that of the process to be swapped in. Once that candidate has been found, the image size is determined and the swap-out operation is performed; when the victim has been swapped out, the disk process is swapped into store.

$$
\begin{array}{l}
doDiskSwap \mathrel{\widehat=} \\
\quad (pdescrs.NextProcessToSwapIn[p?/pid!, rqsz?/sz!] \\
\quad \land (sms.CanAllocateInStore \land swapProcessIntoStore) \\
\quad \lor (pdescrs.HaveSwapoutCandidate \\
\qquad \land (pdescrs.FindSwapoutCandidate[outcand/cand!, mspec/slot!] \\
\qquad\quad \land [mspec : MEMDESC;\ start, end : ADDRESS;\ sz : \mathbb{N} \mid \\
\qquad\qquad start = memstart(mspec) \land sz = memsize(mspec) \\
\qquad\qquad \land\ end = (start + sz) - 1] \\
\qquad\qquad\quad \land (swapCandidateOut[outcand/p?, start/start?, \\
\qquad\qquad\qquad\qquad end/end?, sz/sz?] \mathbin{;} \\
\qquad\qquad\qquad swapProcessIn))) \setminus \{outcand, start, end, sz, mspec\}) \\
\qquad\qquad\qquad\qquad \setminus \{p?, rqsz?\}
\end{array}
$$

As can be seen, this swapper is based upon disk residency time to determine whether swapping should occur. Clearly, if there are no processes on disk, no swapping will occur; only when there are more processes than can be simultaneously maintained in main store does swapping begin. This seems a reasonable way of arranging matters: when there is nothing to swap, the swapper does nothing.

$$waitForNextSwap \mathrel{\widehat=} swapsema.Wait$$

$$
\begin{array}{l}
RunProcess \mathrel{\widehat=} \\
\quad WaitForNextSwap \mathbin{;} \\
\quad (\forall i : 1 \mathrel{..} \infty \bullet \\
\qquad doDiskSwap \mathbin{;} \\
\qquad waitForNextSwap)
\end{array}
$$

Proposition 99. *If the owner of a process' code is swapped out, that process cannot proceed.*

PROOF. By Proposition 54, since *MakeUnready* removes the process from the ready queue and alters its state to pstwaiting. □

Proposition 100. *When a parent process is swapped out, all of its children are blocked.*

PROOF. This is an immediate consequence of the *BlockProcessChildren* propositions. □

The final organisation of this subsystem is shown in Figure 4.4.

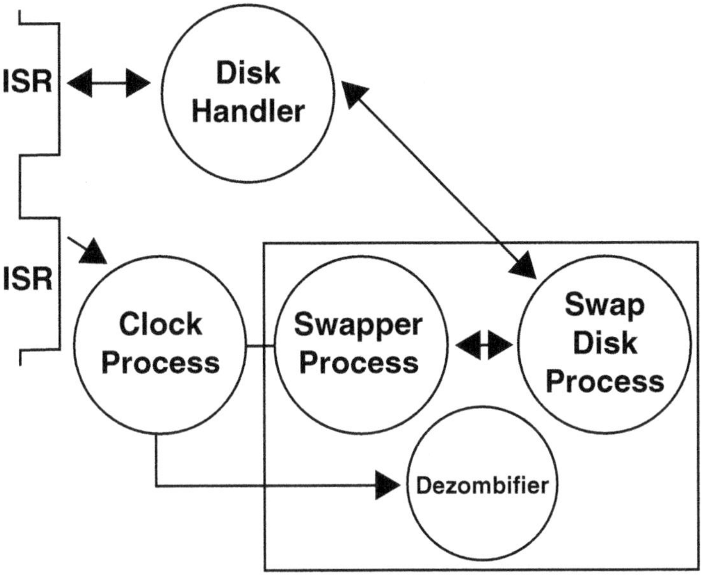

Fig. 4.4. *Interaction between clock, swap and dezombifier processes.*

4.7 Process Creation and Termination

A major issue to be addressed is the following: how are processes created within a system such as this? The answer is that some processes are created at boot time, others when the system is running. Among the latter class are user processes. In this section, mechanisms are defined for creating system and user processes. System processes come in two varieties, so two operations are defined for their creation.

Most system processes never terminate but user processes do, so a primitive is defined to release resources when a user process ends; resource release includes the handling of zombies. For all of the system processes defined in this chapter, termination is an exceptional behaviour.

The operations required to create processes (of all kinds) and to handle the termination of user processes are all collected in the following class.

ProcessCreation

$\upharpoonright$(*INIT*,
CreateUserProcess,
CreateChildUserProcess,
CreateSystemProcess,
CreateDeviceProcess,
TerminateProcess)

$proctab : ProcessTable$
$pdescrs : ProcessStorageDescrs$
$diskrqbuff : SwapRQBuff$
$realstore : REALMAINSTORE$
$lck : Lock$

INIT

$ptab? : ProcessTable$
$dskbf? : SwapRQBuff$
$pdescr? : ProcessStorageDescrs$
$store? : REALMAINSTORE$
$lk? : Lock$

$proctab' = ptab?$
$diskrqbuff' = dskbf?$
$pdescrs' = pdescr?$
$realstore' = store?$
$lck' = lk?$

$createNewPDescr \ldots$

$createAUserProcess \ldots$

$CreateUserProcess \ldots$

$CreateChildUserProcess \ldots$

$CreateSystemProcess \ldots$

$CreateDeviceProcess \ldots$

$writeImageToDisk \mathrel{\widehat{=}} \ldots$

$deleteProcessFromDisk \mathrel{\widehat{=}} \ldots$

$freeProcessStore \mathrel{\widehat{=}} \ldots$

$deleteSKProcess \mathrel{\widehat{=}} \ldots$

$TerminateProcess \mathrel{\widehat{=}} \ldots$

The first operation to define creates a new process descriptor and adds it to the process table. In order to define this operation, the following functions are required:

$mkpstack : \mathbb{N}_1 \rightarrow PSTACK$
$mpdata : \mathbb{N}_1 \rightarrow PDATA$

These functions are intended to simulate the allocation of storage for the classes of structure. A refined specification would fill in these details—for the present, the axiomatic definitions will suffice.

The *CreateNewPDescr* operation just creates a new process descriptor. It is supplied with the basic information required to create one through its arguments. The predicate creates descriptors for the new process' stack and data areas (using the above-declared functions). The identifier of the new process is also supplied as an argument. The schema is somewhat uninteresting from the operating systems viewpoint; however, it does show how an Object-Z entity is dynamically created.

The operation *CreateNewPDescr* is, therefore, as follows:

createNewPDescr

$pid? : APREF$
$kind? : PROCESSKIND$
$prio? : SCHDLVL$
$timequant? : TIME$
$stacksize?, datasize? : \mathbb{N}$
$code? : PCODE$
$mspec? : MEMDESC$
$rqsz? : \mathbb{N}$

$\exists pd : ProcessDescr; stat : PROCSTATUS; stk : PSTACK; data : PDATA \bullet$
$\quad stat = \mathsf{pstnew}$
$\quad \land stk = mkpstack(stacksize?)$
$\quad \land data = mkpdata(datasize?)$
$\quad pd.Init[stat/stat?, kind?/knd?, prio/slev?, timequant?/tq?,$
$\qquad stk/pstack?, data/pdata?, mspec?/mem?, rqsz?/msz?]$
$\quad \land proctab.AddProcessToTable[pd/pd?]$

(The *AddProcessToTable* operation requires no substitution because it expects the process to be identified by a variable *pid?*.)

The user-process creation operation proper is as follows. It creates a process descriptor for the new process, thus enabling it to be represented within the system. As part of this, a test (*proctab.CanGenPId*) is made as to whether the system has reached its maximum number of processes. The schema is complicated by the fact that allocation might have to take place on disk and not in main store. It should be noted that the identifier of the newly created process is returned by this operation; this will be of some importance, as will be seen.

createAUserProcess

$code? : PCODE$
$stacksize?, datasize? : \mathbb{N}$
$prio? : SCHDLVL$
$timequant? : TIME$
$newpid! : APREF$

$\exists p : APREF; rqsz : \mathbb{N}; prio : SCHDLVL;$
$\qquad mspec : MEMDESC; kind : PROCESSKIND; qimage : MEM \mid$
$\qquad\quad kind = \mathsf{ptuserproc} \land prio = userqueue \bullet$
$\quad proctab.CanGenPId \land (proctab.NewPId[p/p!] \land p = newpid!$

$$
\begin{aligned}
&\quad \land\ rqsz = \#code? + stacksize? + datasize?\\
&\quad \land\ ((realstore.RSCanAllocateInStore[rqsz/rqsz?]\\
&\qquad\qquad \land\ realstore.RSAllocateFromHole[rqsz/rqsz?, mspec/mspec!]\\
&\qquad\qquad \land\ createNewPDesc[rqsz/rqsz?, mspec/mspec?]\\
&\qquad\qquad \land\ pdescrs.MakeInStoreProcessSwappable[p/pid?])\\
&\qquad \lor\ (\land\ mspec = mkmspec(0, rqsz)\\
&\qquad\qquad \land\ createNewPDesc[rqsz/rqsz?, mspec/mspec?]\\
&\qquad\qquad \land\ realstore.CreateProcessImage[stacksize?/stksz?,\\
&\qquad\qquad\qquad\qquad datasize?/datasz?, image/image!]\\
&\qquad\qquad \land\ writeImageToDisk[p/pid?, image/image?]\\
&\qquad\qquad \land\ pdescrs.MakeProcessOnDiskSwappable[p/pid?]))\\
&\quad \land\ pdescrs.AddProcessStoreInfo[p/p?, mspec/mdesc?, rqsz/sz?])
\end{aligned}
$$

If there are no free process identifiers and *CanGenPId* fails, an error should be raised. However, for the purposes of clarity, errors are ignored in this book. The case should be noted, however.

In a similar fashion, a creation operation for system and device processes needs to be defined. There are some differences between it and the user-process creation operation. In particular, all system-process identifiers and storage areas can be predefined, so they can be supplied as configuration-time or boot-time parameters. The schema defining the operation is as follows:

createASystemProcess

$$
\begin{aligned}
&kind? : PROCESSKIND\\
&pid? : APREF\\
&code? : PCODE\\
&stacksize?, datasize? : \mathbb{N}\\
&prio? : SCHDLVL\\
&mspec? : MEMDESC\\
\hline
&\exists\, rqsz : \mathbb{N};\ tquant : TIME \mid\\
&\qquad\qquad tquant = \infty \land rqsz = \#code? + stacksize? + datasize? \bullet\\
&\quad realstore.RSAllocateInSTore[rqsz/rqsz?]\\
&\quad \land\ createNewPDEsc[rqsz/rqsz?, tquant/timequant?]
\end{aligned}
$$

There should be no errors raised by calls to this operation.

The following operation writes a new process image to disk. It will be loaded into main store by the swapper at some later stage. This is the operation used above in the definition of the *createAUserProcess* schema.

writeImageToDisk

$$
\begin{aligned}
&pid? : APREF\\
&image? : MEM\\
\hline
&(\exists\, rq : SWAPRQMSG \bullet\\
&\quad rq = NEWSPROC\langle\langle pid?, image?\rangle\rangle \land diskrqbuff.Write[rq/rq?])
\end{aligned}
$$

It is now possible to continue with the definition of the interface operations for the creation of all three kinds of process. The first operation is the one that creates user processes. This differs from the other two schemata in that it requires locking and that it returns a new process identifier.

$$
\begin{array}{l}
\mathit{CreateUserProcess} \mathrel{\widehat{=}} \\
\quad \exists\, pprio : PRIO;\ tquant : TIME \mid \\
\qquad\qquad pprio = userqueue \land tquant = minpquantum \bullet \\
\qquad lck.Lock \\
\qquad \mathbin{⨾}(createAUserProcess[pprio/prio?, tquant/timequant?] \mathbin{⨾} lck.Unlock)
\end{array}
$$

The lock is required because:

- This is an operation that is called when other processes are executing.
- This is an operation that is intended to be called from user processes.

So, it is reasonable to ask, how are processes actually created? In particular, how is the first user process created? Without an initial user process, a process outside the kernel that can call this primitive, how are user processes created? The answer is simple: there is a kernel call that creates the initial user process. The initial process is called the *UrProcess* and is created when the kernel finishes its initialisation. What is required is, then, the following:

$$
\begin{array}{l}
\mathit{CreateUrProcess} \mathrel{\widehat{=}} \\
\quad \exists\, pprio : PRIO;\ tquant : TIME \mid \\
\qquad\qquad pprio = userqueue \land tquant = minpquantum \bullet \\
\qquad createAUserProcess[pprio/prio?, tquant/timequant?]
\end{array}
$$

This operation requires stack and data region sizes to be created: they will be zero or very small. They are not specified because the *UrProcess* might be used for purposes other than simply creating user processes (e.g., it could count them, exchange messages with them, and so on). For this reason, the storage areas are not specified by the existential. The operation also returns a new process identifier (element of *APREF*): it can be stored within the kernel or just ignored.

Child processes are created by the operation that is defined next. It should be noted that the basic operation is still *createAUserProcess*.

$$
\begin{array}{l}
\mathit{CreateChildUserProcess} \mathrel{\widehat{=}} \\
\quad (\exists\, pprio : PRIO;\ tquant : TIME \mid \\
\qquad\qquad pprio = userqueue \land tquant = minpquantum \bullet \\
\qquad lck.Lock \\
\qquad \mathbin{⨾}(createAUserProcess[pprio/prio?, tquant/timequant?] \\
\qquad\quad \land\ proctab.AddChildOfProcess[rqprocid?/parent?, newpid!/child?]) \\
\qquad\quad \mathbin{⨾} lck.Unlock)
\end{array}
$$

Now come the two operations to create the two kinds of system processes. Both operations are based on the *createASystemProcess* operation. This operation performs the same role in the creation of system and device processes as *createAUserProcess* does in the creation of user processes.

A difference between the two following operations and the ones for user processes is that the priorities are different. System and device processes each have their own priority level. They are assigned the appropriate priority by the creation operation.

First, there is the system-process creation operation:

$$\begin{array}{l} \mathit{CreateSystemProcess} \mathrel{\widehat=} \\ \quad (\exists \mathit{kind} : \mathit{PROCESSKIND};\ \mathit{prio} : \mathit{SCHDLVL} \mid \\ \qquad\qquad \mathit{kind} = \mathsf{ptsysproc} \land \mathit{prio} = \mathit{sysprocqueue} \bullet \\ \qquad \mathit{createASystemProcess}[\mathit{kind}/\mathit{kind}?, \mathit{prio}/\mathit{prio}?]) \end{array}$$

Next, there is the operation to create device processes:

$$\begin{array}{l} \mathit{CreateDriverProcess} \mathrel{\widehat=} \\ \quad (\exists \mathit{kind} : \mathit{PROCESSKIND};\ \mathit{prio} : \mathit{SCHDLVL} \mid \\ \qquad\qquad \mathit{kind} = \mathsf{ptsysproc} \land \mathit{prio} = \mathit{sysprocqueue} \bullet \\ \qquad \mathit{createASystemProcess}[\mathit{kind}/\mathit{kind}?, \mathit{prio}/\mathit{prio}?]) \end{array}$$

The storage areas are defined by the kernel-configuration operation, and the code is statically defined as part of the kernel code. The following operations are for use when processes terminate. In the present kernel, user processes are the only ones that can terminate; all the other processes must continue running until the system shuts down.

As noted above, the identifier of the process is also statically allocated. This allows, *inter alia*, the identifiers to be hard-coded into all communications. (This will be of great convenience when IPC is defined in terms of messages, as they are in the next chapter, where a full interface to the entire kernel is defined.)

Next, it is necessary to handle process termination. Processes cannot simply be left to terminate. The resources belonging to a terminating process must be released in an orderly fashion. For this kernel, as it stands, processes can only hold storage as a resource, so this must be released before the process descriptor representing the process is deleted. In addition to releasing store, a process might have unterminated children and must, therefore, become a zombie before it can be killed off completely. The following operations implement the basics (and add a few extra operations to give the reader an idea of some of the other things that might need to be handled during termination).

If a process is on disk when it is terminated (say, because of system termination or because of some error that we have not specified in this chapter), its image must be erased. The operation whose schema follows performs that operation.

$$\begin{array}{|l} \hline \mathit{deleteProcessFromDisk} \\ \hline p? : \mathit{APREF} \\ \hline (\exists \mathit{rq} : \mathit{SWAPRQMSG} \bullet \\ \qquad \mathit{rq} = \mathit{DELSPROC}\langle\!\langle p? \rangle\!\rangle \\ \qquad \land\ \mathit{diskrqbuff}.\mathit{Write}[\mathit{rq}/\mathit{rq}?]) \\ \hline \end{array}$$

When a process terminates, its storage must be freed. The following schema defines what happens. It is really just an interface to *FreeMainstoreBlock*:

$$
\begin{array}{|l}
\hline
\textit{freeProcessStore} \\
p? : APREF \\
descr? : MEMDESC \\
\hline
(\exists\, start : ADDRESS;\ sz : \mathbb{N} \mid \\
\qquad\qquad start = memstart(descr?) \wedge sz = memsize(descr?) \bullet \\
\quad realstore.FreeMainstoreBlock[start/start?, sz/sz?]) \\
\hline
\end{array}
$$

Finally, the full operation for releasing process storage is defined. The way in which storage is released will, at some point, depend upon whether the terminating process owns its code or shares it with some other process. Clearly, if the process owns its code, the store for the code can just be deleted—provided, that is, the process has only terminated children. The basic operation for releasing storage is as follows. It should be noted that there will be some extra work for handling zombie processes.

$$
\begin{array}{l}
releaseProcessStorage \mathrel{\widehat{=}} \\
\quad deleteProcessFromDisk \\
\quad \wedge ((proctab.IsCodeOwner \wedge proctab.DelCodeOwner) \\
\qquad \vee (\exists\, owner : APREF \bullet \\
\qquad\qquad proctab.DelCodeSharer[owner/owner?, p?/sharer?])) \\
\quad \wedge (\exists\, pd : ProcessDescr;\ md : MEMDESC \bullet \\
\qquad proctab.DescrOfProcess[p?/pid?, pd/pd!] \\
\qquad \wedge pd.StoreDescr[md/descr!] \\
\qquad \wedge freeProcessStore[md/descr?]) \\
\quad \wedge pdescrs.RemoveProcessStoreInfo
\end{array}
$$

System and device processes are easier to handle. Their storage can just be deleted. In this system, there are no hierarchical relationships between system and driver processes. The schema defining the operation that releases kernel process storage is the following:

$$
\begin{array}{l}
deleteSKProcess \mathrel{\widehat{=}} \\
\quad releaseProcessStorage \\
\quad \wedge proctab.DelProcess[p?/pid?]
\end{array}
$$

It can be argued that a system shutdown can be performed without freeing the storage that system and device processes occupy. This is messy, so the operation just defined frees the process' space and then deletes its process descriptor. This choice has the consequence that *any* process in this kernel can execute the operation just defined when it terminates.

The operations just defined can be used to define the *TerminateProcess* operation. This operation is defined below.

$$
\begin{array}{l}
\mathit{TerminateProcess} \mathrel{\widehat{=}} \\
\quad \mathit{lck.Lock} \mathbin{;} \\
\quad ((\exists \mathit{pd} : \mathit{ProcessDescr} \bullet \\
\qquad \mathit{proctab.DescrOfProcess}[p?/\mathit{pid}?] \\
\qquad \wedge (\neg\ \mathit{proctab.ProcessHasChildren} \\
\qquad\quad \wedge ((\mathit{proctab.ProcessHasParent} \\
\qquad\qquad \wedge\ \mathit{proctab.ParentOfProcess}[\mathit{parent}/\mathit{parent}!] \\
\qquad\qquad \wedge\ \mathit{proctab.RemoveProcessFromParent} \\
\qquad\qquad\qquad [\mathit{parent}/\mathit{parent}?, p?/\mathit{child}?] \\
\qquad\qquad \wedge\ \mathit{pd.SetStatusToTerminated} \\
\qquad\qquad \wedge\ \mathit{deleteSKProcess}) \setminus \{\mathit{parent}\} \\
\qquad\quad \vee (\mathit{pd.SetStatusToTerminated} \\
\qquad\qquad\quad \wedge\ \mathit{deleteSKProcess})) \\
\qquad \vee (\mathit{proctab.ProcessHasChildren} \\
\qquad\quad \wedge\ \mathit{proctab.MakeZombieProcess}[p?/\mathit{pid}?]))) \\
\quad \wedge\ \mathit{lck.Unlock})
\end{array}
$$

The operation uses a lock instead of a semaphore because, strictly speaking, it belongs to the layer implementing the process abstraction.

The termination operation also has to handle the case in which a parent process terminates before any of its children do. If a parent terminates, its storage will be deallocated but this will also remove its code from main store. Without code, the children cannot execute, so a mechanism must be implemented to prevent the parent's code from being deleted. (If parents and children share data storage, it, too, must be prevented from deallocation.) The zombie mechanism whose operations were defined together with the process table is used to do this.

Basically, when a parent process terminates, a check is made to see if there are any active child processes. If there are no active children, the parent is allowed to terminate normally. Otherwise, the parent is unreadied and placed in a special waiting state (which we refer to, here, as the "zombie" state). When *all* the children of a zombie parent have terminated, the parent can be deallocated (properly terminated). The deallocation is the same as for normal processes; each zombie must have a process descriptor, at least to record the locations and sizes of its storage areas. The only problem is that children can create children: in the model, this requires that the transitive closure of the child relation be used to determine *all* the children of a parent process.

4.8 General Results

This final section contains the proof of a number of propositions that deal with properties of the kernel.

The propositions stated and proved in this section are collected here for convenience.

Proposition 101. *When a process is swapped in, it enters the ready queue.*

PROOF. The predicate of schema *swapProcessIn* contains *MakeReady*[*p*?/*pid*?] as a conjunct in an unconditional location. □

Proposition 102. *When a parent process is swapped out, none of its children appear in the ready queue.*

PROOF. By Proposition 89. □

Proposition 103. *When a parent process is swapped in, its children change state and appear in the ready queue.*

PROOF. The appropriate schema contains an instance of *MakeReady*. □

Proposition 104. *When a device request is made, the current process enters a waiting state and is no longer in the ready queue.*

PROOF. In this kernel, there is really only one good case upon which to make an argument: clock alarms. When a process makes an alarm request, has its context swapped out by the SVC ISR and its state is set to pstwaiting. Furthermore, the ISR calls *MakeUnready* on the requesting process to remove it from the scheduler. The process is held by the clock driver.

Device requests are made via SVCs, so the above will always hold. □

Proposition 105. *When a device completes, the requesting process is returned to the ready queue.*

PROOF. Again, the clock driver is the only example but it is normative. When each process is awakened from its sleeping state (when its alarm clock "rings"), *MakeReady* is called to return the process to the ready queue. The *MakeReady* operation changes the status attribute in the process' descriptor to reflect the fact that it is ready (sets the status to pstready, that is). □

Proposition 106. *While a process is waiting for a device request to complete, it is in neither the ready nor the running state. It is in the waiting state.*

PROOF. Proposition 104 states that the requesting process is unreadied by the SVC ISR. Therefore, it cannot be in the ready state. The ISR also calls the scheduler to execute another process, so the requesting process cannot be executing. □

Proposition 107. *Processes marked as* zombie *cannot be swapped out.*

PROOF. By the schema, *FindSwapoutCandidate*:

$$
\begin{array}{l}
\textit{FindSwapoutCandidate} \\
\hline
p? : APREF \\
cand! : APREF \\
rqsz? : \mathbb{N} \\
slot! : MEMDESC \\
\hline
(\exists p : APREF;\ t : TIME \mid residencytime(p) = t \bullet \\
\quad p \neq p? \land \\
\quad pstatus(p) \notin illegalswapstatus \land \\
\quad pkind(p) \neq \mathsf{ptsysproc} \land \\
\quad pkind(p) \neq \mathsf{ptdevproc} \land \\
\quad pmemsize(p) \geq rqsz? \land \\
\quad (\forall p_1 : APREF \bullet \\
\qquad (p_1 \neq p \land p_1 \neq p? \\
\qquad\qquad \land p_1 \in \operatorname{dom} residencytime \land t \geq residencytime(p_1)) \\
\qquad\quad \Rightarrow p = cand! \land pmem(p) = slot!))
\end{array}
$$

The critical line is $pstatus(p) \notin illegalswapstatus$, where:

$$
\begin{array}{l}
illegalswapstatus = \{\mathsf{pstrunning}, \mathsf{pstwaiting}, \mathsf{pstswappedout}, \\
\qquad\qquad\qquad\qquad \mathsf{pstnew}, \mathsf{pstterm}, \mathsf{pstzombie}\}
\end{array}
$$

Since this line appears as a conjunct, if it is false, it will invalidate the entire predicate. □

Proposition 108. *Processes marked as* zombie *cannot make device requests.*

PROOF. To make a device request, a process must be ready. Processes marked as zombie are not active and are about to terminate. Therefore, they cannot make any requests apart from those that release resources.

□

Corollary 8. *Processes marked as* zombie *cannot be present in device queues.*

PROOF. This follows from the immediately preceding proposition. □

Proposition 109. *Each process is resident in at most one queue at any time.*

PROOF. The possible queues in which a process can reside are:

- one of the ready queue components;
- a device request queue (if appropriate);
- a semaphore queue; or

- the clock's waiting sets.

In the definition of the semaphore *Wait* operation, there is an instance of *MakeUnready*. This operation removes the caller from the ready queue. Furthermore, the operation applies *Enqueue* to the caller to place it in the local queue of processes waiting on the semaphore. Therefore, any process performing a *Wait* operation cannot simultaneously be in a ready queue and the semaphore's queue of waiting processes.

When a process is ready, it is not waiting on the clock or a device or semaphore (by definition). When a process is waiting on a device, it cannot be marked as ready. □

Proposition 110. *Each process is in exactly one state at any time.*

This is the analogue of the informal property that a process is resident in at most one queue at any time.

PROOF. The state of each process (with the exception of the idle process, which is a special case) is represented by its *status* attribute. Inspection of the operations reveals that the value of this attribute is modified appropriately. □

Proposition 111. *The scheduling régime employed by this kernel is fair.*

PROOF. The *fairness* property is interpreted as: no process waits infinitely long before executing. Therefore, it must be shown that the scheduler does not require processes to wait for infinite periods of time.

First, if there are only user-level processes, by Propositions 59 and 64, the user-level queue implements a fair policy.

Clearly, by Proposition 62, all device processes execute as soon as possible.

Similarly, by Proposition 63, all system processes are executed before user-level ones and after all device processes. Indeed, if there are no device processes in the scheduler, system processes are executed in preference to user-level ones.

It can be observed that device and system processes are guaranteed by design either:

- to terminate in a finite time; or
- to block after a relatively short period of execution.

Either way, device and system processes do not execute for infinite periods of time before either terminating or blocking.

The next case to consider is that in which an infinite number of device processes are executed on an infinite number of occasions between the execution of user processes.

The clock process is driven by a periodic interrupt. The handler processes are guaranteed to terminate promptly. However, the swapper processes might have to wait for an infinite time because of a disk fault; when waiting, processes are not in the scheduler's ready queues.

To have a sufficient number of device processes in the scheduler, a user process would have to make repeated requests. This implies that *at least* one user process is executing infinitely often because device and system processes do not usually make device requests (the swapper process can be discounted because of its structure). However, user processes exhaust their time quantum after a finite period of activity and are returned to the back of the user-level ready queue, thus permitting rescheduling. If a process is waiting on a device, it is not in the scheduler's queues.

Finally, there is the case of infinite execution of the idle process. The idle process is only executed when there are no other processes available in the scheduler to execute. Therefore, if the idle process is executing and another process becomes ready, the scheduler will block the idle process in favour of the other process.

In the sense of fairness enunciated at the start, the scheduler is fair. □

Ich weiß nicht, was soll bedeuten,
Daß ich so traurig bin,
Ein Märchen aus alten Zeiten,
Daß kommt mir nicht aus dem Sinn
– Heinrich Heine, Die Lorelei

5

Using Messages in the Swapping Kernel

Hoc volo, sic iubeo, sit pro ratione voluntas
– Juvenal, Satires, 223

5.1 Introduction

The kernel that was the topic of the last chapter employed semaphores as its inter-process communication primitive. In that kernel, semaphores were used both to synchronise processes, as in the case of driver processes being awakened by a semaphore shared with the corresponding ISR. Semaphores were also used to synchronise processes that shared data; for example, the clock driver and alarm requests.

Semaphores have, of course, a long and distinguished history, but they are open to abuse. One must (almost) always ensure that the *Wait* and *Signal* operations come in pairs. Sometimes, as was the case in the last chapter, the two operations are applied to a single semaphore in two processes (as in the driver wake-up method). It is all too easy to make mistakes in their use.

It was decided to base this chapter on a message-passing model. The model adopted is synchronous message passing. The reason for using messages is that they make inter-process interaction more explicit; this is reflected in the organisation of the model. The presence of operations that send or receive messages is much clearer, it seems to us, than the use of semaphores. Message passing can also be integrated more easily with networked services than can a semaphore-based scheme. The choice of synchronous message passing is motivated by two facts:

- It is easier to prove synchronous message-passing systems.
- Synchronous message passing is often assumed by process algebras (of particular relevance to this book is CCS [21]).

As is well known, message passing can be defined in terms of semaphores and semaphores can be implemented using messages. The two mechanisms are, then, of equal power. However, as indicated in the last paragraph, spotting a misuse of message-passing primitives is much easier than with semaphores.

The system modelled in this chapter is similar in many respects to Tanenbaum's MINIX system [30], the precursor of Linux. The reason for this is that we believe that MINIX is one of the superior examples of the method. (Xinu [9] is another clear example but one that does not draw as much from message passing as does MINIX). We are, in any case, more familiar with the MINIX kernel, another contributing reason for using it.

Although this chapter purports to contain the model of a kernel, it actually does rather less. The primitive structures required to support a message-passing model are presented. This model is interrupt-driven, so the model of a generic ISR is also presented. The kernel in full is not modelled because so much of it can be constructed with relative ease. Instead of showing every single case of the use of messages, critical examples (e.g., the clock driver process) are included. The chapter also includes essential modifications to other structures required to model message passing.

The kernel is organised in a way identical to the kernel of the previous chapter (Chapter 4). The only difference between the two is that, whereas the kernel of Chapter 4 used semaphores, the current one uses messages. Figure 4.1 is therefore an adequate depiction of the current kernel.

5.2 Requirements

The requirement is to model an operating system kernel that is based upon the exchange of messages between processes. The kernel should be an example of the interrupt-driven variety. In this class of kernel, all context switches are performed within ISRs and scheduling decisions only take effect, therefore, when the next interrupt is serviced. The scheduler can be called at many places in the system so that the current needs are taken into account; however, any scheduling decision taken outside of an ISR will, quite probably, be overridden by the one made during the activation of the next ISR.

When a process sends a message, it raises an interrupt and uses the ISR associated with it to perform the send operation. This provides a clear interface and usage protocol for messages. It also fits in well with the general requirement that the kernel be interrupt-driven.

The kernel should include a clock process. The periodic nature of the clock will ensure that some scheduling decisions take place; if there is no peripheral activity, kernels such as this will do nothing. The clock should allow user and other processes to request alarm calls.

The kernel should also support low-level storage management, very much as did the kernel in the last chapter. Swapping, in the sense of the last chapter's kernel, can be included if desired.

5.3 Message-Passing Primitives

The model begins with the specification of the message-passing primitives and their support in the kernel.

A type representing all messages is required. It is an atomic type and defined as:

$[MSG]$

It is necessary to determine the source of each message. Processes can receive messages from ISRs and from other processes. The idle process can never originate messages. Therefore, the message source type, *MSGSRC*, is defined as follows:

$MSGSRC == \{\mathsf{hardware}, \mathsf{any}\} \cup APREF$

The element any above is important because processes can state the identity of the process from which they wish to receive their next message. There must be a way to denote the fact that a process can state that it does not care where the next message comes from: this is the role of any. This value is of some importance when specifying the send and receive operations.

The attentive reader will observe that the model that follows does not, as a matter of fact, type check as Z or Object-Z. The reason for this is that the message operators are heavily overloaded; in order not to clutter the model, the type *MSG* is defined at the start and assumed for the message operators. The overloading *could* have been handled in a variety of ways. It was decided not to clutter the model with such details (they can be added with relative ease). The use of overloading is, we believe, akin to the omission of errors in the previous models (and, for that matter, in this one), something that can be added later. For the present, the structures that model message passing are the centre of interest, not the mechanisms of overloading.

The process descriptor structure must be modified to hold messages. It is assumed that messages are stored in some kind of structure before they are read; also, because it is a synchronous message-passing model, sending processes must be suspended if the destination is not ready to receive from them. There is a number of places where the message and the process queues can be located within the kernel. However, it is clear that, wherever they are stored, they should be somewhere that makes their association with source and destination processes clear. The queue structures will, clearly, be shared between every process in the system and must be protected in some fashion. Locks will be used below because:

- The message primitives are implemented at a level below that at which full process abstraction is available (message passing is, in any case, a part of the process abstraction in this kernel).
- Messages are used to wake driver processes. They must be available to ISRs.

- The code segments implementing the message primitives are all relatively short, so locking should not interfere too much with the overall performance of the system. Moreover, messages are, themselves, passed via an interrupt, so interrupts are disabled in any case.

For these reasons, it was decided to place the message structures in the process descriptor of each process. This requires slight modifications to the process descriptor object. The modifications are additions rather than deletions.

The *ProcessDescr* structure is defined as follows. Rather than give the entire class definition again, only those additions are shown below. The variables and operations omitted from the new *ProcessDescr* class definition are denoted by .... The remainder of the class remains the same as it was in the previous chapter. The reader can find the assumed structure in Section 4.4. The types used in the omitted parts of the class are exactly the same as in Section 4.4.

ProcessDescr

$\upharpoonright(INIT, \ldots,$
$SetInMsg, InMsg, SetOutMsg, OutMsg, SetNextMsgSrc, NextMsgSrc,$
$WaitingSenders, AddWaitingSenders)$

$status : PROCSTATUS$
$kind : PROCESSKIND$
$schedlev : SCHDLVL$
$regs : GENREGSET$
$time\_quantum : TIME$
$statwd : STATUSWD$
$ip, memsize : \mathbb{N}$
$stack : PSTACK$
$data : PDATA$
$code : PCODE$
$mem : MEMDESC$
$qinmsgbuff, myoutmsgbuff : MSG$
$nextmsgsrc : MSGSRC$
$waitingsenders : \mathrm{seq}\ MSGSRC$

INIT

$\ldots nextmsgsrc' = \mathsf{any}$
$waitingsenders' = \langle\,\rangle$

...

$SetInMsg \mathrel{\widehat=} \ldots$
$InMsg \mathrel{\widehat=} \ldots$
$SetOutMsg \mathrel{\widehat=} \ldots$

$OutMsg \mathrel{\widehat=} \ldots$

$SetNextMsgSrc \mathrel{\widehat=} \ldots$

$NextMsgSrc \mathrel{\widehat=} \ldots$

$WaitingSenders \mathrel{\widehat=} \ldots$

$AddWaitingSenders \mathrel{\widehat=} \ldots$

The class adds four variables to its state: *inmsgbuff*, *myoutmsgbuff*, *nextmsgsrc* and *waitingsenders*, respectively. These components are interpreted, informally, as:

- *inmsgbuff*: A one-element buffer containing the most recently read message; the process will copy the message into local store when it reads it. Sender processes deposit their messages into this slot when message exchange occurs.
- *myoutmsgbuff*: A one-element buffer containing the next message this process is to send.
- *nextmsgsrc*: The identifier of the process from which this process wants next to receive a message. If this process is prepared to accept a message from any source, this component has the value any.
- *waitingsenders*: A sequence of elements of *MSGSRC*; that is, a queue containing the identifiers of processes that are waiting to send a message to this process.

The *PROCSTATUS* type needs to be extended:

$$
\begin{aligned}
PROCSTATUS ::= {} & \mathsf{pstnew} \\
\mid {} & \mathsf{pstrunning} \\
\mid {} & \mathsf{pstready} \\
\mid {} & \mathsf{pstwaiting} \\
\mid {} & \mathsf{pstwaitingmsg} \\
\mid {} & \mathsf{pstswappedout} \\
\mid {} & \mathsf{pstzombie} \\
\mid {} & \mathsf{pstterm}
\end{aligned}
$$

The additional state, pstwaitingmsg, is added. This is the state of a process that is waiting to receive a message.

The operations required to support these additional components are now defined. They are all quite straightforward.

SetInMsg

$\Delta(inmsgbuff)$
$m? : MSG$

$inmsgbuff' = m?$

This operation is to be executed when a sending process synchronises with this process. By setting *inmsgbuff* to a message, this process has received the message.

The receiving process executes the *InMsg* operation when it wants to read the current message.

$$
\begin{array}{l}
\textit{InMsg} \\
\hline
m! : MSG \\
\hline
m! = inmsg \\
\hline
\end{array}
$$

A sending process places the message it is trying to send in this component of its process table. This component represents a standard location for outbound messages; it is copied by the primitives to the *inmsgbuff* component of the receiver when the message is sent.

$$
\begin{array}{l}
\textit{SetOutMsg} \\
\hline
\Delta(outmsgbuff) \\
m? : MSG \\
\hline
outmsgbuff' = m? \\
\hline
\end{array}
$$

The following operation retrieves the contents of the *outmsgbuff* component of the process descriptor.

$$
\begin{array}{l}
\textit{OutMsg} \\
\hline
m! : MSG \\
\hline
m! = outmsgbuff \\
\hline
\end{array}
$$

This operation is called by a process when it advertises the identifier of the next process from which it wishes to receive a message. If the process is willing to accept a message from any process or ISR, it supplies the value any. If the process wishes to receive from some piece of hardware, it supplies the value hardware (in general, a process will receive only from one piece of hardware, e.g., the clock or a disk driver).

$$
\begin{array}{l}
\textit{SetNextMsgSrc} \\
\hline
msrc? : MSGSRC \\
\hline
nextmsgsrc' = msrc? \\
\hline
\end{array}
$$

This operation retrieves the value stored in *nextmsgsrc*. This is used to determine which process to resume when *waitingsenders* is not empty.

$$
\begin{array}{l}
\textit{NextMsgSrc} \\
\hline
msrc! : MSGSRC \\
\hline
msrc! = nextmsgsrc \\
\hline
\end{array}
$$

As indicated before the schema, a search is made when *waitingsenders* is non-empty to determine whether a process is to send its message to this process.

This operation returns the queue of the identifiers of those processes waiting to send a message to this process.

WaitingSenders

$sndrs! : \mathrm{seq}\ MSGSRC$

$waitingsenders = sndrs!$

When a process is sending a message to this process and this process is unable to accept it, the sender is enqueued here.

AddWaitingSenders

$\Delta(waitingsenders)$
$sndr? : MSGSRC$

$waitingsenders' = waitingsenders \frown \langle sndr? \rangle$

The *ProcessTable* is also the same except for some minor additions. As with the *ProcessDescr* class, the class is presented below in outline form *only* and shows only those components that are required to support message passing. The rest of the definition can be found in Section 4.4 of Chapter 4.

The differences between the *ProcessTable* required by the previous kernel and the present one consist in two operations—*MessageForDriver* and *AddDriverMessage*, which manipulate messages whose destination is a driver process—and in a new table, *drivermsgs*, which maps the identifier of the destination process (which should be the identifier of a driver process) to the next message it is intended to receive.

ProcessTable

$\upharpoonright(INIT,$
$\ldots$
$\quad MessageForDriver,$
$AddDriverMessage)$

$\ldots$
$drivermsgs : APREF \nrightarrow MSG$

$\ldots$
$(\forall p : APREF \mid p \in \mathrm{dom}\ drivermsgs \bullet$
$\quad (\exists pd : ProcessDescr;\ k : PROCESSKIND \bullet$
$\quad\quad DescrOfProcess[p/pid?, pd/pd!]$
$\quad\quad \land pd.ProcessKind[k/knd!]$
$\quad\quad \land k = \mathsf{ptdevproc}))$
$\ldots$

INIT

...

$\mathrm{dom}\, drivermsgs' = \varnothing$

$MessageForDriver \mathrel{\widehat{=}} \ldots$

$AddDriverMessage \mathrel{\widehat{=}} \ldots$

Messages for drivers are handled slightly differently. The support for them is provided by the following pair of operations.

MessageForDriver

$pid? : APREF$

$dmsg! : MSG$

$dmsg! = drivermsgs(pid?)$

AddDriverMessage

$\Delta(drivermsgs)$

$pid? : APREF$

$dmsg? : MSG$

$drivermsgs' = drivermsgs \oplus \{pid? \mapsto dmsg?\}$

Driver messages must be treated differently because the driver might be busy when the message is sent. Messages to drivers are of high priority and must be delivered. Therefore, any messages with a driver as destination are temporarily stored if the driver cannot immediately receive them.

There is a generic message-based ISR that responds to interrupts by creating and sending messages from hardware.

Process context manipulation must be redefined to account for messages. This is an interrupt-driven kernel, so the context switch is a logical place to insert message-handling code.

The *Context* structure is the same as that defined in the last chapter. It is repeated here because it plays a central role in the modelling of the message-passing subsystem. The class is

Context

$\upharpoonright(INIT, SaveState, RestoreState,$
$\quad SwapOut, SwapIn, SwitchContext)$

$ptab : ProcessTable$

$sched : LowLevelScheduler$

$hw : HardwareRegisters$

$$
\begin{array}{l}
\textit{INIT} \\
\hline
ptb? : ProcessTable \\
shd? : LowLevelScheduler \\
hwregs? : HardwareRegisters \\
\hline
ptab' = ProcessTable \\
sched' = LowLevelScheduler \\
hw' = hwregs? \\
\hline
\end{array}
$$

$SaveState \mathrel{\widehat=} \ldots$

$RestoreState \mathrel{\widehat=} \ldots$

$SwapOut \mathrel{\widehat=} \ldots$

$SwapIn \mathrel{\widehat=} \ldots$

$SwitchContext \mathrel{\widehat=} \ldots$

The *SaveState* operation's schema is very much as in the previous kernel. Since it is relatively short, it is repeated here.

The *SaveState* and *RestoreState* operations are intended to be called from within an ISR. The *SaveState* stores the contents of the hardware register in the descriptor of the process referred to by *currentp*. The scheduler, just like the one in Chapter 4, can then be called to select another process to execute (if there are none, the idle process is run); as a part of this selection operation, *currentp* becomes bound to the identifier of the selected process. At the end, *RestoreState* should be called to copy the newly selected process' state to the hardware.

$$
\begin{array}{l}
\textit{SaveState} \\
\hline
(\exists\, cp : IPREF \bullet \\
\quad sched.CurrentProcess[cp/cp!] \\
\quad (\exists\, pd : ProcessDescr \bullet \\
\quad\quad ptab.DescrOfProcess[cp/pid?, pd/pd!] \\
\quad\quad \land (\exists\, regs : GENREGSET;\ stk : PSTACK; \\
\quad\quad\quad\quad\quad ip : \mathbb{N};\ stat : STATUSWD;\ tq : TIME \bullet \\
\quad\quad\quad hw.GetGPRegs[regs/regs!] \land hw.GetStackReg[stk/stk!] \\
\quad\quad\quad \land hw.GetIP[ip/ip!] \land hw.GetStatWd[stat/stwd!] \\
\quad\quad\quad \land sched.GetTimeQuantum[tq/tquant!] \\
\quad\quad\quad \land pd.SetFullContext[regs/pregs?, ip/pip?, \\
\quad\quad\quad\quad\quad stat/pstatwd?, stk/pstack?, tq/ptq?]))) \\
\hline
\end{array}
$$

The current process referred to in the following schema is *not necessarily* the same as the one referred to in the previous schema. Basically, whichever process is referred to by *currentp* is the next to run and its context is switched onto the processor. It should be noticed that the instruction pointer is the last register to be set. This is because it hands the processor to the process that owns it.

$$
\begin{array}{l}
\textit{RestoreState} \\
\hline
(\exists\, cp : IPREF \bullet \\
\quad sched.CurrentProcess[cp/cp!] \\
\quad \land (\exists\, pd : ProcessDescr \bullet \\
\quad\quad ptab.DescrOfProcess[cp/pid?, pd/pd!] \\
\quad\quad \land (\exists\, regs : GENREGSET;\ stk : PSTACK;\ ip : \mathbb{N}; \\
\quad\quad\quad\quad\quad stat : STATUSWD;\ tq : TIME \bullet \\
\quad\quad\quad pd.FullContext[regs/pregs!, ip/pip!, stat/pstatwd!, \\
\quad\quad\quad\quad\quad\quad stk/pstack!, tq/ptq!] \\
\quad\quad\quad \land hw.SetGPRegs[regs/regs?] \\
\quad\quad\quad \land hw.SetStackReg[stk/stk?] \\
\quad\quad\quad \land hw.SetStatWd[stat/stwd?] \\
\quad\quad\quad \land sched.SetTimeQuantum[tq/tquant?] \\
\quad\quad\quad \land hw.SetIP[ip/ip?])))
\end{array}
$$

The general operation for swapping out a context is defined by the next operation:

$$
\begin{array}{l}
SwapOut \mathrel{\widehat=} \\
\quad (\exists\, cp : IPREF;\ pd : ProcessDescr \bullet \\
\quad\quad sched.CurrentProcess[cp/cp!] \\
\quad\quad \land ptab.DescrOfProcess[pd/pd!] \\
\quad\quad \land pd.SetProcessStatusToWaiting \\
\quad \land SaveState \mathbin{⨾} \\
\quad sched.MakeUnready[currentp/pid?])
\end{array}
$$

The operation calls *SaveState* and then unreadies the process referred to by *currentp*.

Similarly, *SwapIn* is the general interface to the operation that swaps a context onto the processor.

$$
\begin{array}{l}
SwapIn \mathrel{\widehat=} \\
\quad (\exists\, cp : IPREF;\ pd : ProcessDescr \bullet \\
\quad\quad sched.CurrentProcess[cp/cp!] \\
\quad\quad \land pd.SetProcessStatusToRunning \\
\quad\quad RestoreState) \\
SwitchContext \mathrel{\widehat=} SwapOut \mathbin{⨾} SwapIn
\end{array}
$$

The low-level scheduler is identical to the one defined in the last chapter (Section 4.5).

This kernel requires a global variable to store clock ticks that might have been missed by drivers that are not waiting when the clock interrupt arrives. This variable is not protected by a lock. It is assumed that there is no need because it can only be accessed and updated under strict conditions (there *can be* no other process active when update and read occur—this is guaranteed by the fact that this kernel executes on a uni-processor; if the kernel were ported to a multi-processor, matters might be a little different).

GlobalVariables

$\upharpoonright(INIT, missed\_ticks)$

$missed\_ticks : \mathbb{N}$

INIT

$missed\_ticks' = 0$

Finally, we come to the core of the model of message passing. This is the generic ISR that sends messages to devices to wake them up. This is the most detailed specification of an ISR that has been given so far in this book. The reason for this is that it is central to the operation of the message-passing mechanism.

The generic ISR is modelled by the *GenericMsgISR* class as follows:

GenericMsgISR

$\upharpoonright(INIT, SendInterruptMsg)$

$did : APREF$
$mm : MsgMgr$
$ctxt : Context$
$busydds : \mathbb{F}\, APREF$
$glovars : GlobalVariables$

INIT

$isrname? : APREF$
$msgmgr? : MsgMgr$
$ctxtops? : Context$
$gvs? : GlobalVariables$

$ctxt' = ctxtops?$
$did' = isrname?$
$mm' = msgmgr?$
$busydds' = \varnothing$
$glovars' = gvs?$

$saveState \mathrel{\widehat=} \ldots$

$restoreState \mathrel{\widehat=} \ldots$

$shouldRunDriver \mathrel{\widehat=} \ldots$

$SendInterruptMsg \mathrel{\widehat=} \ldots$

Before discussing the more interesting points about this ISR, it is worth pointing out that this model, like the others in this book, makes only minimal assumptions about the hardware on which the kernel runs. In particular, it is assumed that ISRs do not dump the hardware registers anywhere when an interrupt occurs. It is merely assumed that, when the interrupt does occur, a context will be current; that context is the interrupted one. It is also the context that might be switched out, should the scheduler so determine.

It is worth remembering that the interrupt mechanisms of any particular processor could differ considerably from this one; it is a reliable assumption that the contents of *currentp* will be unaffected by any interrupt. Where the context is deposited by a particular cpu is a detail that cannot be accounted for by this model. In what follows, it should be assumed that the state save and restore operations are able to locate the registers belonging to the current process; that is, to the process referred to by *currentp*.

$saveState \mathrel{\widehat{=}} ctxt.SaveState$

This operation saves the current context in the descriptor of the process referred to by *currentp* when the interrupt occurs.

$restoreState \mathrel{\widehat{=}} ctxt.RestoreState$

This is the operation that installs a context in the cpu. The process might be different from the one current when the interrupt occurred.

$shouldRunDriver \mathrel{\widehat{=}} sched.IsEmptyDriverQueue \land sched.IdleIsCurrent$

This operation forces the execution of the driver associated with this ISR if the scheduler's queues are empty and the idle process is currently running. The driver will always have a higher priority than the idle process, so the net effect of this operation is to pre-empt the idle process' execution. It is called from the next operation.

The usual form of an ISR using messages is:

(* Current process is running. *)
$SaveState \mathbin{;}$
(* Do some processing and create message. *)
$SendInterruptMsg \mathbin{;}$
(* A new process might be in *currentp*. *)
$RestoreState$
(* Possible new process executing. *)

The following operation sends a message to a process when the interrupt occurs. The process will usually be the driver associated with the interrupt.

At the end of the operation, the *shouldRunDriver* operation is executed, followed by a call to the scheduler.

There is one more operation defined in the class. It is the one defined next. This operation sends messages generated by ISRs. These messages are used to wake the driver process that corresponds to a driver. The problem is that the driver might have been interrupted or could even be blocked when the message is to be sent. For that reason, the schema records the identifier of the destination driver when it is busy. In addition, the operation is used by the clock ISR to send a message to the clock process. Given the above, it is possible that some clock ticks might be missed, so each invocation of the *SendInterruptMsg* operation adds 1 to a "missed tick" counter (called *missed_ticks* and defined in the global variables class, *GlobalVariables*, defined immediately before the *GenericMsgISR* class).

SendInterruptMsg

$$
\begin{array}{l}
\Delta(missedclicks, busydds) \\
driver? : APREF \\
m? : MSG \\
\hline
(\exists\, dpd : ProcessDescr \bullet \\
\quad ptab.DescrOfProcess[did/pid?, dpd/pd!] \\
\quad \land (\neg\, mm.IsWaitingToReceive[dpd/pd?] \\
\qquad \land ((driver? = CLOCKISR \\
\qquad\quad \land glovars.missed\_ticks' = glovars.missed\_ticks + 1) \\
\qquad\quad \lor (busydds' = busydds \cup \{driver?\} \\
\qquad\qquad \land ptab.AddDriverMessage[driver?/pid?, m?/dmsg?])))) \\
\land (\forall\, p : APREF \mid p \in busydds \bullet \\
\quad (\exists\, pd : ProcessDescr \bullet \\
\qquad \land ptab.DescrOfProcess[p/pid?, pd/pd!] \\
\qquad \land (mm.IsWaitingToReceive[pd/pd?] \\
\qquad\quad \land (\exists\, hwid : MSGSRC \mid hwid = \mathsf{hardware} \bullet \\
\qquad\qquad \land msgmgr.SendMessage[hwid/src?, p/dest?]))) \\
\qquad\qquad\quad \Rightarrow p \notin busydds') \\
\land (shouldRunDriver \land sched.ScheduleNext)
\end{array}
$$

We are now in a position to prove some fairly general properties of the message-passing system.

Proposition 112. *The message-passing mechanism is synchronous.*

Proof. By the predicates of *SendMessage* and *RcvMessage*.

If the destination is already waiting when the source sends a message, the message is immediately received. Otherwise, the destination eventually enters a waiting state, during which the message is exchanged. □

Proposition 113. *Unless the destination process terminates (or the system is shut down), every message is delivered to its destination.*

PROOF. There are two cases to consider.
Case 1. If the destination is waiting to receive from either the sender or any process (*IsWaitingToReceive*), the message is immediately copied to the destination (*SetInMsg*). The sender is the current process and continues. The message has been delivered.
Case 2. If the destination is not waiting, or waiting for a message from a source other than the current process or the default source, *any*, the sender is enqueued onto the destination's queue (list) of processes that are waiting to send a message to it; the sender is unreadied so that it cannot be scheduled.

When the destination is ready to receive a message, it selects a process from its waiting list and adds it to the ready queue after the message has been copied from the sender to the destination.

A receivers can specify the source of the message it wants to recevie next (this is the sender's process identifier) or the special value *any*. If *any* is specified, the receiver is willing to accept a message from any process that wishes to send to it.

Provided that the receiver lives long enough and provided that the receiver issues enough requests for any sender, all of the messages sent to the receiver are received. □

Proposition 114. *Every device process receives its interrupt messages in the correct order.*

PROOF. This follows from the previous result. The difference is that Interrupt Service Routines (ISRs) are not processes and cannot, therefore, be suspended. If the associated driver process is not yet ready to receive from the ISR, the message is stored in a special location until it can be delivered when a subsequent interrupt occurs. □

Proposition 115. *Every message sent is received in the correct order.*

PROOF. This follows from the behaviour of synchronous messages. Consider two processes, S and D. Each time S sends a message, m, to D, either S is suspended or the message is delivered; when S is resumed, the message is delivered. If S sends two messages to D, say m_1 and m_2, in that order, then S first attempts to send m_1. If D is waiting to receive, the message is delivered and S continues; if D is not waiting to receive, S is suspended until D is ready. In either case, message m_1 is delivered. Next, S tries to send m_2 and goes through the same sequence of operations and state transitions; m_2 is delivered, however. The order in which the messages are delivered is m_1 followed by m_2. By induction, the order in which messages are received is the same as the order in which they are sent. □

Proposition 116. *Nothing happens in the system unless an interrupt occurs.*

PROOF. The critical part of this is: when do context switches occur?

In this kernel, a context switch occurs in the general ISR code. (In the previous kernel, context switches occurred in ISR code and in the semaphore's *Wait* operation.)

When the generic ISR code runs, the actual context changes. First, the register set is switched out; these registers belong to the process named by the current value of *currentp*. The scheduler then executes and changes the value of *currentp*; there can be more than one scheduling decision in this kernel before the ISR exits, but the important fact is that these intermediate scheduling decisions do not affect the context. Indeed, only the last scheduling decision (assignment to *currentp*) occurring before the ISR terminates is switched onto the processor.

Since context switches change the state of the system and cause different processes to execute, and since context switches occur only in ISRs, the proposition has been proved. □

Proposition 117. *Each interrupt causes a context switch. This alters the currently running process.*

PROOF. The reasoning is similar to that of the previous proposition. □

Finally, there are the operations to support message passing at the level of the process. They are collected into the following object (module). This is really more a matter of convenience than of anything else. The module is called *MsgMgr*, the *Message Manager*.

As far as processes are concerned, this module contains the most important operations. It is also a relatively complex module whose workings are, at first sight, somewhat obscure. Because of this, the operations defined below are described (perhaps painfully) in more detail.

MsgMgr

$\upharpoonright(INIT, IsWaitingToReceive, SendMessage, RcvMessage)$

$ptab : ProcessTable$
$sched : LowLevelScheduler$

INIT

$pt? : ProcessTable$
$schd? : LowLevelScheduler$

$ptab' = pt?$
$sched' = schd?$

$IsWaitingToReceive \mathrel{\widehat{=}} \ldots$

$SendMessage \mathrel{\widehat{=}} \ldots$

$RcvMessage \mathrel{\widehat{=}} \ldots$

$isWaitingForSender \mathrel{\widehat{=}} \ldots$

$enqueueSender \mathrel{\widehat{=}} \ldots$

$canReady \mathrel{\widehat{=}} \ldots$

$haveMsgsWithAppropriateSrc \mathrel{\widehat{=}} \ldots$

$copyMessageToDest \mathrel{\widehat{=}} \ldots$

$RcvMessage \mathrel{\widehat{=}} \ldots$

It *must* be remembered that the operations defined by the *MsgMgr* class are always called from inside an ISR. This has implications for the use of context switches and the identity of callers and destinations.

IsWaitingToReceive

$pd? : ProcessDescr$

$$(\exists\, stat : PROCSTATUS;\ wtgfor : PROCWAITINGON \bullet \\ \quad pd.ProcessStatus[stat/st!] \\ \quad \land\ stat = \mathsf{pstwaitingmsg})$$

isWaitingForSender

$pd? : ProcessDescr$

$sender? : APREF$

$$(\exists\, nxtsrc : MSGSRC \bullet \\ \quad pd?.NextMsgSrc[nxtsrc/msrc!] \\ \quad \land\ (nxtsrc = \mathsf{any} \lor nxtsrc = sender?))$$

enqueueSender

$pd? : ProcessDescr$

$sender? : APREF$

$$pd?.addWaitingSender[sender?/sndr?]$$

If a receiver is waiting for a message from a process and that process is not in its waiters queue, it will be placed in "limbo", i.e., removed from all queues. When the message eventually arrives, it is put onto its ready queue by *SendMessage*.

$$
\begin{array}{l}
\textbf{SendMessage} \\
\hline
dest?, src? : APREF \\
m? : MSG \\
\hline
(\exists\, pd : ProcessDescr \bullet \\
\quad ptab.DescrOfProcess[dest?/pid?, pd/pd!] \\
\quad (\exists\, stat : PROCSTATUS;\ wtgfor : PROCWAITINGON \bullet \\
\quad\quad pd.ProcessStatus[stat/st!] \\
\quad\quad \land ((IsWaitingToReceive[pd/pd?] \\
\quad\quad\quad \land isWaitingForSender \\
\quad\quad\quad\quad \land pd.SetInMsg \\
\quad\quad\quad\quad \land sched.MakeReady[dest?/pid?]) \\
\quad\quad \lor ((pd.addWaitingSender[pd/pd?, src?/sender?] \\
\quad\quad\quad\quad \land sched.MakeUnready[src?/pid?]))))) \\
\hline
\end{array}
$$

The next operation to be defined determines whether the process described by the process descriptor bound to $pd?$ can be moved to a pstready state.

$$
\begin{array}{l}
\textbf{canReady} \\
\hline
pd? : ProcessDescr \\
\hline
(\exists\, stat : PROCSTATUS \bullet \\
\quad pd?.ProcessStatus[stat/st!] \\
\quad \land (stat \neq \mathsf{pstterm} \\
\quad\quad stat \neq \mathsf{zombie})) \\
\hline
\end{array}
$$

$$
\begin{array}{l}
\textbf{haveMsgsWithAppropriateSrc} \\
\hline
pd? : ProcessDescr \\
src? : MSGSRC \\
\hline
(\exists\, waiters : \mathrm{seq}\ APREF \bullet \\
\quad pd?.WaitingSenders[waiters/sndrs!] \\
\quad \land waiters \neq \langle\rangle \\
\quad \land src? \in \mathrm{ran}\ waiters) \\
\hline
\end{array}
$$

$$
\begin{array}{l}
\textbf{copyMessageToDest} \\
\hline
cpd?, spd? : ProcessDescr \\
\hline
(\exists\, m : MSG \bullet \\
\quad spd?.OutMsg[m/m!] \\
\quad \land cpd?.SetInMsg[m/m?]) \\
\hline
\end{array}
$$

Note that if the receiver doesn't have any appropriate senders, it is removed from all queues.

RcvMessage

$$
\begin{array}{l}
caller? : APREF \\
src? : APREF \\
m! : MSG \\
\hline
(\exists cpd : ProcessDescr \bullet \\
\quad ptab.DescrOfProcess[caller?/pid?, cpd/pd!] \\
\quad \wedge (\exists waiters : \mathrm{seq}\, APREF \bullet \\
\quad\quad cpd.WaitingSenders[waiters/sndrs!] \\
\quad\quad \wedge ((haveMsgsWithAppropriateSrc \\
\quad\quad\quad \wedge (\forall p : APREF \mid p \in \mathrm{ran}\, waiters \bullet \\
\quad\quad\quad\quad (\exists spd : ProcessDescr \bullet \\
\quad\quad\quad\quad\quad ptab.DescrOfProcess[p/pid?, spd/pd!] \\
\quad\quad\quad\quad\quad \wedge (src? = \mathsf{any} \vee src? = caller? \\
\quad\quad\quad\quad\quad\quad \wedge copyMessageToDest[cpd/cpd?, spd/spd?] \\
\quad\quad\quad\quad\quad\quad \wedge ((canReady[spd/pd?] \wedge MakeReady[p/pid?]) \\
\quad\quad\quad\quad\quad\quad\quad \vee Skip))))) \\
\quad\quad \vee sched.ScheduleNext)))
\end{array}
$$

The operations defined in this class are not designed to be executed directly. It is intended that they be executed by raising an interrupt. Therefore, any user process must raise an interrupt to send a message and raise another to read a message. When sending the message, the ISR has to retrieve the message and the destination address from somewhere (say, some fixed offset from the top of the calling process' stack). In a similar fashion, when a process performs a receive operation, the parameters must be in a standard location.

In order to make this clear (and to permit the proof of one or two relevant propositions), the two interrupt-handling classes are defined, one each for send and receive. The reader should be aware that they are only defined in outline (this is, in any case, in line with our general policy on hardware matters). In order to define them, it will be necessary to assume an operation:

RaiseInterrupt

$$
\begin{array}{l}
ino? : \mathbb{N} \\
\ldots
\end{array}
$$

This can be assumed to be a hardware operation (i.e., it can be modelled as an operation provided by the *HardwareRegisters* class. The input to the operation, *ino*?, is the number of the interrupt.

Both classes are subclasses of *GenericISR*, thus permitting the save and restore operations on contexts.

The first class is the one that implements the interface for sending messages. Its main operation, *ServiceInterrupt* handles the message by calling the *SendMessage* operation defined in the *MsgMgr* class.

SendISR

$\upharpoonright(INIT, ServiceInterrupt)$

$GenericISR$

$mmgr : MsgMgr$
$sched : LowLevelScheduler$

INIT

$mm? : MsgMgr$
$sch? : LowLevelScheduler$

$mmgr' = mm?$
$sched' = sch?$

$ServiceInterrupt \mathrel{\widehat{=}}$
$\quad \ldots$
$\quad \land\ sched.CurrentProcess[mypid/cp!]$
$\quad \land\ mmgr.SendMessage[mypid/src?, \ldots]$
$\quad \ldots$

The second class is the one that implements the interface for receiving messages. Its main operation, *ServiceInterrupt*, handles the message by calling the *RcvMessage* operation defined in the *MsgMgr* class. When a process is ready to receive a message, it raises an interrupt, thus invoking the *ServiceInterrupt* operation.

ReceiveISR

$\upharpoonright(INIT, ServiceInterrupt)$

$GenericISR$

$mmgr : MsgMgr$

INIT

$mm? : MsgMgr$

$mmgr' = mm?$

$ServiceInterrupt \mathrel{\widehat{=}}$
$\quad \ldots$
$\quad \land\ sched.CurrentProcess[mypid/cp!]$
$\quad \land\ mmgr.RcvMessage[mypid/caller?, \ldots]$
$\quad \ldots$

It will be assumed that there is a mechanism by which kernel processes can send and receive messages. Rather than performing a full system call, this alternative mechanism will be used (it still performs a *RaiseInterrupt*). It will be denoted by *KMsgMgr*.

Proposition 118. *The sender of a message is always the current process.*

PROOF. In order to send a message, the sending process must execute a call that involves *RaiseInterrupt* in order to raise the interrupt associated with message sending. However, the only process that can do this is the one currently referenced by *currentp* since it denotes the only executing process at any time.

Furthermore, the sender, *src?*, for the message to be sent is bound to the value of *currentp* by *sched.CurrentProcess*. □

Proposition 119. *The receiver of a message is always the current process.*

PROOF. By reasoning similar to the first paragraph of the previous proposition (the caller can be the only executing process on a uni-processor machine).

In this case, the *caller?* of the receive operation, *mmgr.RcvMessage*, is bound to the value of *currentp* by *sched.CurrentProcess*. □

Proposition 120. *If a sending process is not one for which the destination is waiting, the sender is removed from the ready queue.*

PROOF. The relevent class is *MsgMgr* and the relevant operation is, clearly, *SendMessage*. The second disjunct of *SendMessage*'s predicate is as follows:

$$
\begin{array}{l}
(\neg\ isWaitingToReceive[pd/pd?] \\
\quad \lor \neg\ isWaitingForSender) \\
\qquad \land\ pd.addWaitingSender[pd/pd?, src?/sender?] \\
\qquad \land\ sched.MakeUnready[src?/pid?]
\end{array}
$$

This conjunct deals with the case in which the destination is either not in the special waiting-for-message state or it is not waiting to receive from the current process (that is, the process referred to by *currentp*). As can be seen, the sender is made unready by the second conjunct. This removes it from the head of the ready queue so that it cannot be rescheduled until the receiver is ready for it. □

Corollary 9. *The sending process is always referred to by currentp.*

PROOF. By Proposition 118. □

Proposition 121. *When a process, d, receives a message, m, from a process, s, m is placed into d's incoming message slot.*

PROOF. The significant line of *mmgr.RcvMessage* is *copyMessageToDest*[...], which expands into:

$$\exists\, m : MSG \bullet \\ \quad spd?.OutMsg[m/m!] \\ \quad \land\ cpd?.SetInMsg[m/m?]$$

where both *cpd?* and *spd* are process descriptors. With one more level of expansion and simplification, this becomes:

$$cpd.inmsgbuff' = spd?.outmsgbuff$$

This is clearly the operation required in the statement of the proposition. □

Proposition 122. *If a process, d, is waiting for a message from any process and there is a waiting sending process, s, the process s will be readied.*

PROOF. By the predicate of *mmgr.RcvMessage*, having copied the message over, if the waiting process can be readied, it is placed in the ready queue by *MakeReady*:

$$copyMessageToDest \\ \quad \land\ ((canReady[spd/pd?] \land MakeReady[p/pid?] \ldots$$

where *spd* is the process descriptor of the message source (*s*ender *p*rocess *d*escriptor) and *p* is its process identifier. □

It is necessary to make message passing easier to use in user-level processes. In support of this, a new class is defined. This new class, called *UserMessages*, exports two operations, *Send* and *Receive*. The exported operations are intended to place the parameters required by the send and receive operations in an easily accessible place and raises the necessary interrupts. The class is defined in outline as:

UserMessages

$\upharpoonright(INIT, Send, Receive)$

$$sendintno, recvintno : \mathbb{N}$$

INIT

$$sendno?, recvno? : \mathbb{N}$$

$$sendintno' = sendno? \\ recvintno' = recvno?$$

$$raiseSendInterrupt \mathrel{\widehat=} RaiseInterrupt[sendintno/intno?, \ldots]$$

$$raiseRcvInterrupt \mathrel{\widehat=} RaiseInterrupt[recvintno/recvno?, \ldots]$$

$$Send \mathrel{\widehat=} raiseSendInterrupt[m?/msg?, \ldots]$$

$$Receive \mathrel{\widehat=} raiseRcvInterrupt[m!/msg!, \ldots]$$

This definition is intended merely to give the reader an idea of what it should look like. The state variables, *sendinto* and *recvintno*, denote the number of the send and receive interrupts, respectively. The second argument to the message-passing operations is the message being sent or received, respectively. In both cases, there will be other parameters (e.g., for receiving, the origin of the message can be specified).

The *UserMessages* class will be instantiated once in the kernel interface library.

Before moving on, it is worth pausing to reflect at a high level on what has been presented so far. The reader should note that the above model is not the only one. The Xinu operating system [9] is also based on message passing but does not integrate messages with interrupts and is, therefore, in our opinion, a somewhat cleaner kernel design. However, many readers will, as a result of reading standard texts, be of the (false) opinion that kernels *must* be interrupt-driven. The kernel of the last chapter was driven, in part, by interrupts and, in part, by semaphores (recall that semaphores cause a context switch and a reschedule); many real-time kernels are also driven only partially by interrupts. Nevertheless, many kernels are organised entirely around interrupts. This approach has merits (uniformity and simplicity) when there are interactive users hammering away at keyboards and torturing hapless mice! The current kernel was intended to show how such kernels *can* be structured, even if that structure becomes somewhat opaque in places; it is not intended as a statement that this is how they *must* be organised. (For a completely different approach, Wirth et al.'s fascinating and elegant OBERON system [35, 36] is recommended).

With these points noted, it is possible to move on to the use of the message-passing subsystem as defined above. The reader will be relieved to learn that what follows is far less convoluted than that which has gone before. Indeed, it is believed that the following models are cleaner and easier to construct and understand than those defined using semaphores.

5.4 Drivers Using Messages

In this section, the semaphore-based drivers from the swapping kernel are presented in an altered form: they now use message passing. Because of its centrality, the clock is the first to be considered. This will be followed by the swapper process, a process that turned out to be quite complex when defined in the last chapter.

Because these driver processes have already been presented in detail in the last chapter, it is only necessary to cover those parts that differ as a result of the use of message passing. This will allow us to ignore the common parts, thus simplifying the exposition.

5.4.1 The Clock

To begin the discussion, the constant *ticklength* is repeated. It is the amount of time represented by a single tick of the clock:

$$ticklength : TIME$$

Next, a message type is defined:

$$CLOCKMSG ::= CLKTICK\langle\langle TIME\rangle\rangle$$

CLOCKISR

$\upharpoonright(INIT, ServiceISR)$

$GenericISR$

$time : TIME$
$msgs : KMsgMgr$

INIT

$tm? : TIME$
$msgman? : MsgMgr$

$msgs' = msgman?time' = tm?$

$ServiceISR \mathrel{\widehat=}$
$\quad time' = time + ticklength\,_{9}^{\circ}$
$\quad (\exists\, dest : APREF;\ m : CLOCKMSG \mid$
$\qquad dest = \mathsf{clock} \land m = CLOCKTICK\langle\langle time\rangle\rangle \bullet$
$\quad SendInterruptMsg[dest/driver?, m/m?])$

Although logically correct, there are pragmatic issues that need to be discussed. There are similar issues with the page-fault mechanism, which is addressed in some detail in the next chapter. Meanwhile, the clock driver process is presented.

The process is essentially the same as the one specified in the last chapter (Section 4.6.3). The difference, of course, is that the process uses message passing to communicate the current time and to record requests for alarms. Alarms themselves are implemented in the same way as in the last chapter.

In order to implement message passing, it is necessary to provide an interface to the internal kernel message manager, *KMsgMgr*, to the clock driver.

ClockDriver

$\upharpoonright(INIT, RunProcess)$

$$
\begin{array}{l}
swaptabs : ProcessStorageDescrs \\
msgman : KMsgMgr \\
lk : Lock \\
now : TIME \\
missing : TIME \\
alarms : APREF \twoheadrightarrow TIME
\end{array}
$$

$INIT \mathrel{\widehat{=}} \ldots$

$addAlarm \mathrel{\widehat{=}} \ldots$

$cancelAlarm \mathrel{\widehat{=}} \ldots$

$processRequests \mathrel{\widehat{=}} \ldots$

$alarmsToCall \mathrel{\widehat{=}} \ldots$

$updateSwapperTimes \mathrel{\widehat{=}} \ldots$

$getNextTick \mathrel{\widehat{=}} \ldots$

$RunProcess \mathrel{\widehat{=}} \ldots$

The initialisation operation is basically the same as in the last chapter. It is necessary, of course, to initialise the reference to the message manager.

The operation to inform the swapper process of the current time is now implemented as a message-passing method. It is defined as follows:

$$
\begin{array}{l}
updateSwapperTimes \mathrel{\widehat{=}} \\
\quad swaptabs.UpdateAllStorageTimes \fatsemi \\
\quad (\exists src, dest : APREF;\ m : MSG \mid src = \mathsf{clock} \land dest = \mathsf{swapper} \bullet \\
\qquad m = SWAPTIME\langle\!\langle now \rangle\!\rangle \\
\qquad \land\ msgman.SendMessage[dest/dest?, src/src, m/m?])
\end{array}
$$

The main point about this operation is that it now constructs a message of type *SWAPTIME* that contains the current time as its payload. The destination to which the message is to be sent is the swapper process.

As noted above, the *SendMessage* operation is implicitly overloaded.

The clock driver obtains the current time from the hardware clock via the clock ISR, which sends a *CLKTICK* message to the driver process. Sending this message is the main point of the ISR.

It should be noted that the entire clock could be implemented within the ISR. This would lead to an implementation that could place reasonable guarantees on alarms and on the time displayed by the clock. The clock in this model, as in the last, is really organised to display the various interactions between components and how they can be modelled formally. For this reason, the optimisation remark below is of importance.

The *CLKTICK* message contains the current time in an appropriate form (probably in terms of milliseconds). The *missed_ticks* global variable is updated just in case any drivers are unable to synchronise.

$$
\begin{array}{l}
getNextTick \mathrel{\widehat=} \\
\quad (\exists\, src, dest : APREF;\ cm : MSG;\ tm : TIME \mid \\
\qquad\qquad src = \mathsf{hardware} \land dest = \mathsf{clock} \bullet \\
\qquad msgmgr.RcvMessage[src/src?, dest/dest?, cm/m!] \\
\qquad \land\ cm = CLKTICK\langle\!\langle tm \rangle\!\rangle \\
\qquad \land\ missing = glovars.missed\_ticks \\
\qquad \land\ now' = now + missed\_ticks + tm \\
\qquad \land\ glovars.missed\_ticks' = glovars.missed\_ticks - missing
\end{array}
$$

This operation is another example of how the receiving process can request the next message from a specific source, in this case from the clock (which happens to be the clock ISR).

The main routine of this process is *RunProcess*. It implements an infinite loop (modelled by quantification over an infinite domain). The main routine is resumed when the ISR sends it a message containing the current time. The routine then disables interrupts and sends a message to update the swapper's tables; it also updates the current process quantum (this is a shared variable in the scheduler, so access needs to be locked). The driver then decodes other messages. If the message is of type *NULLCLKRQ*, it checks the alarms that it needs to call and readies the processes that are waiting for them. Because readying waiting processes alters the scheduler's queues, the scheduler is called to select a process to run after the driver suspends. If the message is an alarm request (denoted by a message of type *AT*), the requesting process is added to the alarm queue in the driver. Otherwise, the message is a *NOTAT* message; this message type is used to cancel a previously requested alarm. In both cases, processes waiting for alarms are readied and the scheduler is called. Since these operations are within the scope of the universal quantifier, the process then waits for the next message from the ISR.

$$
\begin{array}{l}
RunProcess \mathrel{\widehat=} \\
\quad (\forall\, i : 1 \mathinner{\ldotp\ldotp} \infty \bullet \\
\qquad (getNextTick \mathbin{;} \\
\qquad (lk.Lock \mathbin{;} \\
\qquad\qquad (ptab.UpdateCurrentProcessQuanta[now'/tm?] \\
\qquad \land\ updateSwapperTimes[now/tm?] \land sched.UpdateProcessQuantum \\
\qquad \land\ lk.Unlock))) \\
\qquad \lor ((rq? = NULLCLKRQ \land alarmsToCall \land sched.ScheduleNext) \\
\qquad \lor ((rq? = AT\langle\!\langle p, t \rangle\!\rangle \land addAlarm[p/p?, t/tm?]) \\
\qquad \lor (rq? = NOTAT\langle\!\langle p, t \rangle\!\rangle \land cancelAlarm[p/p?, t/tm?]) \\
\qquad\qquad \land\ alarmsToCall \land sched.ScheduleNext)))
\end{array}
$$

The main routine is organised as a three-way branch, or as three disjuncts:

1. The first case handles messages from the clock ISR. This causes swapper and quantum updates inside a lock.
2. The second case receives requests from processes needing an alarm.
3. The final case receives requests from processes wanting to cancel previously requested alarms.

This organisation imposes no priorities on the messages. It is essential for the clock to be able to respond to ISR messages but it must also be available to handle alarm requests and cancellations. If the process waited for clock ticks *and then* waited for requests and cancellations, it would only be able to respond to the latter when a clock tick had awakened the process. It is expected that alarms (and cancellations) will be comparatively rare and that the probability that they will occur at the same time as a clock tick is low. It is also expected that the time taken to execute the clock driver is very much less than the time between clock ticks (if this turns out to be false, the driver can be optimised by splitting it into two processes).

5.5 Swapping Using Messages

In this section, the various processes implementing the swapping process are presented in an altered form: they now use message passing.

The swapper processes defined in this chapter use message passing instead of semaphores and shared variables. It is, therefore, necessary to define message types.

The message types contain information required by the desired operation. The request type (which is assumed to be a sub-type of *MSG*, so over-loading is, again, implicit) is as follows:

$$
\begin{aligned}
SWAPRQMSG ::= &\ NULLSWAP \\
\mid &\ SWAPOUT\langle\langle PREF \times ADDRESS \times ADDRESS\rangle\rangle \\
\mid &\ SWAPIN\langle\langle PREF \times ADDRESS\rangle\rangle \\
\mid &\ NEWSPROC\langle\langle PREF \times RMEM\rangle\rangle \\
\mid &\ DELSPROC\langle\langle PREF\rangle\rangle
\end{aligned}
$$

An extra message type is required to replace the semaphore that indicated that the image-copy operation has been completed (Section 4.6.1). Instead of signalling on a semaphore, the disk process sends a message of the following type to the swapper:

$$SDRPYMSG ::= DONE$$

SWAPDISKDriverProcess

$\upharpoonright(INIT, SwapDriver)$

$dmem : APREF \twoheadrightarrow\!\!\!\!\!\!\shortmid\;\; RMEM$
$sms : SharedMainStore$
$msgmgr : KMsgMgr$

$$
\begin{array}{l}
\underline{\quad INIT \quad} \\
store? : SharedMainStore \\
mm? : KMsgMgr \\
\hline
\mathrm{dom}\ dmem' = \varnothing \\
sms' = store? \\
msgmgr' = mm?
\end{array}
$$

$writeProcessStoreToDisk \mathrel{\widehat=} \ldots$

$readProcessStoreFromDisk \mathrel{\widehat=} \ldots$

$deleteProcessFromDisk \mathrel{\widehat=} \ldots$

$handleRequest \mathrel{\widehat=} \ldots$

$SwapDriver \mathrel{\widehat=} \ldots$

The *SwapDriver* operation is the main one of the process. It is defined as an infinite loop whose body receives messages from swapdisk:

$$
\begin{array}{l}
SwapDriver \mathrel{\widehat=} \\
\quad \forall i : 1 \mathinner{..} \infty \bullet \\
\qquad (\exists\, src, dest : APREF;\ rq : SWAPRQMSG \mid \\
\qquad\qquad\qquad src = \mathsf{swapper} \land dest = \mathsf{swapdisk} \bullet \\
\qquad\quad msgmgr.RcvMessage[src/src?, dest/dest?, rq/m!] \\
\qquad\quad \land (rq \neq NULLSWAP \land handleRequest \\
\qquad\quad \land (\exists\, SDRPYMSG \bullet \\
\qquad\qquad msgmgr.SendMessage[dest/src?, src/dest?, rpy/m?])))
\end{array}
$$

As is common with such processes, the real work is performed by an operation that is not the main entry point. In this case, it is *handleRequest*:

$$
\begin{array}{l}
\underline{\quad handleRequest \quad} \\
rq? : SWAPRQMSG \\
\hline
(\exists\, p : APREF;\ start, end : ADDRESS;\ mem : RMEM \bullet \\
\qquad rq? = SWAPOUT\langle\!\langle p, start, end \rangle\!\rangle \\
\qquad \land\ sms.CopyMainStore[start/start?, end/end?, mem/mseg!] \\
\qquad \land\ writeProcessStoreToDisk[p/p?, mem/ms?]) \\
\quad \lor (\exists\, p : APREF;\ ldpt : ADDRESS;\ mem : RMEM \bullet \\
\qquad rq? = SWAPIN\langle\!\langle p, ldpt \rangle\!\rangle \\
\qquad \land\ readProcessStoreFromDisk[p/p?, mem/ms!] \\
\qquad \land\ sms.WriteMainStore[ldpt/loadpoint?, mem/mseg?]) \\
\quad \lor (\exists\, p : APREF \bullet \\
\qquad rq? = DELSPROC\langle\!\langle p \rangle\!\rangle \land deleteProcessFromDisk[p/p?]) \\
\quad \lor (\exists\, p : APREF;\ img : RMEM \bullet \\
\qquad rq? = NEWSPROC\langle\!\langle p, img \rangle\!\rangle \\
\qquad \land\ writeProcessStoreToDisk[p/p?, img/img?])
\end{array}
$$

The operation uses the type of the latest message to determine which action to take (dispatches, that is, according to message type). Its definition is similar to that in the previous chapter.

The swapper process is now as follows:

$$
\begin{array}{l}
\textit{SwapperProcess} \\
\hline
\upharpoonright(\mathit{INIT}, \mathit{SwapperProcess}) \\
\quad \mathit{pdescrs} : \mathit{ProcessStorageDescrs} \\
\quad \mathit{proctab} : \mathit{ProcessTable} \\
\quad \mathit{sms} : \mathit{SharedMainStore} \\
\quad \mathit{hw} : \mathit{HardwareRegisters} \\
\quad \mathit{realmem} : \mathit{SharedMainStore} \\
\quad \mathit{msgmgr} : \mathit{KMemMgr} \\
\mathit{INIT} \mathrel{\widehat=} \ldots \\
\mathit{requestWriteoutSegment} \mathrel{\widehat=} \ldots \\
\mathit{requestReadinSegment} \mathrel{\widehat=} \ldots \\
\mathit{SwapperProcess} \mathrel{\widehat=} \ldots \\
\mathit{SwapProcessOut} \mathrel{\widehat=} \ldots \\
\mathit{SwapCandidateOut} \mathrel{\widehat=} \ldots \\
\mathit{SwapProcessIn} \mathrel{\widehat=} \ldots \\
\mathit{SwapProcessIntoStore} \mathrel{\widehat=} \ldots \\
\mathit{DoDiskSwap} \mathrel{\widehat=} \ldots
\end{array}
$$

The operations of this process are similar to those in the semaphore-based one. The primary difference is that messages are used instead of the semaphore and shared variable.

The following operation is dispatched when a *SWAPRQMSG* is received. It sends a *SWAPOUT* message to the swap-disk process and waits for the *SDRPYMSG* to synchronise.

$$
\begin{array}{l}
\textit{requestWriteoutSegment} \\
\hline
p? : \mathit{APREF} \\
\mathit{start}?, \mathit{end}? : \mathit{ADDRESS} \\
\hline
(\exists\, \mathit{rq} : \mathit{SWAPRQMSG};\ \mathit{src}, \mathit{dest} : \mathit{APREF} \mid \\
\qquad\qquad\qquad \mathit{src} = \mathsf{swapper} \land \mathit{dest} = \mathsf{swapdisk} \bullet \\
\quad \mathit{rq} = \mathit{SWAPOUT}\langle\!\langle p?, \mathit{start}?, \mathit{end}? \rangle\!\rangle \\
\quad \land\ \mathit{msgmgr}.\mathit{SendMessage}[\mathit{src}/\mathit{src}?, \mathit{dest}/\mathit{dest}?, \mathit{rq}/m?] \\
\quad \land\ (\exists\, \mathit{rpy} : \mathit{SDRPYMSG} \bullet \\
\qquad \mathit{msgmgr}.\mathit{RcvMessage}[\mathit{dest}/\mathit{src}?, \mathit{src}/\mathit{dest}?, \mathit{rpy}/m!] \\
\qquad \land\ \mathit{rpy} = \mathit{DONE}))
\end{array}
$$

Operation *requestReadinSegment* is, again, a message-dispatched one that sends a message to the swap-disk process.

$$
\begin{array}{l}
\textit{requestReadinSegment} \\
\hline
p? : APREF \\
loadpoint? : ADDRESS \\
\hline
(\exists\, rq : SWAPRQMSG;\ src, dest : APREF \mid \\
\qquad\qquad src = \mathsf{swapper} \land dest = \mathsf{swapdisk} \bullet \\
\quad rq = SWAPIN\langle\!\langle p?, loadpoint? \rangle\!\rangle \\
\quad \land\ msgmgr.SendMessage[src/src?, dest/dest?, rq/m?] \\
\quad \land\ (\exists\, rpy : SDRPYMSG \bullet \\
\qquad msgmgr.RcvMessage[dest/src?, src/dest?, rpy/m!] \\
\qquad \land\ rpy = DONE))
\end{array}
$$

This operation, like the previous one, shows how much can be done just by exchanging messages. It also demonstrates how much clearer message-based code is than that based on semaphores.

The final operation of this subsection is the main one of the process. It loops forever waiting for messages to arrive. It then calls *DoDiskSwap* to do the real work.

$$
\begin{array}{l}
SwapperProcess \mathrel{\widehat=} \\
\quad \forall\, i : 1 \mathinner{..} \infty \bullet \\
\qquad (\exists\, src, dest : APREF;\ m : MSG \mid \\
\qquad\qquad m = dest = \mathsf{swapper} \land src = \mathsf{clock} \bullet \\
\qquad\quad msgmgr.RcvMessage[/src?, /dest?, m/m!] \\
\qquad \land\ DoDiskSwap)
\end{array}
$$

5.6 Kernel Interface

Some readers will have been wondering why a proper system interface has not been defined in any of the models so far. The first kernel merits no such interface: it is completely open (and, therefore, somewhat unsafe) so that it can be as fast as possible. The second kernel is based on semaphores and semaphores, as has been noted (probably too) often, are vital but low-level primitives. A semaphore-based kernel interface was considered to be less clear than one based on messages. It falls to this kernel to define a kernel interface (much as it fell to the one in the last chapter to prove that semaphores behave properly) because it can be defined in a relatively clean fashion.

In order to define the kernel interface, it is necessary to be clear as to its purpose. The interface provides those operations that can be executed by user processes when they require kernel operations to be performed for them. The operations that can be performed generally include such things as:

- i.o operations, traditionally file operations.

- read/write requests to communications networks (often expressed in terms of protocol-specific operations).
- requests to alter process priority (e.g., `nice` and `renice` in some versions of Unix.
- process creation and termination—creation requests can also create child processes.
- storage allocation and deallocation requests.

The kernel of this chapter, like the previous one, contains no file system (for reasons given in Chapter 1) and contains no network interface (for reasons of space[1]). Like the last kernel, all that the interface has to support is process creation and termination. In both cases, there is no way to alter the priority of a user process: it just remains a user-level process. The `nice` operation is easy to implement: it is merely the inclusion of a lower-priority queue in the scheduler. There is plenty, in any case, to do with process creation and termination.

The operations to be included in the system calls for this kernel are the ones that mostly constitute the process abstraction as far as user processes are concerned. The primary operations are:

- Create a new process.
- Terminate a process.

The kinds of process that can be created are:

- Create a process that can potentially have children (the usual case).
- Create a child process.

These operations are immediately supported by the primitives defined in the last chapter. This is a process model that is supported by Unix and its derivatives. When creating a child process, a flag must be passed to the parent and child processes so that they can determine the role they play. This is a simple fork-join model. It is worth noting that message passing between processes in this model is relatively simple: it can be arranged for the parent to receive the identifier of the child process (the child can also be passed the identifier of the parent so that two-way communication can be immediately established).

In addition, some systems (e.g., Windows) support a kind of process creation in which the newly created processes do not share code or data; they are autonomous processes. In this kind of process model, the sub-processes are able to communicate with their creator and with each other. To do this, it is necessary to pass process identifiers between the processes so that messages can be exchanged. The management of such matters as termination among autonomous communicating processes is usually left to the programmer. This is what will be done in this chapter.

[1] It would be extremely interesting to model one, but that is for another day.

In order to provide these operations, a set of parameters must be defined for each operation. In each case, the identifier of the newly created process is returned.

It is easy enough to arrange for processes to create other processes. The question arises as to how a completely new process is created (i.e., a process that is not created by another). One way to do this is to have a general root process that receives creation messages; another is to introduce a command interface. Here, the nature of the process is ignored. It is just a process that calls the necessary library routines; no *urprocess* is, therefore, defined. Whatever approach is adopted, a library of operations must be provided to perform all interface operations. A call to a library operation will actually cause the requesting messages to be sent to the kernel (this is, in essence, what the Mach kernel does). The library is the real focus of attention in this section.

The basis of the library is a set of message types. Each type represents an operation to be performed by the kernel. In addition to these messages, there must be one or more processes that are able to receive and interpret these messages. The library will have to access the storage on which each process' code is stored. Since this store is outside the scope of this model, it must be assumed that the library has already accessed this store and obtained basic parameters from it before the following messages are sent to the kernel. (It should be noted that obtaining this data might also involve kernel operations that are not defined here.)

The messages contain the parameters for the operations to be performed. In each case, the message contains the identifier of the sending process. It also contains two natural numbers (elements of $\mathbb{N}$). They denote the size of the data area and the size of the stack area that are, respectively, required by the new process. The last message type just has a process reference as its payload: it is the identifier of the process that is to terminate. The creation message types also contain a first component of type $PCODE$: this is the code of the process that is to be created. The reader should note that what is going on is actually a conflation of the process-creation and code-loading functions. Since no loading operation is specified here (because no external storage media have been specified), it is not possible to specify something like a file name instead of a complete chunk of code[2]. Alternatively, the operations provided here could be seen as being at a level that is lower than the one called by user processes. The purpose of the specification presented here is to show the reader *how* it can be done in principle, not to give the exact details—given message passing, it is a relatively easy process to define an interface between,

[2] In any case, kernels might vary in the way in which code is stored. Some might store it in FLASH; others might, for instance, use a network interface to a SAN. The conventional method is to employ a file store but a database is equally possible. For present purposes—the modelling of kernels that are free from things like file systems—the approach adopted here seems quite reasonable.

say, a file system and the creation operation. This is not done for the reasons given here and elsewhere in this book.

$$
\begin{aligned}
SYSCALLMSG ::= &\ CRTPROC\langle\langle APREF \times PCODE \times \mathbb{N} \times \mathbb{N}\rangle\rangle \\
| &\ CRTCHLD\langle\langle APREF \times PCODE \times \mathbb{N} \times \mathbb{N} \times \mathbb{N}\rangle\rangle \\
| &\ TERMPROC\langle\langle APREF\rangle\rangle \\
\ldots &
\end{aligned}
$$

These messages will generate reply messages. In most cases, if the operation succeeds, the identifier (of type *APREF*) of the newly created process will be returned, together with other information.

$$
\begin{aligned}
SYSRPY ::= &\ NEWPROC\langle\langle APREF \times APREF\rangle\rangle \\
| &\ NEWCHLDPROC\langle\langle APREF \times APREF\rangle\rangle \\
\ldots &
\end{aligned}
$$

The messages contain the identifier of the newly created process and the identifier of the process that made the creation request; in addition, they contain a natural number that is a code returned by the operation.

When creating processes, it is important for the creating process to have the identifier of the process just created. This enables the two processes to communicate using messages; it also allows the creating process to communicate the identifier of the newly created process to other processes; and finally, the newly created process' identifier can be stored in a table by its creating process.

User interface operations are all collected into a single library. As usual, this library will be represented as a class. The class takes as initialisation parameters the following: the identifier of the process to which it is linked (*mypid*), a pointer to the message manager module (*msgman*), and the identifier of the process handling system calls (*kernitfid*).

SysCallLib

$\upharpoonright(INIT, UserCreateProcess, UserCreateChildProcess, TerminateProcess, \ldots)$

$mypid : APREF$
$msgman : UserMsgMgr$
$kernintfid : APREF$

INIT

$me? : APREF$
$mm? : UserMsgMgr$
$intfid? : APREF$

$mypid' = me?$
$msgman' = mm?$
$kernintfid' = intfid?$

$UserCreateProcess \mathrel{\widehat{=}} \ldots$
$UserCreateChildProcess \mathrel{\widehat{=}} \ldots$
$TerminateProcess \mathrel{\widehat{=}} \ldots$
$\ldots$

The class seems a little impoverished, dealing, as it does, only with process creation and termination. In most systems, the kernel also handles I/O requests, often at the level of the file system. The above class is included as an illustration of what can be done and could, without too much effort, be immediately extended to include calls that perform I/O operations at a level just above the drivers; indeed, I/O operations of considerable sophistication could be provided, as well as operations, say, on streams (i.e., sequences of messages). This whole issue has been ignored because we still have not settled the question of which devices are being managed by this system.

In each case, the creation operations just create a message and send it to the kernel. They then wait for a reply that contains the identifier of the newly created process. The operations are defined as follows.

The first operation is the user-level process-creation operation. It requests the creation of a separate process, requiring the code (*code*?), size of data (*datasize*?) and stack (*stacksize*?) areas, respectively, and returns the identifier of the newly created process (*newpid*!).

UserCreateProcess

$code? : PCODE$
$datasize?, stacksize? : \mathbb{N}$
$newpid! : APREF$

$(\exists\, m : SYSCALLMSGm = CRTPROC\langle\!\langle code?, stacksize?, datasize?\rangle\!\rangle \bullet$
$\quad msgman.SendMessage[mypid/src?, kernintfid/dest?, m/m?])$
$\quad \mathbin{;}(\exists\, rpmsg : SYSRPY;\ newid : APREF \bullet$
$\quad\quad msgman.RcvMessage[kernintfid/src!, mypid/dest?, rpmsg/m!]$
$\quad\quad \land\ rpmsg = NEWPROC\langle\!\langle newid\rangle\!\rangle$
$\quad\quad \land\ newpid! = newid)$

The second operation creates a child process. It requests the creation of a process that will share code with the calling process (which should supply a pointer to its own code segment as *code*?); the remainder of the arguments are the same. The identifier of the newly created process is returned as *newpid*!.

UserCreateChildProcess

$code? : PCODE$
$datasize?, stacksize? : \mathbb{N}$
$newpid! : APREF$

$$(\exists m : SYSCALLMSG \mid m = CRTCHLD\langle\langle code?, stacksize?, datasize?\rangle\rangle \bullet$$
$$msgman.SendMessage[mypid/src?, kernintfid/dest?, m/m?])$$
$$\fatsemi(\exists rpmsg : SYSRPY;\ newid : APREF \bullet$$
$$msgman.RcvMessage[kernintfid/src!, mypid/dest?, rpmsg/m!]$$
$$\wedge\ rpmsg = NEWCHLDPROC\langle\langle newid\rangle\rangle$$
$$\wedge\ newpid! = newid)$$

The termination message is sent by the following operation:

TerminateProcess

$mypid? : APREF$

$$(\exists m : SYSCALLMSG \mid m = TERMPROC\langle\langle mypid?\rangle\rangle \bullet$$
$$msgman.SendMessage[mypid/src?, kernintfid/dest?, m/m?])$$

The caller does not wait for a reply. When this message is sent, it is all over as far as the caller is concerned!

The kernel supports a single process that receives the requests sent by *SysCallLib*'s operations. In the version presented here, it is just a loop that receives messages and performs the appropriate operations. This is very simple, of course, and no checking is performed; the process could do other things and engage in message exchange with processes inside the kernel in a more sophisticated version. Again, the point, here, is to demonstrate the principles.

The process is called *KernIntf*. It is defined by the following class:

KernIntf

$\upharpoonright(INIT, RunProcess)$

$mypid : APREF;\ msgman : UserMessages;\ procmgr : ProcessCreation$

INIT

$me? : APREF$
$mm? : UserMessages$
$pcr? : ProcessCreation$

$mypid' = me?$

$msgman' = mm?$

$procmgr' = pcr?$

$$
\begin{array}{l}
RunProcess \mathrel{\widehat=} \\
\quad \forall i : 1 \mathinner{\ldotp\ldotp} \infty \bullet \\
\qquad (\exists m : SYSCALLMSG \bullet \\
\qquad\quad msgmgr.RcvMessage[mypid/caller?, src/src?, m/m!] \\
\qquad\quad \land ((\exists src : APREF;\ stcksz, datasz : \mathbb{N}; \\
\qquad\qquad\qquad code : MEM;\ newpid : APREF \bullet \\
\qquad\qquad m = CRTPROC\langle\langle src, code, stcksz, datasz\rangle\rangle \\
\qquad\qquad \land procmgr.CreateUserProcess \\
\qquad\qquad\qquad [code/code?, stcksz/stacksize?, \\
\qquad\qquad\qquad\quad datasz/datasize?, newpid/newpid!] \\
\qquad\qquad \land (\exists crmsg : SYSRPY \mid crmsg = NEWPROC\langle\langle newpid\rangle\rangle \bullet \\
\qquad\qquad\quad msgmgr.SendMsg[mypid/src?, src/dest?, crmsg/m?])) \\
\qquad\quad \lor (\exists src : APREF;\ stcksz, datasz : \mathbb{N}; \\
\qquad\qquad\qquad code : MEM;\ newpid : APREF \bullet \\
\qquad\qquad m = CRTCHLD\langle\langle src, code, stcksz, datasz\rangle\rangle \\
\qquad\qquad \land procmgr.CreateChildUserProcess \\
\qquad\qquad\quad [code/code?, stcksz/stacksize?, datasz/datasize?, \\
\qquad\qquad\qquad src/rqprocid?, newpid/newpid!] \\
\qquad\qquad \land (\exists retmsg : SYSRPY \mid \\
\qquad\qquad\qquad retmsg = NEWCHLDPROC\langle\langle newpid\rangle\rangle \bullet \\
\qquad\qquad\quad msgmgr.SendMsg[mypid/src?, src/dest?, retmsg/m?])) \\
\qquad\quad \lor (\exists tpid : APREF \bullet \\
\qquad\qquad m = TERMPROC\langle\langle tpidsrc\rangle\rangle \\
\qquad\qquad procmgr.TerminateProcess[tpid/p?]) \\
\qquad\quad \lor \ldots))
\end{array}
$$

The class is incomplete because it does not handle any requests other than user-process creation and termination. With the exception of sending and receiving messages, the kernel does not contain any other services that would be useful to user processes. The message-passing operations are not included because they are obvious given the above definitions.

It is now possible to state an important proposition about the kernel.

Proposition 123. *Only one user process can be in the kernel at any one time.*

PROOF. First, it is necessary to clarify what is meant by the statement of the proposition. What is intended by the proposition is the following: at any one time, it is possible for at least one and at most one system call to be processed. Since system calls are: (i) procedure calls and (ii) requests for the kernel to perform some action (as defined by the system call library, *SysCallLib*, as the sketch-form adopted here is called).

As a procedure call, any system call can belong to a single thread of execution only. This implies that exactly one process at a time can be performing the system call.

The procedures comprising the system-call library all send and receive messages. Therefore, the rest of the proof must be in terms of the properties of the message-passing subsystem.

The message-passing subsystem is driven by interrupts (Proposition 116) and only one interrupt can be serviced at any time (this is an informal property of interrupts). Therefore, while any user process is sending (or receiving) a message, there can be no other process performing the same operation (more generally, there can be no other process performing any operation).

By Proposition 114, messages are processed in the order in which they are received. By inspection of the system-call operations, each is structured as a message send followed by a message receive. Furthermore, message passing is synchronous by Proposition 112.

The system calls are implemented by a process that just waits for messages, services them and returns the reply. Therefore, until a system call has been completed, it is not possible for the caller to proceed. Furthermore, by the organisation of the message-passing subsystem, it is not possible for another user process to proceed until the *KernIntf* process has replied to a message.

The remainder of the kernel will have its operations hidden by syntactic methods and only imported by the *KernIntf* process. □

Corollary 10. *Message passing can be used to implement mutual exclusion.*

PROOF. By Propositions 112 and 114. □

6

Virtual Storage

Nil posse creari de nilo
– Lucretius, De Rerum Natura, I, 155

6.1 Introduction

In this chapter, mechanisms to support virtual storage will be modelled. Virtual storage affords a considerable number of advantages to the operating system designer and user. Virtual storage is allocated in units of a page and pages can be collected into independent segments; virtual storage defines clear boundaries between address spaces so that each process can maintain its own address space. This clearly provides a measure of protection between processes but requires additional methods for exchanging data between processes.

In virtual storage systems, main store is shared between processes in the usual way but is defined as a sequence of *page frames*, each a block of store one page long. The storage belonging to each process is swapped into main store when required and copied to a paging disk (or some other mass-storage device) when not required. Strategies for selecting pages to copy to the paging disk and for determining which page to bring into main store must be defined.

6.2 Outline

The storage system to be designed is to have the following features:

- The virtual store should have four segments: one each for code, data, stack and heap.
- The system uses *demand paging* with reference counting for victim selection.
- Pages can be shared (and unshared) between processes.
- Segments can be shared (and unshared) between processes;

- Storage should be mapped in the sense that disk blocks can be directly mapped to main-store pages and vice versa.
- Message passing will be used for IPC.

The virtual storage system is composed of:

- A page-fault handler. This is invoked when a page fault occurs. It determines the identity of the *logical* page that caused the page fault. It invokes the page-fault driver process and passes to it the identifier of the faulting process and the page number. It unreadies the faulting process.
- A page-fault driver. This takes a message from the page-fault ISR and sends a message to the paging disk to retrieve the page whose reference caused the fault. If there are no free page frames in (main) physical store, it selects a victim page in physical store and sends it to the paging disk. When the faulting page is received from the paging disk, it is copied into a main store page whose physical address is identified with the logical page number in the faulting process. The faulting process is then readied and the driver waits for the next page fault.

The above scheme is sub-optimal. As part of the model, optimisations are suggested, particularly for the interactions between the driver and paging disk.

Even though it is sub-optimal, the above scheme is logically sufficient. It is, therefore, appropriate to concentrate on it as the model for this chapter. This exemplifies the method adopted in this book: capturing the logical functioning of the model is much more important than optimisation. The optimisation included here is introduced as an example of how it can be done without too much of a compromise to the model.

The structure of this kernel is shown in Figure 6.1. Comparison of this figure with the corresponding one in Chapter 4 (Figure 4.1) reveals their similarities. In the current kernel, virtual and not real storage forms the basis of the system. Apart from the need for structures and operations to support virtual storage (the subject of this chapter), the main difference lies in the kernel bootstrapping operations (which are not considered in this book).

6.3 Virtual Storage

In this section, the basic structures required to model a virtual store are introduced.

The following axiomatic definition defines the number of real pages (pages in real store or physical pages) and the size of the page frame. Neither constant is assigned a value, so the specification is of the loose variety.

$$\begin{array}{|l} numrealpages : \mathbb{N} \\ framesize : \mathbb{N} \end{array}$$

The basic virtual address is represented by an atomic type:

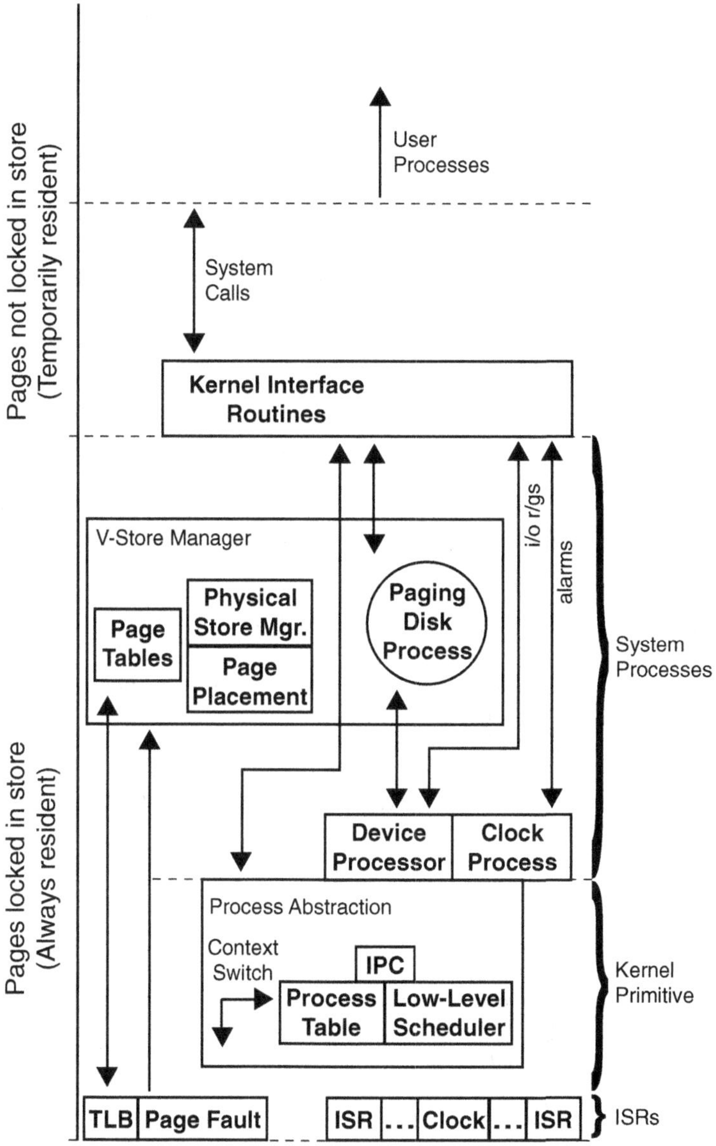

Fig. 6.1. *The layer-by-layer organisation of the kernel, including virtual storage-management modules.*

$[VIRTUALADDRESS]$

$$| \; maxvirtpagespersegment : \mathbb{N}$$

This is the maximum number of virtual pages that a process can own.

$$PAGEOFFSET == 1 \mathinner{..} framesize$$
$$PHYSICALPAGENO == 1 \mathinner{..} numrealpages$$
$$LOGICALPAGENO == \mathbb{N}$$

where:

- *PAGEOFFSET* denotes the offsets into a page;
- *PHYSICALPAGENO* denotes the indices of pages in the main-store page frame;
- *LOGICALPAGENO* denotes the indices of logical pages (pages belonging to a process).

Note that *LOGICALPAGENO* is not bounded above. The reason for this is that the actual number of logical pages that a process can have is a hardware-dependent factor and is not relevant to the current exercise.

For every virtual address, the hardware performs an address translation that maps the virtual address to a logical frame number and an offset into that frame. The signature of this function, *addresstrans*, is:

$$| \; addresstrans : VIRTUALADDRESS \rightarrow DECODEDADDRESS$$

The definition of this function is hardware-specific and is, in any case, not particularly relevant to the current exercise.

The type *DECODEDADDRESS* is defined as:

$$DECODEDADDRESS == LOGICALPAGENO \times FRAMEOFFSET$$

and has the following projections:

$$
\begin{array}{|l}
dlogicalpage : DECODEDADDRESS \rightarrow LOGICALPAGENO \\
dpageoffset : DECODEDADDRESS \rightarrow PAGEOFFSET \\
\hline
\forall addr : DECODEDADDRESS \bullet \\
\quad dlogicalpage(addr) = fst\ addr \\
\quad dpageoffset(addr) = snd\ addr
\end{array}
$$

For a segmented, paged architecture, the address-decoding function can be defined as:

$$| \; saddresstrans : VIRTUALADDRESS \rightarrow SDECODEDADDRESS$$

where:

$$SDECODEDADDRESS == SEGMENT \times LOGICALPAGENO \times PAGEOFFSET$$

and:

$$\begin{array}{l} saddrseg : SDECODEDADDRESS \rightarrow SEGMENT \\ spageno : SDECODEDADDRESS \rightarrow LOGICALPAGENO \\ spagoffset : SDECODEDADDRESS \rightarrow PAGEOFFSET \\ \hline \forall saddr : SDECODEDADDRESS \bullet \\ \quad saddrseg(saddr) = fst\ saddr \\ \quad spageno(saddr) = fst(snd\ saddr) \\ \quad spagoffset(saddr) = snd^2\ saddr \end{array}$$

The *PAGEMAP* translates logical to physical page numbers. *PAGEFRAME* translates physical page numbers to the actual pages in store. Finally, *PAGE* defines a storage page as a vector of *PSU*; the vector has *PAGEOFFSET* elements. (*PSU* is, it will be recalled, the *Primary Storage Unit*, which is assumed to be a byte.)

$$\begin{array}{l} PAGEMAP == LOGICALPAGENO \twoheadrightarrow PHYSICALPAGENO \\ PAGEFRAME == PHYSICALPAGENO \rightarrow PAGE \\ PAGE == PAGEOFFSET \rightarrow PSU \end{array}$$

The empty page is defined as follows:

$$\begin{array}{l} NullPage : PAGE \\ \hline NullPage = (\lambda\, i : PAGEOFFSET \bullet 0) \end{array}$$

It is a page (vector of bytes), *PAGEOFFSET* bytes long, with each byte set to zero.

Pages are associated with a number of flags, including, among others:

- in core (i.e., in main store);
- locked In (i.e., must always remain in main store);
- shared (i.e., shared between at least two processes).

For each process, these properties can be represented by sets (thus simplifying the modelling of pages).

For the remainder of this section, a segmented virtual store will be modelled. The hardware nearest to the model presented here is the Intel X86 series. The reader is warned that this is a *logical* model of segmented, paged storage, not an exact rendition of any existing hardware implementations. Furthermore, details such as the *Translation Lookaside Buffer*, or *TLB*, (the associative store that typically holds a few entries from the current process' page table) are not modelled in detail, the reason being that, as usual, hardware differs considerably in implementation and, in particular, the size of the TLB can vary considerably among MMUs[1].

[1] *Memory Management Unit.*

Although most segmented hardware supports more than four segments per process, for the present, only three segments will be considered. Linux on X86 requires three segments: one each for code, stack and data, although the hardware permits a maximum of 16 segments (the virtual address size is 32 bits). The segment names are as follows:

$$SEGMENT == \{\mathsf{code}, \mathsf{data}, \mathsf{stack}, \mathsf{heap}, \ldots\}$$

A great many programming languages now require heap (dynamic) storage. A problem for dynamic store, as with stacks, is that, quite frequently, it has to be expanded. Within a three-segment organisation, the heap is part of either the stack or data segment; this can limit the maximum size of the heap somewhat on 32-bit machines. To simplify manipulation, it is assumed here that the data segment contains static data (global variables, literal pools, fixed-length buffers and so on) and the heap is given its own segment. The heap can, therefore, grow to the maximum segment size at runtime, as can the stack segment.

$$\begin{array}{l} usedsegment : SEGMENT \\ \hline \forall s : SEGMENT \bullet \\ \quad usedsegment(s) \Leftrightarrow s \in \{\mathsf{code}, \mathsf{data}, \mathsf{stack}, \mathsf{heap}\} \end{array}$$

It is now possible to define per-process page tables.

Each segment is composed of a number of pages. The following function translates a physical address and segment into a logical page number:

$$\begin{array}{l} pages\_in\_segment : APREF \times SEGMENT \times \\ \qquad (APREF \nrightarrow SEGMENT \nrightarrow \mathbb{F}\, LOGICALPAGENO) \rightarrow \\ \qquad\qquad \mathbb{F}\, LOGICALPAGENO \\ \hline \forall p : APREF;\ sg : SEGMENT;\ f : APREF \nrightarrow SEGMENT \nrightarrow \\ \qquad\qquad \mathbb{F}\, LOGICALPAGENO \bullet \\ \quad pages\_in\_segment(p, sg, f) = f(p)(sg) \end{array}$$

The following (inverse) functions mark and unmark pages. They are higher-order functions that take the specification of a page and a page attribute map as arguments and return the modified page attribute map.

$$\begin{array}{l} mark\_page : APREF \times SEGMENT \times LOGICALPAGENO \times \\ \qquad (APREF \nrightarrow SEGMENT \nrightarrow \mathbb{F}\, LOGICALPAGENO) \rightarrow \\ \qquad\qquad (APREF \nrightarrow SEGMENT \nrightarrow \mathbb{F}\, LOGICALPAGENO) \\ unmark\_page : APREF \times SEGMENT \times LOGICALPAGENO \times \\ \qquad (APREF \nrightarrow SEGMENT \nrightarrow \mathbb{F}\, LOGICALPAGENO) \rightarrow \\ \qquad\qquad (APREF \nrightarrow SEGMENT \nrightarrow \mathbb{F}\, LOGICALPAGENO) \\ \hline \forall p : APREF;\ sg : SEGMENT;\ lpno : LOGICALPAGENO; \\ \qquad f : APREF \nrightarrow SEGMENT \nrightarrow \mathbb{F}\, LOGICALPAGENO \bullet \\ \quad mark\_page(p, sg, lpno, f) = f(p) \oplus \{sg \mapsto (f(p)(sg) \cup \{lpno\})\} \\ \quad unmark\_page(p, sg, lpno, f) = f(p) \oplus \{sg \mapsto (f(p)(sg) \setminus \{lpno\})\} \end{array}$$

It is now proved that these two functions are mutual inverses.

Proposition 124. *mark_page and unmark_page*$^{-1}$ *are mutually inverse.*

PROOF. Write $f(p)(sg) = h$, then:

$$\begin{aligned} mark &= f(p) \oplus \{sg \mapsto (h \cup \{lpno\})\} \\ unmark &= f(p) \oplus \{sg \mapsto (h \setminus \{lpno\})\} \end{aligned}$$

Calculating the value of $unmark \circ mark$, we obtain:

$$(f(p) \oplus \{sg \mapsto (h \cup \{lpno\})\}) \oplus \{sg \mapsto (h \setminus \{lpno\})\}$$

Writing $f(p)(sg) = s$, then $mark = s \cup \{lpno\}$ and $unmark = s \setminus \{lpno\}$, so:

$$\begin{aligned} & unmark \circ mark \\ &\quad = (s \setminus \{lpno\}) \circ (s \cup \{lpno\}) \\ &\quad = (s \cup \{lpno\}) \setminus \{lpno\} \\ &\quad = s \end{aligned}$$

Conversely:

$$\begin{aligned} & mark \circ unmark \\ &\quad = (s \cup \{lpno\}) \circ (s \setminus \{lpno\}) \\ &\quad = (s \setminus \{lpno\}) \setminus \{lpno\} \\ &\quad = s \end{aligned}$$

Therefore, *mark* and *unmark* are mutual inverses and the proposition is proved. □

The page table abstraction can be modelled as follows. The variable *freepages* represents those pages in main store that are not allocated to any process. The page table proper is *pagetable.* The variables *executablepages*, *writablepages* and *readablepages* are intended to refer to pages the owner has marked executable (i.e., code pages), read-only (e.g., a constant data segment) and read-write (e.g., a stack). Pages can be shared between processes and some are locked into main store. When a page is locked, it cannot be removed from main store. The kernel's own storage is often marked as locked into main store. It is so locked because a page fault could prevent the kernel from responding in time to a circumstance. It is also necessary to keep track of those pages that are currently in main store: these are referred to as being "in core", hence the name of the variable, *incore.* The *pagecount* counts the number of pages in each segment of each process. There is an *a priori* limit to the number of pages in a segment and *pagecount* is intended to keep track of this and to provide a mechanism for raising an error condition if this limit is exceeded. The final variable, *smap*, is a relation between elements of *PAGESPEC*; it denotes those pages that are shared and it will be explained in more detail below.

There are different ways to organise page tables. The simplest is a linear sequence of page references. As virtual storage sizes increase, simple linear structures do not perform well, so tree-like structures are to be preferred. These trees can be arranged to perform mapping on two or three levels. The model defined here is intended to be suggestive of a tree structure, even though it can also be implemented as a table.

The class that follows defines an abstract data type. It represents the page table type. The type exports a large number of operations and has the most complex invariant in this book.

PageTables

$\upharpoonright$(*INIT*, *HaveFreePages*, *NumberOfFreePages*, *AllocateFreePage*, *MakePageFree*, *PhysicalPageNo*, *InitNewProcessPageTable*, *RemoveProcessFromPageTable*, *AddPageToProcess*, *HasPageInStore*, *IncProcessPageCount*, *DecProcessPageCount*, *LatestPageCount*, *UpdateMainstorePage*, *RemovePageFromProcessTable*, *RemovePageProperties*, *RemovePageFromProcess*, *IsPageInMainStore*, *MarkPageAsIn*, *MarkPageAsOut*, *IsSharedPage*, *MarkPageAsShared*, *UnsharePage*, *IsLockedPage*, *LockPage*, *UnlockPage*, *MakePageReadable*, *MakePageNotReadable*, *MakePageExecutable*, *IsPageExecutable*, *MakePageNotExecutable*, *MakePageWritable*, *IsPageWritable*, *MakePageNotWritable*)

$freepages : \mathbb{F}\, PHYSICALPAGENO$
$pagetable : APREF \twoheadrightarrow SEGMENT \twoheadrightarrow PAGEMAP$
$executablepages, writablepages, readablepages,$
$sharedpages, lockedpages,$
$incore : APREF \twoheadrightarrow SEGMENT \twoheadrightarrow \mathbb{F}\, LOGICALPAGENO$
$pagecount : APREF \twoheadrightarrow SEGMENT \twoheadrightarrow \mathbb{N}$
$smap : PAGESPEC \leftrightarrow PAGESPEC$

InvPageTables

$0 \leq \#freepages \leq numrealpages$

$\mathrm{dom}\, incore \subseteq \mathrm{dom}\, pagetable$
$\mathrm{dom}\, sharedpages \subseteq \mathrm{dom}\, pagetable$
$\mathrm{dom}\, lockedpages \subseteq \mathrm{dom}\, pagetable$
$\mathrm{dom}\, pagecount = \mathrm{dom}\, pagetable$
$\mathrm{dom}\, executablepages = \mathrm{dom}\, pagetable$
$\mathrm{dom}\, writablepages = \mathrm{dom}\, pagetable$
$\mathrm{dom}\, readablepages = \mathrm{dom}\, pagetable$

$$
\begin{array}{|l}
\hline
\textit{INIT} \\
\hline
\mathit{freepages}' = \varnothing \\
\operatorname{dom} \mathit{pagetable}' = \varnothing \\
\operatorname{dom} \mathit{sharedpages}' = \varnothing \\
\operatorname{dom} \mathit{lockedpages}' = \varnothing \\
\operatorname{dom} \mathit{incore}' = \varnothing \\
\operatorname{dom} \mathit{pagecount}' = \varnothing \\
\operatorname{dom} \mathit{executablepages}' = \varnothing \\
\operatorname{dom} \mathit{writablepages}' = \varnothing \\
\operatorname{dom} \mathit{readablepages}' = \varnothing \\
\operatorname{dom} \mathit{smap} = \varnothing \\
\hline
\end{array}
$$

$HaveFreePages \mathrel{\widehat=} \ldots$

$NumberOfFreePages \mathrel{\widehat=} \ldots$

$AllocateFreePage \mathrel{\widehat=} \ldots$

$MakePageFree \mathrel{\widehat=} \ldots$

$PhysicalPageNo \mathrel{\widehat=} \ldots$

$InitNewProcessPageTable \mathrel{\widehat=} \ldots$

$RemoveProcessFromPageTable \mathrel{\widehat=} \ldots$

$AddPageToProcess \mathrel{\widehat=} \ldots$

$HasPageInStore \mathrel{\widehat=} \ldots IncProcessPageCount \mathrel{\widehat=} \ldots$

$DecProcessPageCount \mathrel{\widehat=} \ldots$

$LatestPageCount \mathrel{\widehat=} \ldots$

$UpdateMainstorePage \mathrel{\widehat=} \ldots$

$RemovePageFromPageTable \mathrel{\widehat=} \ldots$

$RemovePageProperties \mathrel{\widehat=} \ldots$

$RemovePageFromProcess \mathrel{\widehat=} \ldots$

$IsPageInMainStore \mathrel{\widehat=} \ldots$

$MarkPageAsIn \mathrel{\widehat=} \ldots$

$MarkPageAsOut \mathrel{\widehat=} \ldots$

$IsSharedPage \mathrel{\widehat=} \ldots$

$MarkPageAsShared \mathrel{\widehat=} \ldots$

$UnsharePage \mathrel{\widehat=} \ldots$

$IsLockedPage \mathrel{\widehat=} \ldots$

$LockPage \mathrel{\widehat=} \ldots$

$UnlockPage \mathrel{\widehat=} \ldots$

$MakePageReadable \mathrel{\widehat=} \ldots$

$MakePageNotReadable \mathrel{\widehat=} \ldots$

$MakePageExecutable \mathrel{\widehat=} \ldots$

$IsPageExecutable \mathrel{\widehat=} \ldots$

$MakePageNotExecutable \mathrel{\widehat=} \ldots$

$MakePageWritable \mathrel{\widehat=} \ldots$

$IsPageWritable \mathrel{\widehat=} \ldots$

$MakePageNotWritable \mathrel{\widehat=} \ldots$

It will be noted that the invariant is partially stated in the class definition. The remainder is specified by the *InvPageTables* schema defined below after the other operations have been defined. This will bring the invariant closer to some of the proofs in which it is required.

The following schema represents the test that there are pages in main store (physical pages) that are free.

HaveFreePages

$freepages \neq \varnothing$

NumberOfFreePages

$np! : \mathbb{N}$

$np! = \#freepages$

The following operation models the allocation of a free page to a process. It removes the page denoted by *ppno*! from the set of free pages, *freepages*.

AllocateFreePage

$\Delta(freepages)$
$ppno! : PHYSICALPAGENO$

$ppno! \in freepages$
$freepages' = freepages \setminus \{ppno!\}$

Proposition 125. *AllocateFreePage implies that* $\#freepages' = freepages + 1$.

PROOF.

$$
\begin{aligned}
&\#freepages' \\
&\quad = \#(freepages \setminus \{ppno?\}) \\
&\quad = \#freepages - \#\{ppno?\} \\
&\quad = \#freepages - 1
\end{aligned}
$$

□

Proposition 126. *If freepages* $= n$, $AllocateFreePage^n$ *implies* $freepages' = 0$.

PROOF. By induction, using the last proposition. □

The next operation returns a page to the set of free pages.

$$
\begin{array}{|l}
\hline
MakePageFree \\
\Delta(freepages) \\
ppno? : PHYSICALPAGENO \\
\hline
freepages' = freepages \cup \{ppno?\} \\
\hline
\end{array}
$$

Proposition 127. *MakePageFree implies that* $\#freepages' = \#freepages - 1$.

PROOF.

$$
\begin{aligned}
&\#freepages' \\
&\quad = \#(freepages \cup \{ppno?\}) \\
&\quad = \#freepages + \#\{ppno?\} \\
&\quad = \#freepages + 1
\end{aligned}
$$

□

Proposition 128. $AllocateFreePage[p/ppno!] \mathbin{;} MakePageFree[p/ppnp?]$ *implies that* $freepages' = freepages$.

PROOF. The sequential composition can be written as:

$$
\begin{aligned}
&\exists\, freepages'' : \mathbb{F}\, PHYSICALPAGENO \mid freepages'' = freepages \setminus \{ppno!\} \bullet \\
&\quad freepages' = freepages \cup \{ppno?\}
\end{aligned}
$$

Renaming and simplifying:

$$freepages' = (freepages \setminus \{p\}) \cup \{p\}$$

Then:

$$
\begin{aligned}
&(freepages \setminus \{p\}) \cup \{p\} \\
&\quad = (freepages \setminus \{p\}) \cup \{p\} \\
&\quad = freepages \cup (\{p\} \setminus \{p\}) \\
&\quad = freepages \cup \varnothing \\
&\quad = freepages
\end{aligned}
$$

□

The *PhysicalPageNo* operation contains a use of the *pagetable* variable. This variable is a higher-order function. Its use might appear a little odd.

Essentially, to obtain the physical page number corresponding to a logical page number, the process has to locate the segment in which the page occurs and then translate the logical page number.

$$
\begin{array}{|l}
\textit{PhysicalPageNo} \\
\hline
p? : APREF \\
sg? : SEGMENT \\
lpno? : LOGICALPAGENO \\
ppgno! : PHYSICALPAGENO \\
\hline
ppgno! = pagetable(p?)(sg?)(lpno?)
\end{array}
$$

When a process is allocated, it is given an entry in the page table. The following schema models this operation. It just adds the process' identifier as a key in the subtables and sets everything to zero (empty or $\varnothing$).

$$
\begin{array}{|l}
\textit{InitNewProcessPageTable} \\
\hline
\Delta(incore, sharedpages, lockedpages) \\
p? : APREF \\
\hline
pagetable' = pagetable \cup \{p? \mapsto \mathsf{code} \mapsto \varnothing\} \cup \{p? \mapsto \mathsf{data} \mapsto \varnothing\} \\
\qquad\qquad \cup\{p? \mapsto \mathsf{stack} \mapsto \varnothing\} \cup \{p? \mapsto \mathsf{heap} \mapsto \varnothing\} \\
incore' = (incore \cup \{p? \mapsto \mathsf{code} \mapsto \varnothing\}) \cup (\{p? \mapsto \mathsf{data} \mapsto \varnothing\} \\
\qquad\qquad (\cup\{p? \mapsto \mathsf{stack} \mapsto \varnothing\}(\cup\{p? \mapsto \mathsf{heap} \mapsto \varnothing\}))) \\
sharedpages' = (sharedpages \cup \{p? \mapsto \mathsf{code} \mapsto \varnothing\}) \cup (\{p? \mapsto \mathsf{data} \mapsto \varnothing\} \\
\qquad\qquad (\cup\{p? \mapsto \mathsf{stack} \mapsto \varnothing\}(\cup\{p? \mapsto \mathsf{heap} \mapsto \varnothing\}))) \\
lockedpages' = (lockedpages \cup \{p? \mapsto \mathsf{code} \mapsto \varnothing\}) \cup (\{p? \mapsto \mathsf{data} \mapsto \varnothing\} \\
\qquad\qquad (\cup\{p? \mapsto \mathsf{stack} \mapsto \varnothing\}(\cup\{p? \mapsto \mathsf{heap} \mapsto \varnothing\})))
\end{array}
$$

Proposition 129. *InitNewProcessPageTable implies that the new process has no pages.*

PROOF. For a process to have pages, it must have at least one page in *at least one* segment. However, for a process, p, and all segments, sg, $pagetable'(p)(sg) = \varnothing$. □

Corollary 11. *InitNewProcessPageTable implies that the new process has no in-core pages.*

PROOF. Similar to the above. □

Similar results can be proved for all other page attributes, e.g., locked pages.

Conversely, when a process terminates or is killed, its storage is returned to the free pool and all of the information associated with it in the page tables is removed. The following schema models this operation:

$$
\begin{array}{|l}
\hline
\textit{RemoveProcessFromPageTable} \\
\Delta(pagetable, incore, sharedpages, lockedpages) \\
p? : APREF \\
\hline
pagetable' = \{p?\} \lhd pagetable \\
incore' = \{p?\} \lhd incore \\
sharedpages' = \{p?\} \lhd sharedpages \\
lockedpages' = \{p?\} \lhd lockedpages \\
\hline
\end{array}
$$

Proposition 130. *The predicate of RemoveProcessFromPageTable implies that $p \notin \mathrm{dom}\, pagetable$.*

PROOF. The first line of the predicate is $pagetable' = \{p?\} \lhd pagetable$. Taking domains:

$$
\begin{aligned}
\mathrm{dom}\, pagetable' &= \mathrm{dom}(\{p?\} \lhd pagetable) \\
&= (\mathrm{dom}\, pagetable) \setminus \{p?\}
\end{aligned}
$$

□

Proposition 131. *For any $p : APREF$, $RemoveProcessFromPageTable[p/p?]$ implies that the process:*

1. *is no longer incore, swappable, etc.; and*
2. *is no longer in the page table for its owning process.*

PROOF. Each conjunct of the predicate employs the domain subtraction operation ($\lhd$) to remove p from the domain of each function. This implies that p is removed from each table. □

Propositions about page attributes can be proved. They follow the pattern of the last proposition.

When the storage image of a process is augmented by the addition of fresh pages, the following operation is the basic one used to extend the process' page table entry. Each page is specified as a process reference, a segment and a logical page number; in addition, the physical page number of the page to be added is also included. Since the process and segment are already present in the table, the logical to physical page number mapping is added to the table at the specified point.

$$
\begin{array}{|l}
\hline
\textit{AddPageToProcess} \\
\Delta(pagetable) \\
p? : APREF \\
sg? : SEGMENT \\
lpno? : LOGICALPAGENO \\
ppno? : PHYSICALPAGENO \\
\hline
pagetable' = pagetable(p?)(sg?) \cup \{lpno? \mapsto ppno?\} \\
\hline
\end{array}
$$

There now follows a predicate that returns true when the process specified by $p?$ has at least one page in main store:

HasPageInStore

$p? : APREF$

$p? \in \operatorname{dom} incore$

The per-segment page count is incremented by the following schema:

IncProcessPageCount

$\Delta(pagecount)$
$p? : APREF$
$sg? : SEGMENT$

$pagecount' = pagecount(p?) \oplus \{sg? \mapsto pagecount(p?)(sg?) + 1\}$

The counter is decremented by the following schema:

DecProcessPageCount

$\Delta(pagecount)$
$p? : APREF$

$pagecount' = pagecount(p?) \oplus \{p? \mapsto pagecount(p?)(sg?) - 1\}$

When a page is added to a segment, the page count is incremented. When a page is removed from a segment, the page count is decremented. The current value of the page count is obtained by the following schema:

LatestPageCount

$p? : APREF$
$sg? : SEGMENT$
$lpno! : LOGICALPAGENO$

$lpno! = pagecount(p?)(sg?)$

If the logical to physical page mapping is changed, the following schema performs the update in the page table.

UpdateMainStorePage

$\Delta(pagetable)$
$p? : APREF$
$sg? : SEGMENT$
$lpno? : LOGICALPAGENO$
$ppno? : PHYSICALPAGENO$

$pagetable' = pagetable(p?)(sg?) \oplus \{lpno? \mapsto ppno?\}$

When a page is removed from the page table, the entry representing it must be removed. The removal operation is defined as follows:

$$
\begin{array}{l}
\textit{RemovePageFromPageTable} \\
\hline
\Delta(pagetable) \\
p? : APREF \\
sg? : SEGMENT \\
lpno? : LOGICALPAGENO \\
\hline
(\exists\, pmap : PAGEMAP \mid pmap = pagetable(p?)(sg?) \bullet \\
\quad pagetable' = pagetable(p?) \oplus \{sg? \mapsto (\{lpno?\} \lhd pmap)\})
\end{array}
$$

The removal of a page also requires the removal of the attributes of that page. The attributes are removed using the *unmark_page* function (when a page is allocated, the attributes it possesses are marked using the *mark_page* function).

$$
\begin{array}{l}
\textit{RemovePageProperties} \\
\hline
\Delta(executablepages, readablepages, writablepages, sharedpages, \\
\quad lockedpages, incore) \\
p? : APREF \\
sg? : SEGMENT \\
lpno? : LOGICALPAGENO \\
\hline
executablepages' = unmark\_page(p?, sg?, lpno?, executablepages) \\
readablepages' = unmark\_page(p?, sg?, lpno?, readablepages) \\
writablepages' = unmark\_page(p?, sg?, lpno?, writablepages) \\
sharedpages' = unmark\_page(p?, sg?, lpno?, sharedpages) \\
lockedpages' = unmark\_page(p?, sg?, lpno?, lockedpages) \\
incore' = unmark\_page(p?, sg?, lpno?, incore)
\end{array}
$$

Finally, the high-level operation to remove a page from a process is defined as follows:

$$
\begin{array}{l}
RemovePageFromProcess \mathrel{\widehat{=}} \\
\quad RemovePageFromProcessTable \mathbin{;} \\
\quad MakePageFree \mathbin{;} \\
\quad RemovePageProperties
\end{array}
$$

It is possible to determine whether a page is in main store by determining whether it is in the *incore* attribute. The following schema defines this predicate. Note that it uses the *pages_in_segment* function.

$$
\begin{array}{l}
\textit{IsPageInMainStore} \\
\hline
p? : APREF \\
sg? : SEGMENT \\
lpno? : LOGICALPAGENO \\
\hline
lpno? \in pages\_in\_segment(p?, sg?, incore)
\end{array}
$$

Proposition 132. *IsPageInMainStore iff lpno? is an in-core page.*

Proof. The predicate states that $lpno? \in pages\_in\_segment(p?, sg?, incore)$. By the definition of *pages_in_segment*,

$$lpno? \in incore(p?)(sg?)$$

□

Proposition 133. *IsSharedPage iff lpno? is a shared page; that is, iff lpno? is an element of sharedpages(p)(sg), for some p and sg.*

Proof. Similar to the previous proof. □

Proposition 134. *IsLockedPage iff lpno? is a locked page; that is, iff lpno? is an element of lockedpages.*

Proof. Similar to the previous proof. □

When a page is swapped into main store, it is marked as being "in". The following schema performs this marking. The schema that immediately follows marks pages as "out" (i.e., as not being main-store resident).

MarkPageAsIn

$\Delta(incore)$
$p? : APREF$
$sg? : SEGMENT$
$lpno? : LOGICALPAGENO$

$incore' = mark\_page(p?, sg?, lpno?, incore)$

Proposition 135. *MarkPageAsIn implies that lpno? is an element of incore'.*

Proof. As can be seen from the schema, the predicate is $incore' = mark\_page(p?, sg?, lpno?, incore)$. Substituting the definition of *mark_page*:

$$\begin{aligned} incore' &= mark\_page(p?, sg?, lpno?, incore) \\ &= incore(p?) \oplus \{sg \mapsto (incore(p?)(sg) \cup \{lpno?\})\} \end{aligned}$$

□

Proposition 136. *For fixed arguments, $p : APREF$, $s : SEGMENT$ and $l : LOGICALPAGENO$, $MarkPageAsIn[p/p?, s/sg?, l/lpno?]^n$ has the same effect as $MarkPageAsIn[p/p?, s/sg?, l/lpno?]$.*

PROOF. This proposition is to be taken as:

$$MarkPageAsIn^n \Leftrightarrow MarkPageAsIn$$

The proposition is proved by substitution from the following general property of sets:

$$(S \cup \{x\}) \cup \{x\} = S \cup \{x\}$$

(i.e., the absorbtive law of set union). □

$$
\begin{array}{l}
\textit{MarkPageAsOut} \\
\hline
\Delta(incore) \\
p? : APREF \\
sg? : SEGMENT \\
lpno? : LOGICALPAGENO \\
\hline
incore' = unmark\_page(p?, sg?, lpno?, incore)
\end{array}
$$

Proposition 137. *MarkPageAsOut is satisfied iff lpno? is not an element of incore.*

PROOF. Substituting the definition of *unmark_page* into the predicate of the schema *MarkPageAsOut*:

$$
\begin{aligned}
incore' &= unmark\_page(p?, sg?, lpno?, incore) \\
&= incore(p?) \oplus \{sg? \mapsto (incore(p?)(sg?) \setminus \{lpno?\})\}
\end{aligned}
$$

□

Proposition 138. *For fixed arguments,* $p : APREF$, $s : SEGMENT$ *and* $l : LOGICALPAGENO$, $MarkPageAsOut[p/p?, s/sg?, l/lpno?]^n$ *has the same effect as* $MarkPageAsOut[p/p?, s/sg?, l/lpno?]$.

PROOF. The statement of the proposition is to be taken as:

$$MarkPageAsOut^n \Leftrightarrow MarkPageAsOut$$

The proposition is proved by substitution from the following general property of sets:

$$(S \setminus \{x\}) \setminus \{x\} = S \setminus \{x\}$$

□

Proposition 139. $MarkPageAsIn[l/lpno?] \fatsemi MarkPageAsOut[l/lpno?]$ *implies that* $incore' = incore$.

PROOF. Writing out the predicates and performing the obvious substitutions:

$$unmark\_page(p?, sg?, lpno?, mark\_page(p?, sg?, lpno?, incore))$$

The result follows from the fact that *unmark_page* and *mark_page* are mutually inverse. □

Proposition 140. *$MarkPageAsOut[l/lpno?] \mathbin{;} MarkPageAsIn[l/lpno?]$ implies that $incore = incore'$.*

PROOF. Writing out the predicates and performing the obvious substitutions:

$$mark\_page(p?, sg?, lpno?, unmark\_page(p?, sg?, lpno?, incore))$$

The result follows from the fact that *unmark_page* and *mark_page* are mutually inverse. □

The next few schemata set and unset attributes in pages. The attributes are represented by the various tables in the *PageTables* class, such as *sharedpages*, *readable* and *locked*. The schemata naturally fall into three sets: one to perform a test and one to set the attribute and one to unset it. The schemata in each of these sets have the same structure. That structure is the obvious one and is quite simple. For these reasons, the schemata will not be described in English: the formal notation can stand on its own.

IsSharedPage

$p? : APREF$
$sg? : SEGMENT$
$lpno? : LOGICALPAGENO$

$lpno? \in pages\_in\_segment(p?, sg?, sharedpages)$

MarkPageAsShared

$\Delta(sharedpages)$
$p? : APREF$
$sg? : SEGMENT$
$lpno? : LOGICALPAGENO$

$sharedpages' = mark\_page(p?, sg?, lpno?, sharedpages)$

UnsharePage

$\Delta(sharedpages)$
$p? : APREF$
$sg? : SEGMENT$
$lpno? : LOGICALPAGENO$

$sharedpages' = unmark\_page(p?, sg?, lpno?, sharedpages)$

IsLockedPage

$p? : APREF$
$sg? : SEGMENT$
$lpno? : LOGICALPAGENO$

$lpno? \in pages\_in\_segment(p?, sg?, lockedpages)$

LockPage

$\Delta(lockedpages)$
$p? : APREF$
$sg? : SEGMENT$
$lpno? : LOGICALPAGENO$

$lockedpages' = mark\_page(p?, sg?, lpno?, lockedpages)$

UnlockPage

$\Delta(lockedpages)$
$p? : APREF$
$sg? : SEGMENT$
$lpno? : LOGICALPAGENO$

$lockedpages' = unmark\_page(p?, sg?, lpno?, lockedpages)$

MakePageReadable

$\Delta(readablepages)$
$p? : APREF$
$sg? : SEGMENT$
$lpno? : LOGICALPAGENO$

$readablepages' = mark\_page(p?, sg?, lpno?, readablepages)$

IsPageReadable

$p? : APREF$
$sg? : SEGMENT$
$lpno? : LOGICALPAGENO$

$lpno? \in readablepages(p?)(sg?)$

MakePageNotReadable

$\Delta(readablepages)$
$p? : APREF$
$sg? : SEGMENT$
$lpno? : LOGICALPAGENO$

$readablepages' = unmark\_page(p?, sg?, lpno?, readablepages)$

MakePageExecutable

$\Delta(executablepages)$
$p? : APREF$
$sg? : SEGMENT$
$lpno? : LOGICALPAGENO$

$executablepages' = mark\_page(p?, sg?, lpno, executablepages)$

IsPageExecutable

$p? : APREF$
$sg? : SEGMENT$
$lpno? : LOGICALPAGENO$

$lpno? \in executablepages(p?)(sg?)$

MakePageNotExecutable

$\Delta(executablepages)$
$p? : APREF$
$sg? : SEGMENT$
$lpno? : LOGICALPAGENO$

$executablepages' = unmark\_page(p?, sg?, lpno, executablepages)$

MakePageWritable

$\Delta(writablepages)$
$p? : APREF$
$sg? : SEGMENT$
$lpno? : LOGICALPAGENO$

$writablepages' = mark\_page(p?, sg?, lpno, writablepages)$

IsPageWritable

$p? : APREF$
$sg? : SEGMENT$
$lpno? : LOGICALPAGENO$

$lpno? \in writablepages(p?)(sg?)$

MakePageNotWritable

$\Delta(writablepages)$
$p? : APREF$
$sg? : SEGMENT$
$lpno? : LOGICALPAGENO$

$writablepages' = unmark\_page(p?, sg?, lpno, writablepages)$

Finally, we come to *InvPageTables*, the page table invariant. As can be seen, this is quite an invariant. Because of its length, it was not possible to make it fit within the Object-Z class box without obfuscating the specification even more. The invariant is mostly concerned with ensuring that each page is represented correctly. For example, no free page has a corresponding page-table entry; all locked pages are always in main store, and so on. Readers can inspect the various clauses of the invariant for themselves.

This invariant clearly demonstrates the need for higher-order logics!

$$
\begin{array}{l}
InvPageTables \mathrel{\widehat=} \\
(\forall ppno : PHYSICALPAGENO \bullet \\
\quad ppno \in freepages \Leftrightarrow \\
\quad \neg\,(\exists p : APREF;\ sg : SEGMENT;\ lpno : LOGICALPAGENO \bullet \\
\qquad\qquad ppno = pagetables(p)(sg)(lpno)))
\end{array}
$$

$$
\begin{array}{l}
(\forall p : APREF;\ sg : SEGMENT \bullet \\
\quad p \in \mathrm{dom}\, lockedpages \land sg \in \mathrm{dom}\, lockedpages(p) \Rightarrow \\
\qquad lockedpages(p)(sg) \subseteq incore(p)(sg))
\end{array}
$$

$$
\begin{array}{l}
(\forall p : APREF;\ sg : SEGMENT \bullet \\
\quad p \in \mathrm{dom}\, incore \land sg \in \mathrm{dom}\, incore(p) \Rightarrow \\
\qquad incore(p)(sg) \subseteq \mathrm{dom}\, pagetable(p)(sg))
\end{array}
$$

$$
\begin{array}{l}
(\forall p : APREF;\ sg : SEGMENT \bullet \\
\quad p \in \mathrm{dom}\, sharedpages \land sg \in \mathrm{dom}\, sharedpages(p) \Rightarrow \\
\qquad sharedpages(p)(sg) \subseteq \mathrm{dom}\, pagetable(p)(sg))
\end{array}
$$

$$
\begin{array}{l}
(\forall p : APREF;\ sg : SEGMENT \bullet \\
\quad p \in \mathrm{dom}\, lockedpages \land sg \in \mathrm{dom}\, lockedpages(p) \Rightarrow \\
\qquad lockedpages(p)(sg) \subseteq \mathrm{dom}\, pagetable(p)(sg))
\end{array}
$$

These clauses are intended to represent the disjointness of segments.

$$
\begin{array}{l}
(\forall p : APREF \mid p \in \mathrm{dom}\, pagetable \bullet \\
\quad (\forall sg_1, sg_2 : SEGMENT \bullet \\
\qquad (sg_1 \neq sg_2 \land \\
\qquad\qquad sg_1 \in \mathrm{dom}\, pagetable(p) \land sg_2 \in \mathrm{dom}\, pagetable(p)) \Rightarrow \\
\qquad\qquad\quad \mathrm{dom}\, pagetable(p)(sg_1) \cap \mathrm{dom}\, pagetable(p)(sg_2) = \varnothing))
\end{array}
$$

$$
\begin{array}{l}
(\forall p : APREF \mid p \in \mathrm{dom}\, pagetable \bullet \\
\quad (\forall sg_1, sg_2 : SEGMENT \bullet \\
\qquad (sg_1 \neq sg_2 \land \\
\qquad\qquad sg_1 \in \mathrm{dom}\, pagetable(p) \land sg_2 \in \mathrm{dom}\, pagetable(p)) \Rightarrow \\
\qquad\qquad\quad \mathrm{ran}\, pagetable(p)(sg_1) \cap \mathrm{ran}\, pagetable(p)(sg_2) = \varnothing))
\end{array}
$$

$$
\begin{array}{l}
(\forall p : APREF;\ sg : SEGMENT \bullet \\
\quad p \in \mathrm{dom}\, executablepages \land sg \in executablepages(p) \Rightarrow \\
\qquad executablepages(p)(sg) \subseteq \mathrm{dom}\, pagetable(p)(sg))
\end{array}
$$

$$
\begin{array}{l}
(\forall p : APREF;\ sg : SEGMENT \bullet \\
\quad p \in \mathrm{dom}\, readablepages \land sg \in readtablepages(p) \Rightarrow \\
\qquad readablepages(p)(sg) \subseteq \mathrm{dom}\, pagetable(p)(sg))
\end{array}
$$

$$(\forall p : APREF;\ sg : SEGMENT \bullet \\ \quad p \in \mathrm{dom}\, writablepages \land sg \in writablepages(p) \Rightarrow \\ \qquad writablepages(p)(sg) \subseteq \mathrm{dom}\, pagetable(p)(sg))$$

$$(\forall pg : PAGESPEC \mid pg \in \mathrm{dom}\, pmap \bullet \\ \quad (\exists p : APREF;\ sg : SEGMENT;\ lpno : LOGICALPAGENO \bullet \\ \qquad (p = pgspecpref(pg) \land \\ \qquad\quad sg = pgspecseg(pg) \land \\ \qquad\quad lpno = pgspeclpno(pg)) \Rightarrow \\ \qquad (p \in \mathrm{dom}\, pagetable \land sg \in \mathrm{dom}\, pagetable(p) \land \\ \qquad\quad lpno \in \mathrm{dom}\, pagetable(p)(sg))))$$

$$(\forall pg : PAGESPEC \mid pg \in \mathrm{ran}\, pmap \bullet \\ \quad (\exists p : APREF;\ sg : SEGMENT;\ lpno : LOGICALPAGENO \bullet \\ \qquad (p = pgspecpref(pg) \land \\ \qquad\quad sg = pgspecseg(pg) \land \\ \qquad\quad lpno = pgspeclpno(pg)) \Rightarrow \\ \qquad (p \in \mathrm{dom}\, pagetable \land sg \in \mathrm{dom}\, pagetable(p) \land \\ \qquad\quad lpno \in \mathrm{dom}\, pagetable(p)(sg))))$$

Proposition 141.

$$InitNewProcessPageTable \Rightarrow \\ \quad (\forall s : SEGMENT \bullet \\ \qquad incore = \varnothing \\ \qquad \land\ sharedpages = \varnothing \\ \qquad \land\ lockedpages = \varnothing)$$

PROOF. By the predicate of the *InitNewProcessPageTable* predicate. □

Proposition 142. $\forall p : PAGE \bullet locked(p) \Leftrightarrow \neg\ swappable(p)$

PROOF. The class invariant states that $\mathrm{dom}\, lockedpages \subseteq \mathrm{dom}\, pagetables$. This permits us to infer that:

$$\forall p : LOGICALPAGENO \bullet \\ \quad p \in lockedpages(p)(sg) \\ \qquad \Rightarrow p \in (\mathrm{dom}\, pagetable(p)(sg))$$

for all $p \in IPREF$ and $sg \in SEGMENT$.

Again, by the same invariant:

$$lockedpages(p)(sg) \subseteq incore(p)(sg)$$

(again, for all p and sg as above).

These two formulæ ensure that every locked page exists in store.

$\Rightarrow$: By the predicate of *PageFaultDriver.findVictimLogicalPage* (q.v.):

$pagetable(p)(sg)(l) = v! \wedge l \notin lockedpage$

First, applying the substitutions $[p/p!, sg/sg!, l/l!]$, then simplifying and rearranging the predicate of *findVictimLogicalPage*, we obtain:

$$\begin{array}{l} p \in \mathrm{dom}\, pagetable \\ \quad \wedge\, sg \in \mathrm{dom}\, pagetable(p) \\ \quad \wedge\, l \in \mathrm{dom}\, pagetable(p)(sg) \\ \quad \wedge\, l \notin lockedpages(p)(sg) \\ \quad \wedge\, pagetable(p)(sg)(l) = v \end{array}$$

where v is the page to be swapped.

From this, it can be concluded that v cannot be a locked page.

$\Leftarrow$. If a page is not swappable, it is locked. Consider the set of swappable pages, S, by the definition of *findVictimLogicalPage* in the substitution instance above, any logical page, l, cannot be in S because $l \notin lockedpages(p)(sg)$. □

Proposition 143. *For all processes,* $lockedpages \cap swappablepages = \varnothing$.

PROOF. This follows immediately from the previous proposition. □

Proposition 144. *If a page, p, is in main store, IsPageInMainStore[p/...] is satisfied.*

PROOF. By the predicate of *IsPageInMainStore*, it is clear that

$$lpno? \in pages\_in\_segment(p, s, incore)$$

and that: $pages\_in\_segment(x, y, f) = f(x)(y)$; its range is a set.

For a page actually to be in store, the following must be satisfied:

$$\begin{array}{l} IsPageInMainstore \text{ implies that:} \\ \quad (\exists\, p : IPREF;\ sg : SEGMENT;\ pp : PHYSICALPAGENO \bullet \\ \qquad pagetable(p)(sg)(lpno) = pp \\ \qquad \wedge\, pp \notin freepages) \end{array}$$

By the invariant:

$$\begin{array}{l} \forall\, pp : PHYSICALPAGENO \bullet \\ \quad pp \in freepages \Rightarrow \\ \qquad \neg\,(\exists\, p : IPREF;\ sg : SEGMENT;\ l : LOGICALPAGENO \bullet \\ \qquad\quad pp = pagetables(p)(sg)(l) \end{array}$$

By the application of contraposition $(p \Rightarrow q \Leftrightarrow q \Rightarrow p)$, the result is obtained.

It is now necessary to show that

$$\begin{array}{l} l \in incore(p)(sg) \Rightarrow \\ \quad (\exists\, pp_1 : PHYSICALPAGENO \bullet pagetables(p)(sg)(l) = pp) \end{array}$$

This can be done using the invariant. □

Proposition 145. *If a page, p, is locked, IsPageInMainStore[p/...] is satisfied.*

PROOF. Similar to that of the last proof. □

Proposition 146. *A locked page can never be swapped out.*

PROOF. By Propositions 142 and 144. □

Proposition 147. *A free page is in the process table of no process.*

PROOF. Assume that process, p, has segment, s, in which the logical page number, l is mapped to physical page, n. In this case, $n = pagetable(p)(s)(l)$.
By the class invariant,

$$\begin{array}{l} \forall\, ppno : PHYSICALPAGENO \bullet \\ \quad ppno \in freepages \Leftrightarrow \\ \qquad \neg\,(\exists\, pid : APREF;\ sg : SEGMENT;\ lpn : LOGICALPAGENO \bullet \\ \qquad\quad ppno = pagetable(pid)(sg)(lpn)) \end{array}$$

By Universal Instantiation:

$$\begin{array}{l} n \in freepages \Leftrightarrow \\ \quad \neg\,(\exists\, pid : APREF;\ sg : SEGMENT;\ lpn : LOGICALPAGENO \bullet \\ \qquad n = pagetable(pid)(sg)(lpn)) \end{array}$$

From the assumption that $n \in freepages$, a contradiction ensues. Therefore,

$$n \notin freepages$$

as required. □

Proposition 148. *A free page is not incore.*

PROOF. The invariant states that:

$$\begin{array}{l} \forall\, ppno : PHYSICALPAGENO \bullet \\ \quad ppno \in freepages \Leftrightarrow \\ \qquad \neg\,(\exists\, pid : APREF;\ sg : SEGMENT;\ lpn : LOGICALPAGENO \bullet \\ \qquad\quad ppno = pagetable(pid)(sg)(lpn)) \end{array}$$

This implies that:

$$freepages \cap pagetable(p)(sg) = \varnothing \qquad (6.1)$$

The class invariant also states that:

$$p \in \mathrm{dom}\, incore \wedge sg \in \mathrm{dom}\, incore(p) \Rightarrow \\ \quad incore(p)(sg) \subseteq \mathrm{dom}\, pagetable(p)(sg)$$

This and equation (6.1) above imply that $pg \in freepages \Leftrightarrow \neg\, incore(p)(sg)(l)$ for all values of p, sg and l. □

Proposition 149. *A free page is not:*

1. *executable;*
2. *readable; or*
3. *writable.*

PROOF. Since a free page is not in the page table, it follows from the invariant that it cannot have any of these attributes. □

6.3.1 The Paging Disk Process

The paging disk holds pages while they are not in main store. The virtual store software copies pages to and from the paging disk to implement the swapping process that underlies the huge address space illusion. Pages are swapped out of store when they are not required and swapped back in when their owning process (or one of them, if the page is shared) refers to them.

The paging disk is part of the subsystem whose design will be discussed in the next few subsections. The subsystem's organisation and interactions are shown in Figure 6.2. It consists of an ISR, a handler or driver process and the paging disk proper.

The paging disk is assumed to be infinite in size. It is represented by a mapping from logical to physical pages (*pagemap*). In addition, the disk has an interface to the message-passing system (*msgmgr*) so that it can communicate with the other processes in the storage-management subsystem. For most of the time, *pagemap* will be the focus of attention.

The class defining the paging disk process is now defined.

PagingDiskProcess

$\upharpoonright(INIT, PageIsOnDisk, StorePageOnDisk, RetrievePageFromDisk,$
$\quad RemoveProcessFromPagingDisk, OnPageRequest)$

$pagemap : APREF \twoheadrightarrow SEGMENT \twoheadrightarrow LOGICALPAGENO \twoheadrightarrow PAGE$
$msgmgr : MsgMgr$

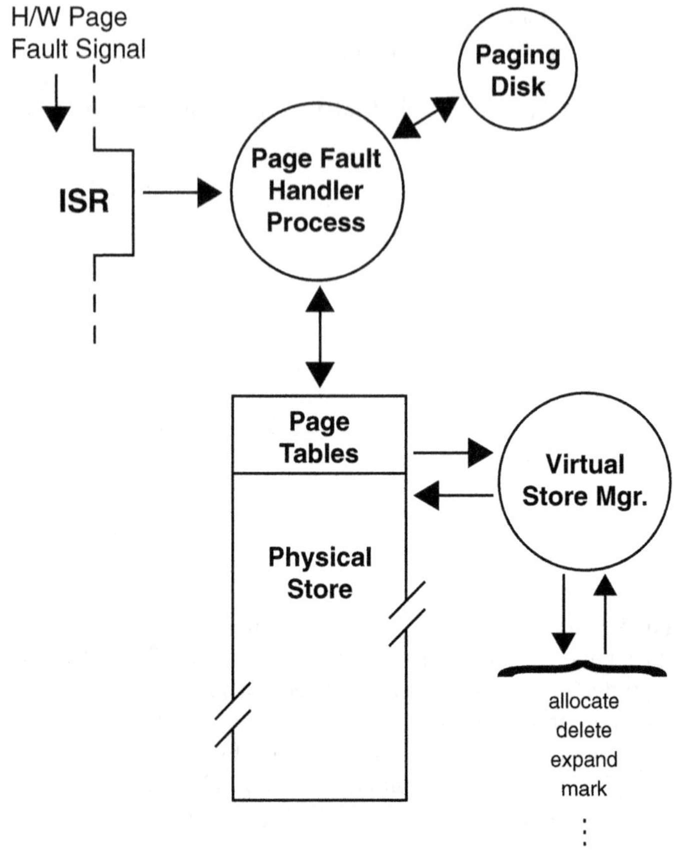

Fig. 6.2. *Interactions between virtual storage components.*

INIT

$msgs? : MsgMgr$

$msgmgr' = msgs?$

$\mathrm{dom}\ pagemap' = \varnothing$

$PageIsOnDisk \mathrel{\widehat=} \ldots$

$StorePageOnDisk \mathrel{\widehat=} \ldots$

$RetrievePageFromDisk \mathrel{\widehat=} \ldots$

$RemoveProcessFromPagingDisk \mathrel{\widehat=} \ldots$

$OnPageRequest \mathrel{\widehat=} \ldots$

The following schema is a predicate that is true when the specified page of the specified process is present on the paging disk.

$$
\begin{array}{|l}
\textit{PageIsOnDisk} \\
\hline
p? : APREF \\
sg? : SEGMENT \\
lpno? : LOGICALPAGENO \\
\hline
p? \in \operatorname{dom} pagemap \\
sg? \in \operatorname{dom} pagemap(p?) \\
lpno? \in \operatorname{dom} pagemap(p?)(sg?) \\
\hline
\end{array}
$$

Pages are stored on the paging disk when they are swapped out of main store. In order for a page to be stored on the paging disk, it must be placed there and indexed properly. The *StorePageOnDisk* operation does this:

$$
\begin{array}{|l}
\textit{StorePageOnDisk} \\
\hline
\Delta(pagemap) \\
p? : APREF \\
sg? : SEGMENT \\
lpno? : LOGICALPAGENO \\
pg? : PAGE \\
\hline
pagemap' = pagemap \oplus \{p? \mapsto \{sg? \mapsto \{lpno? \mapsto pg?\}\}\} \\
\hline
\end{array}
$$

Proposition 150. *If a logical page image is already on disk and that page is swapped out again, that page image is overwritten.*

PROOF. Let $p = pagemap(p?)(sg?)(lpno?)$ and let $pg?$ be $pagemap'(p?)(sg?)(lpno?)$, so $\{p? \mapsto \{sg? \mapsto \{lpno? \mapsto p\}\}\} \in pagemap$ and $\{p? \mapsto \{sg? \mapsto \{lpno? \mapsto pg?\}\}\} \in pagemap'$. Therefore, $pagemap' = pagemap \oplus \{p? \mapsto \{sg? \mapsto \{lpno? \mapsto pg?\}\}\}$. □

When its owning process references a page that is not in main store, the paging disk is instructed to retrieve it from the paging disk. The following schema models the basics of this operation:

$$
\begin{array}{|l}
\textit{RetrievePageFromDisk} \\
\hline
p? : APREF \\
sg? : SEGMENT \\
lpno? : LOGICALPAGENO \\
pg! : PAGE \\
\hline
pg! = pagemap(p?)(sg?)(lpno?) \\
\hline
\end{array}
$$

When a process terminates (or is terminated), all of its pages must be removed from the paging disk. In our model, this can be done very easily by removing the process' identifier from the domain of the *pagemap*. This removes all references to the process from the map—hence, removes the process from the disk, thus:

$$
\begin{array}{l}
\textit{RemoveProcessFromPagingDisk} \\
\hline
\Delta(pagemap) \\
p? : APREF \\
\hline
pagemap' = \{p?\} \ntriangleleft pagemap
\end{array}
$$

Proposition 151. *The predicate of RemoveProcessFromPagingDisk implies that* $p \notin \operatorname{dom} pagemap'$.

PROOF.

$$
\begin{aligned}
&\operatorname{dom} pagemap' \\
&\quad = \operatorname{dom}(\{p?\} \ntriangleleft pagemap) \\
&\quad = (\operatorname{dom} pagemap) \setminus \{p?\}
\end{aligned}
$$

Therefore, $p \notin \operatorname{dom} pagemap'$. □

The paging disk has an operation that handles requests for operations. The operations it can perform are:

- store a page (*STOPG*);
- retrieve a page (*GTPG*), and
- delete a process from the paging disk (*DELPROCPG*).

The following schema defines this operation. The schema uses overloaded calls to the *RcvMessage* primitive provided by the message-handling subsystem. The operation is really just a large "or" based on the type of message that has just been received. The infinite loop is modelled by the exterior universal quantifier. While the operation has nothing to do, it waits for the next message to arrive.

$$
\begin{array}{l}
\textit{OnPageRequest} \\
\hline
\forall i : 1 \ldots \infty \bullet \\
\quad (\exists pdsk, fhandler : APREF;\ msg : MSG \mid \\
\qquad\qquad pdsk = \mathsf{pagedisk} \land fhandler = \mathsf{faultdrvr} \bullet \\
\qquad msgmgr.RcvMessage[pdsk/caller?, fhandler/src?, msg/m!] \\
\qquad \land (\exists p : APREF;\ sg : SEGMENT; \\
\qquad\qquad\qquad lpno : LOGICALPAGENO;\ pg : PAGE \bullet \\
\qquad\quad (msg = STOPG\langle\langle p, sg, lpno, pg\rangle\rangle \\
\qquad\qquad \land storePageOnDisk[p/p?, sg/sg?, lpno/lpno?, pg/pg?])
\end{array}
$$

$$\begin{array}{l}\quad \lor\ (msg = GTPG\langle\!\langle p, sg, lpno\rangle\!\rangle \\ \qquad \land\ retrievePageFromDisk[p/p?, sg/sg?, lpno/lpno?, pg/pg!] \\ \qquad \land\ (\exists\, rpmsg : MSG \mid rpmsg = ISPG\langle\!\langle p, sg, lpno, pg\rangle\!\rangle \bullet \\ \qquad\qquad msgmgr.SendMessage \\ \qquad\qquad\qquad [fhandler/dest?, pdsk/src?, rpmsg/m?])) \\ \quad \lor\ (msg = DELPROCPG\langle\!\langle p\rangle\!\rangle \\ \qquad \land\ removeProcessFromPagingDisk[p/p?])))\end{array}$$

Proposition 152. *OnPageRequest does what it should.*

PROOF. The proof is by cases on message type. It is assumed that all domains are correct so the error cases ignored in the schema need not be introduced here.
Case 1. $m = STOPG$, a request to store a page on disk. The predicate of $storePageOnDisk[p/p?, sg/sg?, lpno/lpno?, pg/pg?]$ states:

$$pagemap' = pagemap \oplus \{p \mapsto \{sg \mapsto \{lpno \mapsto pg\}\}\}$$

so $pg? = pagemap'(p)(sg)(lpno)$.
Case 2. $m = GTPG$, a request to retrieve a page. This follows from the fact that $retrievePageOnDisk$ is a function: $pg? = pagemap(p?)(sg?)(lpno?)$.
Case 3. $m = DELPROCPG$. By the last proposition. □

6.3.2 Placement: Demand Paging and LRU

There are significant issues to be resolved in the design of the virtual storage mechanism. This section deals with the general area of *placement*. Placement is concerned with where pages are to be removed and included in main store. When a process refers to a page that is not currently in main store, it generates a *page fault*, an asynchronous signal that interrupts the process and leads to the satisfaction of the reference. To do this, the support software has to identify the page that is being referenced and then identify a page in physical store that can be swapped out to the paging disk, thus making space for the referenced page to be copied into main store. The issue is slightly complicated by the fact that the system might have some free physical pages in main store (indeed, it might be a policy to keep a block of such pages in reserve). For present purposes, it is assumed that *all* physical pages are allocated and that there is no pool of free pages kept in reserve.

The placement algorithm just outlined is *demand paging*. This is the most common approach. It is documented, like many other possible approaches, in most textbooks on operating systems (e.g., [26, 11, 29, 5]). Demand paging is the most commonly used approach and makes reasonable assumptions about the hardware and the software. It just assumes that the hardware can detect a page fault and that the software can find the referenced page and locate the page in main store where it can be stored; if there is no free page, demand paging assumes that there is a way to find a victim page that can be swapped out to make space.

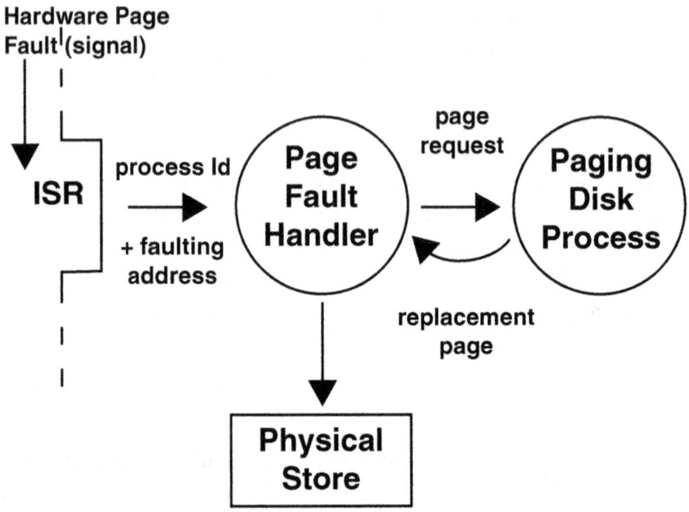

Fig. 6.3. *Process organisation for handling page faults.*

The placement algorithm used in this model has two identifiable aspects:

1. finding pages to remove;
2. pages to include.

The second aspect is solved for us: the pages to include are always those that cause a page fault when referenced by a process.

The general organisation of the processes that handle page faults is shown in Figure 6.3.

6.3.3 On Page Fault

A page fault occurs when a virtual address refers to a page that is not in physical store.

When such a fault occurs, it is assumed that the address causing the fault is available to the operating system. From this address, it is assumed that the following parameters can be computed: segment number, logical page number and the offset into the logical page. In addition, it is assumed that the identity of the process causing the page fault can be determined (it should be the currently running process on a uni-processor; in the terms of this book, the value of *currentp*).

The case in which a page is shared is identical. The important thing to remember is that the owning process causes the page fault. If a page is shared, the important thing is for the page not to be swapped in (or out) more than once. That is why the *smap* is included in *PageTables*. When a shared page is to be swapped, this map is consulted. If the page is shared, it should already be in store. Conversely, if a page is to be swapped in, the *smap* should be consulted for the same reason.

In this section, a page that is to be removed from main store in order to free a page frame will be called a *victim* or *candidate*. It is always a restriction on victim page selection that victim pages are never locked into main store.

To define a relatively simple candidate-finding algorithm, it is necessary to associate page frames in main store with a bit that is set by the hardware whenever a page is referenced and a counter. The counter is a single byte whose value is computed by rotating the reference bit into the top bit of the counter byte and or-ing the counter with the contents of the counter shifted down one bit (ignoring the bottom bit). The types of the reference bit and the counter are:

$BIT == \{0, 1\}$
$N_{256} == 0 \,..\, 255$

(N_{256} is just the naturals $0 \,..\, 2^{16} - 1$—i.e., a 16-bit unsigned.)

The computation of the counter value forms part of the predicate of schema *ComputeHitCounts*. To define a swap-out procedure, it is necessary to extend *PageFrames* a little so that information on page usage is represented.

The class and its operations are relatively straightforward.

PageFrames

$\upharpoonright(INIT, GetPage, OverwritePhysicalPage, ClearRefBitsAndCounter,$
$\quad ComputeHitCounts, IsVictim, VictimPhysicalPageNo)$

$frames : PAGEFRAME$
$refbit : PHYSICALPAGENO \twoheadrightarrow\!\!\!\!\!\!\!\shortmid\;\; BIT$
$count : PHYSICALPAGENO \twoheadrightarrow\!\!\!\!\!\!\!\shortmid\;\; N_{256}$

$\mathrm{dom}\, count = \mathrm{dom}\, refbit$
$\mathrm{dom}\, count \subseteq \mathrm{dom}\, frames$
$\#\, \mathrm{dom}\, refbit = numrealpages$

$$\begin{array}{l}
\underline{\quad INIT \quad} \\
(\forall i : 1 \mathinner{\ldotp\ldotp} numrealpages \bullet \\
\quad frames'(i) = NullPage) \\
\mathrm{dom}\, refbit' = \varnothing \\
\mathrm{dom}\, count' = \varnothing
\end{array}$$

$GetPage \mathrel{\widehat{=}} \ldots$

$OverwritePhysicalPage \mathrel{\widehat{=}} \ldots$

$ClearRefBitsAndCounter \mathrel{\widehat{=}} \ldots$

$ComputeHitCounts \mathrel{\widehat{=}} \ldots$

$IsVictim \mathrel{\widehat{=}} \ldots$

$VictimPhysicalPageNo \mathrel{\widehat{=}} \ldots$

$$\begin{array}{l}
\underline{\quad GetPage \quad} \\
pageno? : PHYSICALPAGENO \\
fr! : PAGE \\
\hline
1 \leq pageno? \leq numrealpages \\
fr! = frames(pageno?)
\end{array}$$

This operation retrieves a page.

The infix function *after* is required by the next schema. Its definition is repeated for convenience:

$$\begin{array}{l}
after : \mathrm{seq}\, X \times \mathbb{N} \rightarrow \mathrm{seq}\, X \\
\hline
\forall m : \mathrm{seq}\, X;\ offset : \mathbb{N} \bullet \\
\quad \mathrm{dom}(m\ after\ offset) = (1 \mathinner{\ldotp\ldotp} \#m - offset) \land \\
\quad (\forall n : \mathbb{N} \bullet \\
\qquad (n + offset) \in \mathrm{dom}\, m \Rightarrow (m\ after\ offset)(n) = m(n + offset))
\end{array}$$

$$\begin{array}{l}
\underline{\quad OverwritePhysicalPage \quad} \\
\Delta(frames) \\
pageno? : PHYSICALPAGENO \\
pg? : PAGE \\
\hline
frames' = frames \oplus \{pageno? \mapsto pg?\}
\end{array}$$

This operation overwrites a page in main store. The input *pageno?* is the index of the page frame in main store and *pg?* is a page full of data.

Proposition 153. *The predicate of the substitution instance of the predicate $OverwritePhysicalPage[p/pageno?, pg/pg?]$ replaces the page indexed by p in frames by the page, pg and only that page.*

PROOF. The $\oplus$ operation can be defined as:

$$(f \oplus g)(x) = \begin{cases} f(x), x \in \operatorname{dom} f \\ g(x), \text{ otherwise.} \end{cases}$$

Then, for the predicate of the schema:

$$(frames \oplus \{pageno? \mapsto pg?\})(x) = \begin{cases} frames(x), x \in \operatorname{dom} frames \\ \{pageno? \mapsto pg?\}, x = pageno? \end{cases}$$

□

The clearing of the reference bits and reference counter in a physical page is defined as follows:

ClearRefBitsAndCounter

$\Delta(refbit, count)$
$ppno? : PHYSICALPAGENO$

$refbit' = refbit \oplus \{ppno? \mapsto 0\}$
$count' = count \oplus \{ppno? \mapsto 0\}$

The following operation computes the hit count for each page. That is, it computes the number of times the page has been referenced since it was copied into main store. It must be performed on a cyclic basis but this model does not specify how the cycle is implemented—hardware is the optimal way to compute such counts because the counter must be updated on each reference. The computation operation is defined by the following schema:

ComputeHitCounts

$\Delta(pcount, count)$

$(\forall i : PHYSICALPAGENO \mid i \in \operatorname{dom} frames \bullet$
$\quad (\exists\, pcount : N_{265} \bullet$
$\quad\quad pcount = (count(i)/2)_{\mathrm{mod}\ 256} + refbit(i) * 2^7 \wedge$
$\quad\quad count' = count \oplus \{i \mapsto pcount\}))$

The lowest *count* value is chosen as the victim:

IsVictim

$(\exists j : PHYSICALPAGENO \mid j \in \operatorname{dom} count \bullet$
$\quad count(j) = min(\operatorname{ran} count))$

The physical page of the victim must be obtained:

VictimPhysicalPageNo

$victim! : PHYSICALPAGENO$

$(\exists : PHYSICALPAGENO \mid i \in \operatorname{dom} count \bullet$
$\quad \wedge\ count(i) = min(\operatorname{ran} count)$
$\quad \wedge\ i = victim!)$

This algorithm is not foolproof but is a reasonable, hardware-independent choice. There are many alternative algorithms in the literature (see, for example, [26] or [29]) but the best will always be determined by the hardware on which the operating system runs. The assumption made here is a minimal one (because many processors implement reference bits in page frames); some machines might provide reference counters directly, while others might record the time of the last reference to each page. It is to be hoped that, in the future, victim determination will be considerably simplified by more co-operative hardware.

The *FindVictim* operation is "safe" in the sense that it will always find an in-core page. The reason for this is that the pageout operation defined above requires that $\neg$ *HaveFreePages* be true before *FindVictim* is called. This ensures that none of the candidate pages is in *freepages*.

$$
\begin{array}{lcl}
PGMSG & ::= & DELPROCPG\langle\langle AREF\rangle\rangle \\
 & | & GETPG\langle\langle AREF \times SEGMENT \times LOGICALPAGENO\rangle\rangle \\
 & | & STOPG\langle\langle AREF \times SEGMENT \times LOGICALPAGENO \times PAGE\rangle\rangle \\
 & | & ISPG\langle\langle AREF \times SEGMENT \times LOGICALPAGENO \times PAGE\rangle\rangle
\end{array}
$$

$$FMSG ::= BADADDR\langle\langle APREF \times SEGMENT \times LOGICALPAGENO\rangle\rangle$$

The ISR handling page faults can be defined as the following class. It is a subclass of the message-based ISR defined in the last chapter.

PageFaultISR

$\upharpoonright(INIT,$
$\quad OnPageInterrupt)$

GenericMsgISR

$sched : LowLevelScheduler$
$ptab : ProcessTable$

INIT

$schd? : LowLevelScheduler$
$pt? : ProcessTable$

$sched' = schd? \wedge ptab' = pt?$

OnPageInterrupt

$intaddr? : VIRTUALADDRESS$

$$
\begin{array}{l}
\exists fmsg : FMSG;\ cp : APREF;\ sg : SEGMENT; \\
\qquad lpno : LOGICALPAGENO;\ offset : \mathbb{N} \bullet \\
\quad saddrseq(saddresstrans(intaddr?)) = sg \\
\quad \land\ spageno(saddresstrans(intaddr?)) = lpno \\
\quad \land\ sched.CurrentProcess[cp/cp!] \land sched.MakeUnready[cp/pid?] \\
\quad \land\ ptab.DescrOfProcess[cp/pid?, pd/pd!] \\
\quad \land\ ctxt.SwapOut \land fmsg = BADADDR\langle\!\langle cp, sg, lpno\rangle\!\rangle \\
\quad \land\ (\exists\ isrid, fdvrid : APREF \mid fdrvrd = \mathsf{faultdrvr} \bullet \\
\qquad SendInterruptMsg[fdrvrid/driver?, fmsg/m?]) \\
\quad ⨟\, ctxt.SwapIn
\end{array}
$$

When a page fault occurs, the ISR sends a message to the *PageFault-Driver*. The page-fault handler or driver is defined by the following class. As is usual with classes that represent processes, it exports only one operation, here *DoOnPageFault*. The driver also contains routines that access page tables but they are not exported. The idea is that once the driver starts, it has exclusive access to these data structures. The data structures, however, still need to be protected by locks.

PageFaultDriver

$\upharpoonright(INIT, DoOnPageFault)$

$$
\begin{array}{l}
sched : LowLevelScheduler;\ ptab : ProcessTable;\ pts : PageTables; \\
vsm : VStoreManager;\ pfs : PageFrames;\ msgman : MsgMgr \\
lck : Lock
\end{array}
$$

INIT

$$
\begin{array}{l}
mmgr? : MsgMgr;\ sch? : LowLevelScheduler \\
ptb? : ProcessTable;\ pgtabs? : PageTables \\
vstoreman? : VStoreManager \\
pgfrms? : PageFrames;\ lk? : Lock
\end{array}
$$

$$
\begin{array}{l}
msgman' = mmgr? \land sched' = sch? \land ptab' = ptb? \land pts' = pgtabs? \\
vsm' = VStoreManager \land pfs' = PageFrames \land lck' = lk?
\end{array}
$$

$$\begin{array}{l}
findVictimLogicalPage \mathrel{\widehat=} \ldots \\
haveVictim \mathrel{\widehat=} \ldots \\
findVictimPage \mathrel{\widehat=} \ldots \\
swapPageToDisk \mathrel{\widehat=} \ldots \\
retrievePageFromDisk \mathrel{\widehat=} \ldots \\
storePageOnDisk \mathrel{\widehat=} \ldots \\
genOnPageFault \mathrel{\widehat=} \ldots \\
onPageFault \mathrel{\widehat=} \ldots \\
DoOnPageFault \mathrel{\widehat=} \ldots
\end{array}$$

The driver has to find a victim page to swap out when there is no free store. The following schema defines this operation. The page to be swapped out is a *logical* one at this point. The schema maps the logical page to a physical page by another operation.

findVictimLogicalPage

$$\begin{array}{l}
p! : APREF \\
sg! : SEGMENT \\
lpno! : LOGICALPAGENO \\
victim! : PHYSICALPAGENO \\
\hline
(\exists\, p : APREF;\ s : SEGMENT;\ l : LOGICALPAGENO \mid \\
\qquad\qquad p \in \operatorname{dom} pagetable \land sg \in \operatorname{dom} pagetable(p) \\
\qquad\qquad\qquad \land\ l \in \operatorname{dom} pagetable(p)(sg) \bullet \\
\qquad pagetable(p)(sg)(l) = victim! \\
\qquad \land\ l \notin lockedpages(p)(sg) \\
\qquad \land\ p! = p \land sg! = s \land lpno! = l)
\end{array}$$

The schema simplifies to:

$$\begin{array}{l}
p! : APREF \\
sg! : SEGMENT \\
lpno! : LOGICALPAGENO \\
victim! : PHYSICALPAGENO \\
\hline
p! \in \operatorname{dom} pagetable \land sg! \in \operatorname{dom} pagetable(p!) \\
\land\ lpno! \in \operatorname{dom} pagetable(p!)(sg!) \\
\land\ pagetable(p!)(sg!)(lpno!) = victim! \\
\land\ lpno! =\notin lockedpages(p!)(sg!)
\end{array}$$

The following definition is a synonym. It tests the reference count of the victim physical page.

$haveVictim \mathrel{\widehat=} pfs.IsVictim$

The final operation to locate a victim page is the following:

$$
\begin{array}{l}
findVictimPage \mathrel{\widehat=} \\
\quad pfs.VictimPhysicalPageNo[victim/victim!] \\
\qquad \wedge\ findVictimLogicalPage[victim/victim!]
\end{array}
$$

It is possible to put a few properties of pages on a more formal basis.

Proposition 154. *Locked pages can never be victims.*

PROOF. The predicate of *findVictimLogicalPage* contains the conjunct $l \notin lockedpages(p)(sg)$, where l is the logical page number, p the process identifier and sg the segment. □

Proposition 155. *Free pages can never be victims.*

PROOF. Since free pages do not belong to any process (by Proposition 147), they do not appear in *pagetable*, so they cannot be victims because the quantifier in *findVictimLogicalPage* ranges over *pagetable*. □

Proposition 156. *Faulting processes are not ready.*

PROOF. The predicate of *PageFaultISR.OnPageInterrupt* contains a reference to the schema *MakeUnready*[*cp*/*pid*?], where *cp* is the current process (i.e., $cp = currentp$), so it is the process that caused the page fault. The action of *MakeUnready* is to remove the process from the ready queue. □

Proposition 157. *A faulting process cannot be executed.*

PROOF. By the previous proposition, *MakeUnready* implies that *cp* cannot be scheduled until it is returned to the ready queue. Furthermore, *cp* cannot continue because *OnPageInterrupt* calls *sched.ScheduleNext* to select the next process to run; this process will not be *cp*. □

Corollary 12. *A faulting process is blocked.*

PROOF. This is a consequence of the previous proposition. □

The operation that actually swaps a page from main store to the paging disk is the following:

$$
\begin{array}{l}
swapPageToDisk \mathrel{\widehat=} \\
\quad (findVictimPage[p/p!, sg/sg!, lpno/lpno!] \\
\qquad \land\ vsm.MarkSharedLogicalPageAsOut[p/p?, sg/sg?, lpno/lpno?] \\
\qquad \land\ pfs.GetPage[pg/fr!, victim!/pageno?] \\
\qquad \land\ storePageOnDisk[p/p?, sg/sg?, lpno/lpno?, pg/pg?] \\
\qquad \land\ pfs.ClearRefBitsAndCounter[victim!/ppno?] \\
\qquad \land\ storePageOnDisk[p/p?, sg/sg?, lpno/lpno?, pg/pg?]) \\
\qquad\qquad\qquad \setminus \{pg, p, sg, lpno\} \\
\qquad \lor\ Skip
\end{array}
$$

First, a victim is located and then marked as being not in store (*MarkShared-LogicalPageAsOut*). The page causing the page fault is then demanded from the paging disk and the victim is sent to the disk for temporary storage. The reference bits of the victim are then cleared so that the referenced page (the one causing the page fault) can be written to it by *storePageOnDisk*.

The next operation to be defined is *retrievePageFromDisk*. This operation, as its name suggests, communicates with the paging disk process to locate the page that is to be brought into main store as a consequence of the last page fault. The reference to this page is, in fact, the one that caused the page fault that resulted in the current execution of the *PageFaultDriver*. It is defined by the following schema:

$$
\begin{array}{|l}
\textit{retrievePageFromDisk} \\
\hline
p? : APREF \\
sg? : SEGMENT \\
lpno? : LOGICALPAGENO \\
pg! : PAGE \\
\hline
(\exists\, msg_o, msg_r : PGMSG;\ src, dest : APREF;\ pg : PAGE \mid \\
\qquad\qquad msg_o = GETPG\langle\!\langle p?, sg?, lpno? \rangle\!\rangle \\
\qquad\qquad\quad \land\ src = \mathsf{faultdrvr} \land dest = \mathsf{pagedsk} \bullet \\
\quad msgman.SendMsg[src/src?, dest/dest?, msg_o/m?] \mathbin{;} \\
\quad (msgman.RcvMsg[src/caller?, dest/src?, msg_r/m!] \\
\qquad \land\ msg_r = ISPG\langle\!\langle p?, sg?, lpno?, pg! \rangle\!\rangle)) \\
\hline
\end{array}
$$

As can be seen, this operation mostly handles messages. First, the operation sends a message requesting that the paging disk retrieve the page specified by the parameters $p?$, $sg?$ and $lpno?$; the page is represented by $pg!$. The page, $pg!$, is returned by the paging disk in the *ISPG* message.

The operation to store a page on disk is similar to the last one.

$$
\begin{array}{|l}
\hline
\textit{storePageOnDisk} \\
p? : APREF \\
sg? : SEGMENT \\
lpno? : LOGICALPAGENO \\
pg? : PAGE \\
\hline
(\exists\, msg : PGMSG;\ src, dest : APREF \mid \\
\qquad\qquad msg = STOPG\langle\!\langle p?, sg?, lpno?, pg? \rangle\!\rangle \\
\qquad\qquad \wedge\ src = \mathsf{faultdrvr} \wedge dest = \mathsf{pagedsk} \bullet \\
\qquad mgman.SendMsg[src/src?, dest/dest?, msg/m?]) \\
\hline
\end{array}
$$

Again, this operation mostly deals with messages. In this case, it contains just one send operation. Messages of type *STOPG* request the paging disk to store the page specified by *p*?, *sg*?, *lpno*?; the page is denoted by *pg*?.

The next operation schema defines the operation performed whenever a page fault occurs. This is a complex operation and its definition reflects this complexity. The operation receives a message containing a specification of the page that was referenced and determined (by the hardware) not to be in main store (it is on the paging disk).

The operation determines whether there are any page frames free in main store. If there are, the page is located on the disk and copied into a free page. To do this, the operation sends a message to the paging-disk process requesting the page.

If there are no page frames free in main store, a page that is currently resident in main store must be placed on the paging disk in order to make space for the one that caused the page fault. A suitable resident page must have a low frequency of reference in the near future and the reference-count mechanism specified above is used to determine which it is. This page is referred to as the *victim* in the following schema and the schemata defining the selection operations. It should be noted that the victim can belong to any process whatsoever but cannot be a page that is locked into main store. The victim is swapped to the paging disk and the one causing the page fault is retrieved and copied into the newly vacated page frame in main store. The process that caused the page fault is then returned to the ready queue and the ISR waits for the next page fault.

It should be noted that the victim can never be a locked page (i.e., a page locked into main store). This condition is imposed so that, in particular, pages locked into main store by the kernel (typically pages containing kernel code and data structures) cannot be swapped out. It is undesirable to swap kernel pages out of store because they might be involved in the operation currently being performed or they might be ISRs and the scheduler.

In some kernel designs, it is possible for kernel processes to be stored in swappable pages. In such a case, the pages would contain data or processes that are considered to be of lower importance, implementing operations that the kernel can afford to have blocked while some of their store is paged out.

The kernel assumed here is the one modelled in Chapter 4 and extended in Chapter 5. It is assumed that *all* of its components (data structures and processes) must be stored in pages that are locked into main store (and hence are stored in pages with the *locked* attribute). (The schemata defined above for victim selection, it should be noted, depend on the commutativity of conjunction.)

$$
\begin{array}{l}
genOnPageFault \mathrel{\widehat=} \\
\quad \exists fmsg : FMSG;\ me, src : APREF \mid me = \mathsf{faultdrvr} \land src = \mathsf{hardware} \bullet \\
\qquad msgman.RcvMessage[me/dest?, src/src?, fmsg/m!] \\
\qquad \land (\exists fpid : APREF;\ fs : SEGMENT; \\
\qquad\qquad\qquad flpno : LOGICALPAGENO; \\
\qquad\qquad\qquad destpg : PHYSICALPAGENO \bullet \\
\qquad\quad fmsg = BADADDR\langle\langle fpid, fs, flpno\rangle\rangle \\
\qquad\quad ((lck.Lock \mathbin{⨟} \\
\qquad\qquad ((pts.HaveFreePages \land pts.AllocateFreePage[destpg/ppno!]) \\
\qquad\qquad\quad \lor (haveVictim \\
\qquad\qquad\qquad \land swapPageToDisk[destpg/victim!])) \land lck.Unlock) \mathbin{⨟} \\
\qquad\quad (lck.Lock \mathbin{⨟} \\
\qquad\qquad (retrievePageFromDisk[fpid/p?, fs/sg?, flpno/lpno?, page/pg!] \\
\qquad\qquad\quad \land pfs.ClearRefBitsAndCounter[destpg/ppno?] \\
\qquad\qquad\quad \land pfs.OverwritePhysicalPage[page/pg?, destpg/pageno?] \\
\qquad\qquad\quad \land vsm.MarkSharedLogicalPageAsIn \\
\qquad\qquad\qquad\qquad [fpid/p?, fs/sg?, flpno/lpno?] \\
\qquad\qquad\quad \land lck.Unlock))) \\
\qquad\quad \mathbin{⨟}(lck.Lock \land sched.MakeReady[fpid/pid?]) \\
\qquad\quad \mathbin{⨟}(sched.ScheduleNext \land lck.Unlock))
\end{array}
$$

$$onPageFault \mathrel{\widehat=} (sched.CurrentProcess[p/cp!] \land genOnPageFault[p/p?]) \setminus \{p\}$$

$$
\begin{array}{l}
DoOnPageFault \mathrel{\widehat=} \\
\quad (\forall i : 1 \mathinner{\ldotp\ldotp} \infty \bullet onPageFault)
\end{array}
$$

Proposition 158. *If there are free pages, no page is swapped out.*

PROOF. The first component of the sequential composition in the predicate of *genOnPageFault* contains the disjunction:

$$
\begin{array}{l}
(pts.HaveFreePages \land pts.AllocateFreePage[destpg/ppno!]) \\
\quad \lor (haveVictim \land swapPageToDisk[destpg/victim!])
\end{array}
$$

If there are free pages, *pts.HaveFreePages* is satisfied. □

Proposition 159. *If there are no free pages in main store, a page is swapped out.*

PROOF. There is a disjunction in the composition forming *genOnPageFault*'s predicate:

$$(pts.HaveFreePages \wedge pts.AllocateFreePage[destpg/ppno!]) \\ \vee (haveVictim \wedge swapPageToDisk[destpg/victim!])$$

In this case, if *HaveFreePages* is not satisfied, $\neg$ *HaveFreePages* must be satisfied and the swap is performed by *swapPageToDisk*. □

Proposition 160. *If a process fault occurs, the referenced page overwrites a page frame freed by swapping out.*

PROOF. In the previous proposition, the page *destpg* was swapped out to disk. In the second component of *genOnPageFault*'s predicate, the operation *overWritePhysicalPage* occurs as a conjunct. By $p \wedge q \vdash p$, the result follows. □

Proposition 161. *If a process faults, the referenced page is brought into store.*

PROOF. The operation *retrievePageFromDisk*[*page*/*pg*] is a conjunct in the second component of the sequential composition in *genOnPageFault*. □

Proposition 162. *If there are free pages, only a free page is written when a page is swapped in.*

PROOF. The predicate of schema *genOnPageFault* contains the following references:

$$pts.AllocateFreePage[destpg/ppno!] \ldots \fatsemi \ldots \\ retrievePageFromDisk[\ldots, page/pg!] \\ pts.OverwritePhysicalPage[page/pg?, destpg/pageno?]$$

The physical page, *destpg*, is the one overwritten by *page* in the operation defined by the schema *OverwritePhysicalPage*. The physical page *destpg* is allocated by *AllocateFreePage*. The page retrieved from disk is *page*; it is retrieved by *retrievePageFromDisk* (the omitted arguments refer to the faulting process). □

Proposition 163. *Newly swapped pages have a zero reference count.*

PROOF. Reference bits are cleared in the newly swapped page by *genOnPageFault*. The second component of this schema's predicate contains, as one of its conjuncts, a reference to *pfs.ClearRefBitsAndCounter*[*destpg*/*ppno*?]—this is a reference to a page frame in physical store, not to the contents (which is denoted by *page* in this predicate). □

Proposition 164. *Newly swapped pages cannot be victims unless all pages are victims.*

PROOF. By the victim-finding operation, the victim has the minimum reference count:

IsVictim

$$(\exists j : PHYSICALPAGENO \mid j \in \mathrm{dom}\, count \bullet$$
$$\quad count(j) = min(\mathrm{ran}\, count))$$

A newly swapped-in page—one that has not yet been referenced—has a reference count of 0. To become a victim, there must be no page with reference count > 0. □

A process can be waiting on a device when one of its pages is chosen to swap out. If the driver copies data and puts it into buffers associated with the waiting process' PCBs, there is only the issue of swapping out the page. However, there is no way *a priori* of knowing whether the page just swapped out will cause a page fault when its owning process next executes. Since swapping out does not affect the operations of the device upon which the victim is waiting, it would appear valid just to pick any process that is not locked into main store.

The reader should note that, logically, this is perfectly adequate. As a proposed implementation, this is unlikely to work well. It is to be expected that page faults will be relatively frequent. The paging disk has a latency time that must be taken into account. When a user process causes a page fault, it must be blocked. Clearly, processes ought to be blocked for the shortest possible time. The paging disk, however, serialises requests. All of this suggests that the swapin/swapout operations should be as fast as possible.

Moreover, the specification, as it stands, allows the ISR to respond to as many interrupts as it can but it must also wait for the driver. The driver's message input is restricted to one immediate and one outstanding message. This suggests that the ISR should enqueue messages on the driver and immediately halt.

Furthermore, the context of the faulting process *must* be swapped immediately. This is because it cannot progress and *must* be taken off the processor: alternatives are hard to discern. Removal of the current process' registers from the processor by the ISR is, therefore, justified.

Similarly, the page disk can lose requests if it just hangs between instructing a disk search and reading the result.

This subsystem also shows limitations with the message-passing régime. Because a synchronous method has been adopted, the sender must wait if the receiver is not in a state to receive. This has the implication that, should the page disk process not be ready to accept another request (either because it is waiting for the disk or because it is processing another request), the page-fault handler will have to wait. This has the implication that the page-fault handler might miss an interrupt.

The presence of operations to add and remove process data from the paging disk complicates matters also. Luckily, these requests will be less common than simple page faults.

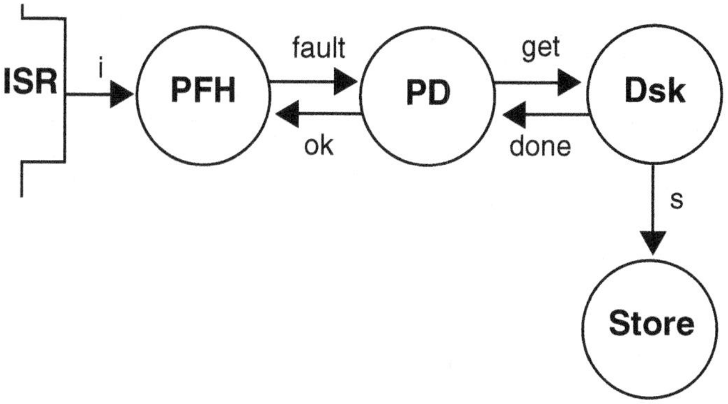

Fig. 6.4. *The actual specification.*

The question for us is the following. Even though the classes and operations presented above provide an adequate logical model, they do not take into consideration all of the pragmatic issues. Should the specification be altered to reflect these pragmatic issues?

The specification presented above can be represented diagrammatically as in Figure 6.4. As can be seen, the virtual-storage management subsystem consists of the ISR (denoted by *ISR* in the figure), the page-fault driver process (denoted by *PFH* in the figure) and the disk driver process (denoted by *PD*). The figure also includes the disk itself (denoted by *DSK*) and main store (denoted by *STORE*).

The arrows in Figure 6.4 denote the messages or interactions between processes. The arrow labelled *i* denotes the message sent by the *ISR* to the page-fault handling process (*PFH*). This message contains the specification of the page that was referenced (in terms of the segment and logical page numbers) and the process (its *APREF*). The page-fault handler sends a *fault* message containing the same information to the paging-disk process, which in turn sends a request (as a *get* message) to the disk proper (actually, to the disk driver). The disk driver retrieves the page and sends a *done* message to the paging-disk handler. The *done* message denotes the fact that the retrieval has been successfully completed.

Once the page has been written to main store, process *PD* sends an *ok* message to the page-fault handler process, *PFH*. On reception of the *ok*, the page-fault handler can wait for another page fault.

It is clearly a highly desirable property for any optimisations of the basic (logical) specification to behave in the same way as the specification. It is also highly desirable, given the present context, to be able to demonstrate this in a formal way. In order to achieve this, the proposed optimisations are formalised as CCS [21] processes so that they can be manipulated in formally sound ways. CCS is chosen as the representation because we are interested in the interactions between the component processes of the subsystem, not in the specification of the components. The processes to be modelled do not have properties suggestive of the use of the π-calculus (e.g., mobility), so CCS appears sufficient.

The subsystem can be represented in CCS as the following set of equations:

$$
\begin{aligned}
&ISR = \bar{i}.ISR\\
&PFH = i.\overline{fault}.ok.PFH\\
&PD = fault.\overline{get}.done.\bar{s}.\overline{ok}.PD\\
&DSK = get.\overline{done}.DSK\\
&STORE = s.STORE
\end{aligned}
$$

The overall arrangement is represented in CCS as:

$$VM_1 = (ISR \mid PFH \mid PD \mid DSK \mid STORE) \setminus \{i, fault, ok, get, done, s\}$$

(Note that actions are hidden using the \ operation.)

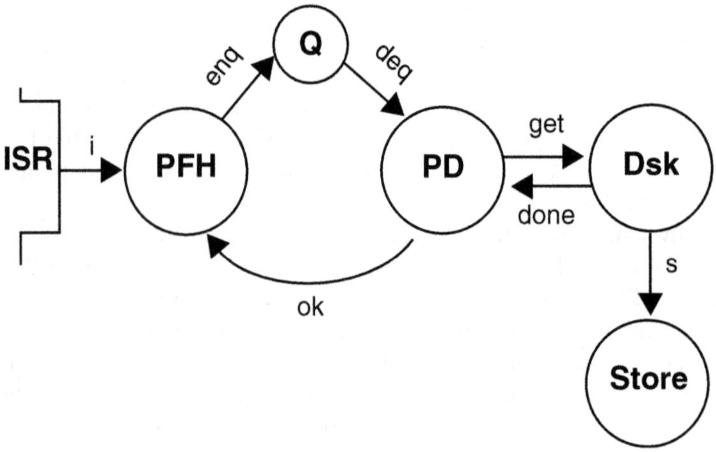

Fig. 6.5. *The specification using a queue.*

As noted above, the design depicted in Figure 6.4 and represented by the above set of CCS process definitions can be optimised. An obvious optimisation is to introduce a queue of requests between the page-fault handler (*PFH*) and paging-disk process (*PD*). This is the arrangement shown in Figure 6.5. In the arrangement shown in Figure 6.5, the *PFH* process places new requests into the queue. The queue is represented by the process named Q in the figure, and the operation of sending a request (really, just an enqueuing of the request) is represented by the *enq* arrow. Requests are removed from the queue process by the paging-disk process, *PD*, in exactly the same way they are in the first case. When the page has been copied to main store, the paging-disk process, *PD* sends the page-fault handling process an *ok* to inform it that: (i) the copy has been performed and that (ii) the process that caused the fault just rectified can now be unblocked.

The argument is that this second version can process more page faults per unit time. In this case, *PFH* does not now wait for the page fault to be rectified before it can wait for a new fault. Instead, it passes the request to the rest of the subsystem and then immediately blocks on the page-fault ISR.

This second arrangement can be represented by CCS processes as follows:

$$VM_2 = init.(ISR \mid PFH_2 \mid Q \mid PD_2 \mid DSK \mid STORE) \\ \backslash\{i, enq, deq, ok, done, get, init, s\}$$

For the definition of this subsystem, processes *ISR*, *DSK* and *STORE* remain as in the first case. The remaining processes must be redefined as follows (the subscripts will be explained below).

$$PFH_2 = i.\overline{enq}.ok.PFH_2$$
$$Q = init.enq.Q_1$$
$$Q_1 = enq.Q1 + deq.Q1$$
$$PD_2 = deq.\overline{get}.done.\overline{s}.\overline{ok}.PD_2$$

It is clearly necessary to distinguish between the two versions of *PFH* that have been defined at this point (a third version will be added shortly). For this reason, subscripts were introduced into the specifications.

It would be useful for these two specifications (models) to be equivalent in some sense. One important sense is that they should be *observationally* equivalent; another is that they should be *bisimilar*.

The property of *observational* equivalence of two processes is very much the intuitive one: two processes are observationally equivalent when they cannot be distinguished by an external observer. In other words, the externally visible events that can be perceived by an external observer are determined by the observer to be the same in content and in order, no matter which process is observed. In the cases of VM_1 and VM_2, the externally observable events are restricted by the hiding operator (\, as in Z).

Bisimilarity is an equivalence that also takes hidden actions into account. (Those readers unfamiliar with the concept should consult [21], Chapter 4, for an extended treatment.)

It is now possible to engage in formal reasoning about processes VM_1 and VM_2 and to prove two important propositions about them. Rather than engaging in a hand proof, it was considered interesting to employ automation. To this end, the *Concurrency Workbench of the New Century* [8] was employed as a tool. The Concurrency Workbench can be used to determine a number of properties of CCS and CSP processes, including observational equivalence and bisimilarity. The propositions concerning equivalence of the various versions of the subsystem were all proved using the Concurrency Workbench.

Proposition 165. *Processes VM_1 and VM_2 are observationally equivalent.*

PROOF. Both VM_1 and VM_2 were encoded in the input format required by the CWB and tested using the `eq -S obseq` command. The CWB system determined that VM_1 and VM_2 are observationally equivalent. □

Matters are different when it comes to bisimilarity. Processes VM_1 and VM_2 are quite dissimilar in internal structure, and process VM_2 offers an initial event, *init*, which initialises the process Q, so it does not appear, at first sight, that VM_1 and VM_2 will be bisimilar. Indeed, this is the case.

Proposition 166. *Processes VM_1 and VM_2 are not bisimilar.*

PROOF. Both VM_1 and VM_2 were encoded in the input format required by the CWB and tested using the `eq -S bisim` command. The CWB system determined that VM_1 and VM_2 are *not* bisimilar. □

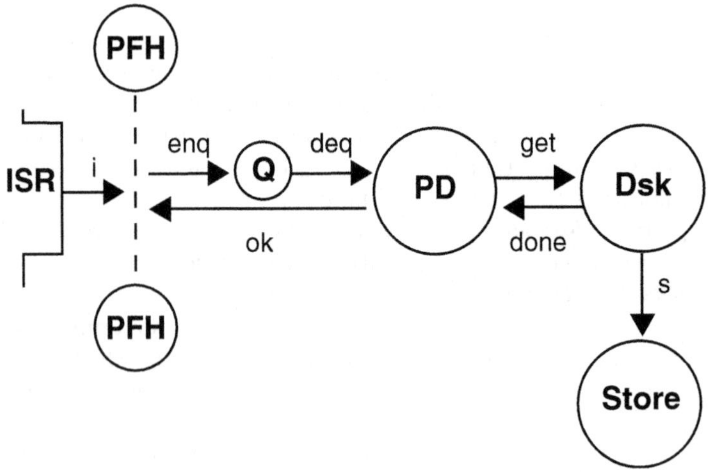

Fig. 6.6. *The ideal specification.*

Finally, a second specification can be considered as an optimisation. This version retains the queue of requests. It differs from the second version by creating a new instance of the *PFH* process whenever an interrupt occurs. This instance performs the same operations as in the second version but immediately terminates (denoted by the **0** process). The CCS specification is (again, subscripts are used to differentiate between versions):

$$VM_3 = init.(ISR \mid PFH_3 \mid Q \mid PD_2 \mid DSK \mid STORE) \\ \backslash\{i, enq, deq, ok, done, get, init, s\}$$

where:

$$PFH_3 = i.(\overline{enq}.ok.\mathbf{0} \mid PFH_3)$$

Proposition 167. *VM_3 is observationally equivalent to VM_2.*

PROOF. The models for VM_2 and VM_3 were encoded in the input format required by the CWB and tested using the `eq -S obseq` command. The CWB system determined that the two are observationally equivalent. □

Proposition 168. *VM_3 is bisimilar to VM_2.*

PROOF. The models for VM_2 and VM_3 were encoded in the input format required by the CWB and tested using the `eq -S obseq` command. The CWB system determined that the two are bisimilar. □

From the first of these two results, we have the next proposition:

Proposition 169. *Process VM_1 is observationally equivalent to VM_3.*

PROOF. By transitivity of the equivalence relation.

(In addition, the encoded models were tested using the `eq -S obseq` command and the result was supported by the CWB system.) □

We also have the following negative result:

Proposition 170. *Processes VM_1 and VM_3 are not bisimilar.*

PROOF. Again, by transitivity of the equivalence relation, this result would be expected.

(In addition, the CWB supported the claim.) □

The last result is not as bad as it sounds. All that is required, here, is that the processes *appear* the same as far as an external observer is concerned. This property is sufficient for the optimisations to be considered equivalent to the original. However, as far as this chapter is concerned, the first version (the one referred to as the "logical" one) is the one that will be adopted.

With this discussion of optimisation out of the way, it is possible to return to the main theme. There follow some propositions dealing with the properties of the logical (Z) model of page faults.

Proposition 171. *Page faults are serviced in the order in which they occur.*

PROOF. There are two senses:

1. Each page fault is an interrupt. In this sense, page faults are all processed in order.
2. In this sense, page faults are serviced by the handler process which services messages sent to it by the page-fault ISR. These messages are sent in temporal (sequence) order. That the handler process operates upon these messages in the correct order is a consequence of the correctness of the message-passing operations.

Each sense implies the statement of the proposition. □

Proposition 172. *Every page fault is serviced eventually (i.e., the page-fault mechanism is fair).*

PROOF. This proof follows from the correctness and fairness of the message-passing operations. □

Proposition 173. *The process causing a page fault has status* pstrunning *when the fault occurs.*

PROOF. Only a running process can cause a page fault.

ISRs cannot cause faults and are not processes.

Drivers are processes. They are locked into main store and, by hypothesis, have all the necessary pages locked in as well.

Only user processes have pages that can be swapped into and out of store.

The fact that a fault occurs implies that the process causing it is executing. There can only be one process that is executing at any one time (i.e., is the current value of *currentp* and has a status of pstrunnning).

This is a property of the scheduler. □

Proposition 174. *A faulting process is not unblocked until the replacement page has been swapped into physical store.*

PROOF. The service operation *onPageFault* first blocks the faulting process by setting its state to pstwaiting and by removing it from the processor. Therefore *blockFaultingProcess* implies that $currentp \neq currentp' \wedge regs' \neq regs$, where *regs* are hardware registers.

The *onPageFault* operation takes the form of a sequential composition. There are three operations, thus forming a composition of the form $S_1 \fatsemi S_2 \fatsemi S_3$. The second operation is itself a sequential composition; it is this part that performs the swapping operation. The operations are, therefore, totally ordered and the order can be directly related to temporal succession.

Let p denote the faulting process and let the composition be written as above. Then post $S_1 \Rightarrow status(p) = \mathsf{pstwaiting}$ and $currentp' \neq p$.

Throughout S_2, $status(p) = \mathsf{pstwaiting}$.

Finally, pre $S_3 \Rightarrow status(p) = \mathsf{pstwaiting}$, while post $S_3 \Rightarrow status(p) = \mathsf{pstready}$ but $p \neq currentp$—the last being a property of *MakeReady*. It cannot be true because the service process is currently executing. □

Proposition 175. *If a process, p, causes a page fault, that process is not marked ready until the faulting page has been replaced.*

PROOF. This follows immediately from the previous proposition. □

Proposition 176. *After execution of onPageFault, there is exactly one physical page in physical store such that its logical page number in the faulting process maps to the physical page number.*

PROOF. The proposition implies that exactly one page is mapped into store by the page-fault mechanism.

Inspection of the predicate of *genOnPageFault* shows that only one page is introduced into store. □

Proposition 177. *If a process, p, causes a page fault, after that page fault has been serviced, the process is in the ready state.*

PROOF. That the process is blocked follows immediately from Proposition 174. The predicate of *genOnPageFault* contains an instance of *MakeReady* with the faulting process' identifier substituted for the formal parameter. □

Proposition 178. *If a page fault occurs and a page has to be swapped out, that page can be a page belonging to the faulting process or to another process.*

PROOF. There are two cases to consider.
Case 1: If there are free pages, they are consumed. Only the faulting process is affected.
Case 2: *HaveVictim* is called. The victim page is found in physical store by *pfs.IsVictim*:

$$
\begin{array}{l}
findVictimPage \mathrel{\widehat=} \\
\quad pfs.VictimPhysicalPageNo \\
\quad \wedge\ findVictimLogicalPage
\end{array}
$$

The value returned by *pfs.VictimPhysicalPageNo* is just a physical page number with the minimum reference count. The schema does not mention processes. Similarly, *findVictimLogicalPage* in this class merely looks up the owning process' identifier and the segment in which the page occurs, as well as the logical page number of the victim page. Therefore, the page can belong to *any* process in the system except those that lock it into main store (i.e., driver and system processes). □

Proposition 179. *If a page fault occurs and there are free pages in physical store, physical store is updated; only one process is affected.*

PROOF. By the definition of *onPageFault*:

$pts.HaveFreePages \wedge pts.AllocateFreePages$

implies that the physical page to which the swapped-in page is to be written is a free page in physical store.

In this case, only the faulting process is affected. This is because the allocation of a free page relates only to a single process (the process to which the allocation is made). □

Corollary 13. *If a page fault occurs within process, p, and if a page has to be swapped out, only a maximum of two processes are affected by the page-swapping operation.*

PROOF. Immediate from the two preceding propositions. □

Proposition 180. *No pages are swapped unless a page fault occurs.*

PROOF. Inspection reveals that only the schema *genOnPageFault* references the relevant operations. □

6.3.4 Extending Process Storage

Processes very often make requests to increase the amount of store they use. In a paged system, this allocates more pages to the process. This subsection contains a model of the operations. It is rarer for processes to release store but operations to perform this task are defined as well. Pages can also be shared between processes. Sharing can be used, for example, to implement message passing in virtual storage.

First, there is a problem with allocating pages. The problem is that there might not be a physical page free when the request is made. The system could block the requesting process until there is a free page in the page frame. This solution does not fit well with the aims of virtual storage. Furthermore, when a new process is created, it will request a set of pages to hold code, data, stack, etc. It would not be acceptable to block the creation of the process until main store becomes free.

A solution more in keeping with the purpose of virtual store is the following. When a request is made to allocate a new page and there is no free physical store, an empty page is allocated on disk. The new page is allocated to the requesting process and can be written to by the requesting process.

This mechanism can be used when allocating extra pages to a new page.

Therefore, it is necessary to begin with the definition of the empty page:

$$pageofzeroes : PAGE$$
$$\forall pg : PAGE \bullet$$
$$\quad pageofzeroes = \lambda\, i : 1 \mathrel{..} framesize \bullet 0$$

This structure need not be held in a page; it can be produced in a loop and then set into a buffer of smaller size. Ideally, it should be packed into a single disk block and then handed to the paging disk. This needs to be iterated n times, where n is the number of disk blocks per page.

Pages must be specified when allocating, deallocating and sharing. A method for specifying them is required. Here, we define a structure for this purpose:

$$PAGESPEC == APREF \times SEGMENT \times LOGICALPAGENO$$

The following axiomatic definitions are required to manipulate objects of type *PAGESPEC*. The relation that comes first will be discussed below. The rest of the definitions are: the constructor function (*mkpgspec*) and the accessor functions (*pgspecpref* to obtain the process reference, *pgspecseg* to obtain the segment name and *pgspeclpno* to obtain the logical page number).

$$smap : PAGESPEC \leftrightarrow PAGESPEC$$
$$mkpgspec : APREF \times SEGMENT \times LOGICALPAGENO \rightarrow PAGESPEC$$
$$pgspecpref : PAGESPEC \rightarrow APREF$$
$$pgspecseg : PAGESPEC \rightarrow SEGMENT$$
$$pgspeclpno : PAGESPEC \rightarrow LOGICALPAGENO$$

$$\forall p : APREF;\ sg : SEGMENT;\ lpno : LOGICALPAGENO \bullet$$
$$\quad mkpgspec(p, sg, lpno) = (p, sg, lpno)$$

$$\forall ps : PAGESPEC \bullet$$
$$\quad pgspecpref(ps) = fst\ ps$$
$$\quad pgspecseg(ps) = fst(snd\ ps)$$
$$\quad pgspeclpno(ps) = snd^2\ ps$$

The type *PAGESPEC* could have been defined earlier in this chapter. If it had been used there, it would have been necessary to represent and manipulate all virtual addresses and page references in terms of *PAGESPEC*. Although this seems attractive, we believe, it would have complicated the specifications somewhat.

The axiomatic definitions begin with *smap*, a relationship between elements of *PAGESPEC*. This is the *sharing map*. When two elements of *PAGESPEC* are in the relation, the processes mentioned as the first component of *PAGESPEC* share the page referred to by the two *PAGESPEC*s.

The operations to allocate, deallocate and share pages are collected into a (somewhat *ad hoc*) class called *VStoreManager*. The class is intended to act as a component in an interface library. It exports most of its operations so that a wide variety of combined operations can be defined elsewhere.

VStoreManager

$\upharpoonright$(*INIT*, *AddNewMainStorePageToProcess*, *AddNewVirtualPageToProcess*, *CanAddPageToProcess*, *MarkLogicalPageAsShared*, *UnshareLogicalPage*, *WithdrawLogicalPage*, *SharedLogicalPageSharers*, *IsSharedLogicalPage*, *RemoveSharedLogicalPageOwner*, *RemoveLogicalPageSharer*, *RawShareLogicalPageBetweenProcesses*, *ShareLogicalPageBetweenProcesses*, *ReturnSharedLogicalPageToOwner*, *MarkSharedLogicalPageAsIn*, *MarkSharedLogicalPageAsOut*, *ShareLogicalSegment*, *ReleaseSharedSegment*, *ReleaseSegmentPagesExcept*, *CanReleaseSegment*, *SharedPagesInSegment*, *CanReleaseProcessVStore*)

$pts : PageTables$
$pfs : PageFrames$

INIT

$pgtabs? : PageTables$
$pfrms? : PageFrames$

$pts' = pgtabs?$
$pfs' = pfrms?$

$makeEmptyPage \mathrel{\widehat=} \ldots$

$AddNewMainStorePageToProcess \mathrel{\widehat=} \ldots$

$AddNewVirtualPageToProcess \mathrel{\widehat=} \ldots$

$CanAddPageToProcess \mathrel{\widehat=} \ldots$

$MarkLogicalPageAsShared \mathrel{\widehat=} \ldots$

$UnshareLogicalPage \mathrel{\widehat=} \ldots$

$WithdrawLogicalPage \mathrel{\widehat=} \ldots$

$SharedLogicalPageSharers \mathrel{\widehat=} \ldots$

$IsSharedLogicalPage \mathrel{\widehat=} \ldots$

$RemoveSharedLogicalPageOwner \mathrel{\widehat=} \ldots$

$RemoveLogicalPageSharer \mathrel{\widehat=} \ldots$

$RawShareLogicalPageBetweenProcesses \mathrel{\widehat=} \ldots$

$ShareLogicalPageBetweenProcesses \mathrel{\widehat=} \ldots$

$ReturnSharedLogicalPageToOwner \mathrel{\widehat=} \ldots$

$MarkSharedLogicalPageAsIn \mathrel{\widehat=} \ldots$

$MarkSharedLogicalPageAsOut \mathrel{\widehat=} \ldots$

$ShareLogicalSegment \mathrel{\widehat=} \ldots$

$ReleaseSharedSegment \mathrel{\widehat=} \ldots$

$ReleaseSegmentPagesExcept \mathrel{\widehat=} \ldots$

$CanReleaseSegment \mathrel{\widehat=} \ldots$

$SharedPagesInSegment \mathrel{\widehat=} \ldots$

$CanReleaseProcessVStore \mathrel{\widehat=} \ldots$

First, there is the hidden operation that creates an empty page:

makeEmptyPage

$pg! : PAGE$

$pg! = pageofzeroes$

This operation is hidden because it is somewhat undesirable for everyone to manipulate the buffers in which new pages are created.

The operation to add a new page to a process using a free-store page is the following. It allocates the page frame, adds the page to the process and clears the page (this is not really necessary but is a nice feature[2]).

$$
\begin{aligned}
&AddNewMainStorePageToProcess \mathrel{\widehat=}\\
&\quad (pts.HaveFreePages\\
&\qquad \land ((pts.IncProcessPageCount \fatsemi (pts.LatestPageCount)\\
&\qquad\qquad \land pts.AllocateFreePage[ppno/ppno!]\\
&\qquad \land pts.AddPageToProcess[ppno/ppno?, lpno!/lpno?]\\
&\qquad \land makeEmptyPage[pg/pg?]\\
&\qquad \land pts.OverwritePhysicalPage[ppno/pageno?, pg/pg?]) \setminus \{ppno\}))
\end{aligned}
$$

Note that an alternative is to swap out a process' page. The approach here is simpler but much more profligate.

The next operation uses the prceding one and then stores the page that it has created on the paging disk. Note that the operation requires there to be *at least* one page-sized buffer in the kernel.

$$
\begin{aligned}
&AddNewVirtualPageToProcess \mathrel{\widehat=}\\
&\quad (\neg\ pts.HaveFreePages\\
&\qquad \land (pts.IncProcessPageCount \fatsemi pts.LatestPageCount)
\end{aligned}
$$

[2] Some (civilised?) operating systems perform this operation on allocating new store.

$$\wedge\ (\exists\ ppno : PHYSICALPAGENO \mid ppno = 1 \bullet$$
$$(makeEmptyPage[pg/pg!]$$
$$\wedge\ pts.AddPageToProcess[lpno!/lpno?, ppno/ppno?]$$
$$\wedge\ StorePageOnDisk[lpno!/lpno?, pg/pg?])) \setminus \{pg\})$$

The value assigned to *ppno* is purely arbitrary. The physical page number of any page on disk has no relevance because it will be mapped to another physical page when swapped into main store.

The following is a predicate. It is true if the process denoted by $p?$ can add at least one page to its $sg?$ segment.

CanAddPageToSegment

$p? : APREF$
$sg? : SEGMENT$

$pagecount(p?)(sg?) < maxvirtpagespersegment$

The actual operation to add a new page to a process is the following. Depending upon the state of main store, it adds the page directly or stores it on the paging disk:

$$AddNewPageToProcess \mathrel{\widehat=}$$
$$AddNewMainStorePageToProcess$$
$$\vee\ AddNewVirtualPageToProcess$$

Proposition 181. *If there are no free pages in main store, a zero page is created for it and written to disk.*

PROOF. The schema *AddNewPageToProcess* is a disjunction:

$$AddNewMainStorePageToProcess$$
$$\vee\ AddNewVirtualPageToProcess$$

Each disjunct is a conjunction, with mutually exclusive first conjuncts.

In particular, *AddNewVirtualPageToProcess* has a predicate containing the following:

$$makeEmptyPage[pg/pg!]$$
$$\wedge$$
$$\ldots$$
$$\wedge\ StorePageOnDisk[\ldots, pg/pg?]$$

The existential quantifier can be removed by the one-point rule, and the result follows from the fact that $p \wedge q \Rightarrow p$. □

Proposition 182. *If there are free pages, one is allocated for the newly created page.*

PROOF. Similar to the above, concentrating on *AddNewMainStorePage*. Again, $p \land q \vdash p$ allows the conclusion to be drawn. □

The only thing that needs to be done when a page is shared between processes is to mark it. This is actually done by adding *PAGESPEC*s to the sharing map, *smap*:

$$
\begin{array}{l}
\textit{MarkLogicalPageAsShared} \\
\hline
\Delta(smap) \\
ownproc?, shareproc? : APREF \\
ownseg?, shareseg? : SEGMENT \\
ownlp?, sharelp? : LOGICALPAGENO \\
\hline
(\exists\, atrip1, atrip2 : PAGESPEC \bullet \\
\quad atrip1 = mkpgspec(ownproc?, ownseg?, ownlp?) \land \\
\quad atrip2 = mkpgspec(shareproc?, shareseg?, sharelp?) \land \\
\quad smap' = smap \cup \{(atrip1, atrip2)\}) \\
\hline
\end{array}
$$

Unsharing, conversely, removes a pair of *PAGESPEC*s from the sharing map:

$$
\begin{array}{l}
\textit{UnshareLogicalPage} \\
\hline
\Delta(smap) \\
p? : APREF \\
sg? : SEGMENT \\
lpno? : LOGICALPAGENO \\
\hline
smap' = smap \rhd \{mkpgspec(p?, sg?, lpno?)\} \\
\hline
\end{array}
$$

If a page is to be reallocated or removed from store completely, it must be totally removed from the sharing map:

$$
\begin{array}{l}
\textit{WithdrawLogicalPage} \\
\hline
\Delta(smap) \\
owner? : APREF \\
sg? : SEGMENT \\
lpno? : LOGICALPAGENO \\
\hline
smap' = \{mkpgspec(p?, sg?, lpno?)\} \lhd smap \\
\hline
\end{array}
$$

In later operations, it is necessary to determine which processes share a given page. The page is specified by the *sg*? and *lpno*? inputs and is shared by (at least) *p*?. The relational image is used to determine all the sharing *PAGESPEC*s and, hence, the sharing processes:

$$
\begin{array}{|l}
\hline
\textit{SharedLogicalPageSharers} \\
p? : APREF \\
sg? : SEGMENT \\
lpno? : LOGICALPAGENO \\
srs! : \mathbb{F}\, PAGESPEC \\
\hline
smap(\!|\ \{mkpgspec(p?, sg?, lpno?)\}\ |\!) = srs! \\
\hline
\end{array}
$$

The following predicate is true when the page is shared:

$$
\begin{array}{|l}
\hline
\textit{IsSharedLogicalPage} \\
p? : APREF \\
sg? : SEGMENT \\
lpno? : LOGICALPAGENO \\
\hline
mkpgspec(p?, sg?, lpno?) \in \operatorname{dom} smap \\
\hline
\end{array}
$$

When, say, a process terminates or when a shared page used to contain a message is withdrawn, the entry for the page must also be removed from the *smap*:

$$
\begin{array}{|l}
\hline
\textit{RemoveSharedLogicalPageOwner} \\
\Delta(smap) \\
p? : APREF \\
sg? : SEGMENT \\
lpno? : LOGICALPAGENO \\
\hline
smap' = \{mkpgspec(p?, sg?, lpno?)\} \lhd smap \\
\hline
\end{array}
$$

All the sharers of a page can be removed from the *smap* by the following operation:

$$
\begin{array}{|l}
\hline
\textit{RemoveLogicalPageSharers} \\
\Delta(smap) \\
p? : APREF \\
sg? : SEGMENT \\
lpno? : LOGICALPAGENO \\
\hline
(\exists\, pg : PAGESPEC \mid pg = mkpgspec(p?, sg?, lpno?) \bullet \\
\qquad smap' = smap \rhd \{pg\}) \\
\hline
\end{array}
$$

The following pair of schemata define the operation of sharing a page between two processes. The first schema defines the actual operation, while the second hides the logical page number. The first operation returns the logical page number of the shared page. Sometimes, this is something of a nuisance, so the second operation hides it. The first operation returns the page number so that it can be referenced, while the second hides the page number—approximately, it states that a page is shared but does not say which it is.

$$
\begin{array}{l}
\mathit{RawShareLogicalPageBetweenProcesses} \mathrel{\widehat=} \\
\quad ((pts.\mathit{IsLockedPage}[owner?/p?, ownseg?/sg?, ownlpno?/lpno?] \\
\qquad \land pts.\mathit{LockPage}[sharer?/p?, shareseg?/sg?, sharelpno!/lpno?]) \\
\qquad \lor \mathit{Skip}) \\
\quad \land \mathit{NextNewLogicalPageNo}[sharer?/p?, shareseg?/sg?, sharelpno!/lpno!] \\
\quad \land (pts.\mathit{PhysicalPageNo}[owner?/p?, ownseg?/sg?, ownlpno?/lpno?, ppno/ppno!] \\
\quad \land pts.\mathit{AddPageToProcess}[sharer?/p?, shareseg?/sg?, sharelpno!/lpno?] \\
\quad \land \mathit{MarkLogicalPageAsShared}[owner?/ownproc?, sharer?/shareproc?, \\
\qquad\qquad\qquad ownlpno?/ownlp?, sharelpno!/sharelp?] \\
\quad \land pts.\mathit{MarkPageAsShared}[owner?/p?, ownseg?/sg?, ownlpno?/lpno?] \\
\quad \land pts.\mathit{MarkPageAsShared}[sharer?/p?, shareseg?/sg?, sharelpno!/lpno?]) \\
\qquad\qquad \setminus \{ppno\}
\end{array}
$$

$$
\begin{array}{l}
\mathit{ShareLogicalPageBetweenProcesses} \mathrel{\widehat=} \\
\quad \mathit{RawShareLogicalPageBetweenProcesses} \setminus \{sharelpno!\}
\end{array}
$$

Proposition 183. *If two processes share a page, pg, that page will appear in the page tables of both processes.*

PROOF. The page, *pg*?, is already assumed to be in the page table of its owning process. The predicate of *RawShareLogicalPageBetweenProcesses* contains, $\mathit{AddPageToProcess}[sharer?/p?, shareseg?/sg?, sharelpno!/lpno?]$ as a conjunct. The arguments to the substitution refer to the process that is receiving the page. □

Proposition 184. *Schema RawShareLogicalPageBetweenProcesses makes the specified page shared by both processes.*

PROOF. The predicate of *RawShareLogicalPageBetweenProcesses* contains an instance of *MarkPageAsShared*, which simplifies to $smap' = smap \cup \{(s_1, s_2)\}$, where:

$$s_1 = mkpgspec(ownproc?, ownseg?, ownlp?)$$

and

$$s_2 = mkpgspec(sharer?, shareseg?, sharelp?)$$

and where *ownproc*? is the owning process identifier and *sharer*? is the identifier of the process being granted the right to share the page; *ownseg*? and *shareseg*? denote the segments in the owner and sharer's spaces; *ownlp*? is the owning process' logical page number and *sharelp*? is the logical page number in the sharer's segment. □

The only point to note about the following is that it unshares the page in question.

$ReturnSharedLogicalPageToOwner \mathrel{\widehat{=}}$
$\quad pts.IsSharedPage[sharer?/p?, shareseg?/sg/, sharelpno?/lpno?]$
$\quad \wedge\ pts.UnsharePage[sharer?/p?, shareseg?/sg/, sharelpno?/lpno?]$
$\quad \wedge\ pts.RemovePageProperties[sharer?/p?, shareseg?/sg/, sharelpno?/lpno?]$
$\quad \wedge\ pts.RemovePageFromPageTable[sharer?/p?, shareseg?/sg/,$
$\qquad\qquad sharelpno?/lpno?]$
$\quad \wedge\ UnshareLogicalPage[sharer?/p?, shareseg?/sg/, sharelpno?/lpno?]$

Proposition 185. *When a page is unshared, it is removed from one of the page tables.*

PROOF. The operation *ReturnSharedLogicalPageToOwner* removes the specified page from one of the processes that share it. This is done by removing the page from the sharer's page table by means of the conjunct:

$$RemovePageFromPageTable[sharer?/p?, shareseg?/sg/, sharelpno?/lpno?]$$

The other conjuncts cancel various attributes, in particular:

$$RemovePageProperties[sharer?/p?, shareseg?/sg/, sharelpno?/lpno?]$$

as well as removing the information that this page is shared by process *sharer?* by

$$UnsharePage[sharer?/p?, shareseg?/sg/, sharelpno?/lpno?]$$

and

$$UnshareLogicalPage[sharer?/p?, shareseg?/sg/, sharelpno?/lpno?]$$

. □

Shared pages need to be marked as in-store or out-of-store. The following pair of schemata define these operations. They will be used below.

$MarkSharedLogicalPageAsIn \mathrel{\widehat{=}}$
$\quad (pts.IsSharedPage \wedge pts.MarkPageAsIn \wedge$
$\qquad (SharedLogicalPageSharers[srs/srs!] \wedge$
$\qquad\quad (\forall pg : PAGESPEC \mid pg \in srs \bullet$
$\qquad\qquad (\exists p : APREF;\ sg : SEGMENT;$
$\qquad\qquad\qquad lpno : LOGICALPAGENO \bullet$
$\qquad\qquad\quad pg = mkpgspec(p, sg, lpno)$
$\qquad\qquad\quad \wedge\ pts.MarkPageAsIn[p/p?, sg/sg?, lpno/lpno?])) \setminus \{srs\}))$
$\quad \vee\ pts.MarkPageAsIn$

Proposition 186. *When a shared page is swapped in, it is swapped in in all sharing processes.*

PROOF. The operation *MarkPageAsIn* occurs within the scope of the universal quantifier which ranges over elements of *PAGESPEC*. These elements are exactly those describing the shared page in processes other than its owner. The operation *MarkPageAsIn* is applied to each descriptor (rather, to its components) and records the fact that the page has entered main store in each of the processes that share it. □

$$
\begin{array}{l}
MarkSharedLogicalPageAsOut \mathrel{\widehat=} \\
\quad (pts.IsSharedPage \land pts.MarkPageAsOut \land \\
\qquad (SharedLogicalPageSharers[srs/srs!] \land \\
\qquad (\forall pg : PAGESPEC \mid pg \in srs \bullet \\
\qquad\quad (\exists p : APREF;\ sg : SEGMENT; \\
\qquad\qquad\qquad\qquad lpno : LOGICALPAGENO \bullet \\
\qquad\qquad mkpgspec(p, sg, lpno) = pg \land \\
\qquad\qquad pts.MarkPageAsOut[p/p?, sg/sg?, lpno/lpno?]))) \setminus \{srs\}) \\
\quad \lor\ pts.MarkPageAsOut
\end{array}
$$

Proposition 187. *When a shared page is swapped out, it is swapped out in all sharing processes.*

PROOF. The operation defined by *MarkPageAsOut* occurs inside the universal quantifier in the predicate of *MarkSharedLogicalPageAsOut*. The universal quantifier ranges over all *PAGESPEC* elements that are related to the process in which the page is being swapped. Each element of type *PAGESPEC* records the process identifier, segment identifier and logical page number of the page in the sharing processes. Therefore, the quantifier ranges over all descriptions of the page in the sharing process. The *MarkPageAsOut* operation is applied to each of these descriptions, thus recording the fact that the page is swapped out. □

Proposition 188. *The operation MarkSharedLogicalPageAsOut can never mark a locked page as swapped out.*

PROOF. By Proposition 154. □

The next schema defines the operation to return the next logical page number available in the segment, *sg?*, of process *p?* (both *sg?* and *p?* are inputs to the schema):

$$
\begin{array}{l}
NextNewLogicalPageNo \mathrel{\widehat=} \\
\quad pts.IncProcessPageCount \fatsemi pts.LatestPageCount
\end{array}
$$

Segments, too, can be shared and unshared. This is not obviously useful but it turns out to be. When a process spawns a copy of itself or creates a child process that shares its code, the code segment can be shared if it is read-only (executable is a page attribute defined at the start of this chapter).

$$
\begin{array}{|l}
\hline
\textit{ShareLogicalSegment} \\
ownerp?, sharerp? : APREF \\
ownerseg?, sharerseg? : SEGMENT \\
\hline
(\forall lpno : LOGICALPAGENO \mid \\
\qquad\qquad lpno \in \mathrm{dom}\, pagetable(ownerp?)(sharerp?) \bullet \\
\quad (\exists ps_1, ps_2 : PAGESPEC;\ nxtlpno : LOGICALPAGENO \bullet \\
\qquad ps_1 = mkpgspec(ownerp?, ownerseg?, lpno) \land \\
\qquad NextNewLogicalPageNo[nxtlpno/lpno!] \land \\
\qquad\qquad ps_2 = mkpgspec(sharerp?, sharerseg?, nxtlpno) \land \\
\qquad\qquad pts.MarkPageAsShared \\
\qquad\qquad\qquad [ownerp?/p?, ownerseg?/sg?, lpno/lpno?] \land \\
\qquad\qquad MarkLogicalPageAsShared[lpno/ownlp?, nxtlpno/sharelp?])) \\
\hline
\end{array}
$$

$$
\begin{array}{|l}
\hline
\textit{ReleaseSharedSegment} \\
\Delta(pagetable) \\
p? : APREF \\
sg? : SEGMENT \\
\hline
pagetable' = \{sg?\} \lhd pagetable(p?) \\
\hline
\end{array}
$$

$$
\begin{array}{|l}
\hline
\textit{ReleaseSegmentPagesExcept} \\
\Delta(pagetable) \\
p? : APREF \\
sg? : SEGMENT \\
except? : \mathbb{F}\, LOGICALPAGENO \\
\hline
pagetable' = ((\mathrm{dom}\, pagetable(p?)(sg?)) \setminus except?) \lhd pagetable(p?)(sg?) \\
\hline
\end{array}
$$

Proposition 189. *If $except? = \varnothing$, ReleaseSegmentPagesExcept implies that the predicate of CanReleaseSegment is satisfied.*

PROOF. The predicate of *CanReleaseSegment* is:

$$\mathrm{dom}\, pagetable(p?)(sg?) = \varnothing$$

The predicate of *ReleaseSegmentPagesExcept* is:

$$pagetable' = ((\mathrm{dom}\, pagetable(p?)(sg?)) \setminus except?) \lhd pagetable(p?)(sg?)$$

Clearly (substituting $\varnothing$ for $except?$):

$$((\mathrm{dom}\, pagetable(p?)(sg?)) \setminus \varnothing) = \mathrm{dom}\, pagetable(p?)(sg?)$$

Now,

$$(\mathrm{dom}\, pagetable(p?)(sg?)) \lhd pagetable(p?)(sg?)$$

implies that $\mathrm{dom}\, pagetable(p?)(sg?) = \varnothing$. This is the predicate of the schema *CanReleaseSegment*, so we are done. □

Proposition 190. *A process whose entire virtual store has been released has no pages.*

PROOF. Obvious given the last proposition and the appropriate schema definitions. □

The following predicate is true if and only if all the pages in the segment, $sg?$, have been deallocated:

CanReleaseSegment

$p? : APREF$
$sg? : SEGMENT$

$\mathrm{dom}\, pagetable(p?)(sg?) = \varnothing$

The pages in the segment $sg?$ of process $p?$ are output by this operation. They constitute the set $sps!$.

SharedPagesInSegment

$p? : APREF$
$sg? : SEGMENT$
$sps! : \mathbb{F}\, LOGICALPAGENO$

$$\begin{array}{l}
(\exists\, psg : \mathbb{F}\, PAGESPEC \bullet \\
\quad psg = smap(\!|\, \{pg : PAGESPEC;\ lpn : LOGICALPAGENO \mid \\
\qquad\qquad pg = mkpgspec(p?, sg?, lpn) \bullet pg\} \,|\!) \land \\
\quad sps! = \{pg : PAGESPEC;\ lpno : LOGICALPAGENO \mid pg \in pgs \bullet \\
\qquad\qquad pgspeclpno(pg)\})
\end{array}$$

If a process has no pages and no segments (i.e., they have never been allocated or all have been released), its page-table entry can be released. The following schema is a predicate defining this case.

CanReleaseProcessVStore

$p? : APREF$

$\mathrm{dom}\, pagetable(p?) = \varnothing$

6.4 Using Virtual Storage

6.4.1 Introduction

This section is a rather looser model than the above model. It deals with the user-level interfaces. The view of virtual store is that much more akin to real store: a linear sequence of locations that can be addressed sequentially. The concept of the virtual address is also different in this section. Instead of a complex structure, virtual addresses are considered atomic, in effect a subset of $\mathbb{N}$.

6.4.2 Virtual Addresses

From this point on, the specification will deal with virtual addresses, not with logical and physical pages. This amounts to the user process' perspective on virtual storage. Hitherto, all the perspective has been that of the virtual storage constructor, so virtual address spaces and virtual addresses have been seen in terms of their components.

For this reason, the following is required:

$$AllocatePageReturningStartVAddress \mathrel{\widehat{=}} \\ \quad AddNewPageToProcess \wedge ComputePageVAddress[lpno!/lpno?]$$

This operation, as its name suggests, allocates a page and returns the virtual address of its start. (Note that the logical page number is *not* hidden; there are reasons for this, as will be seen.)

The following operation computes a virtual address from its components.

ComputePageVAddress

$lpno? : LOGICALPAGENO$
$vaddr! : \mathbb{N}$

$vaddr! = (lpno? - pgallocstart) * framesize$

Here, *pgallocstart* is a constant. It is defined as:

$pgallocstart : \mathbb{N}$

$pgallocstart = 0$

The value of 0 is completely arbitrary, as is now explained.

Some systems map a virtual copy of the operating system onto the virtual address space of each user space (and some privileged processes, too). The virtual copy can be allocated at the start or at the end of virtual store. The allocation of user-process pages has to respect this. Furthermore, there are special addresses that are reserved by the hardware; for example, interrupt vectors, device buffers and status words. These, too, must be allocated somewhere that cannot be directly accessed by untrusted user processes.

The *pgallocstrt* represents the logical page number of the first page in a segment that can be allocated.

For clarity and simplicity, it will be assumed here that the operating system virtual copy is located entirely in its own space. Transfer to the operating system from user space is achieved by a "mode switch". The mode switch activates additional instructions, for example those manipulating interrupts and the translation lookaside buffer. How the mode switch is performed is outside the scope of this book for reasons already given. Mode switches are, though, very common on hardware that supports virtual store.

In some systems, there are parts of the kernel space that are shared between all processes in the system. These pages are pre-allocated and added to the process image when it is created. Because they are pre-allocated, the allocation of user pages in that process must be allocated at some page whose logical page number is greater than zero. The constant *pgallocstart* denotes this offset.

Usually, the offset is used in the data segment only. For simplicity, the offset is here set to 0. Moreover, it is uniformly applied to all segments (since it is 0, this does not hurt).

The one hard constraint on virtual store is that some *physical* pages must never be allocated to user space. These are the pages that hold the device registers and other special addresses just mentioned.

Virtual-store pages are frequently marked as:

- execute only (which implies read-only);
- read-only;
- read-write.

Sometimes, pages are marked write-only. This is unusual for user pages but could be common if device buffers are mapped to virtual store pages.

The operations required to mark pages alter the attributes defined at the start of this chapter. The operations are relatively simple to define and are also intuitively clear. They are operations belonging to the class defined below in Section 6.5.2; in the meanwhile, they are presented without comment.

$$\begin{array}{l} \mathit{MarkPageAsReadOnly} \mathrel{\widehat=} \\ \quad \mathit{MakePageReadable} \land \\ \quad ((\mathit{IsPageWritable} \land \mathit{MakePageNotWritable}) \lor \\ \quad ((\mathit{IsPageExecutable} \land \mathit{MakePageNotExecutable}))) \end{array}$$

$$\begin{array}{l} \mathit{MarkPageAsReadWrite} \mathrel{\widehat=} \\ \quad (\mathit{MakePageReadable} \mathbin{⨾} \mathit{MakePageWritable}) \land \\ \quad (\mathit{IsPageExecutable} \land \mathit{MakePageNotExecutable}) \end{array}$$

$$\begin{array}{l} \mathit{MarkPageAsCode} \mathrel{\widehat=} \\ \quad (\mathit{MakePageExecutable} \mathbin{⨾} \mathit{MakePageReadable}) \land \\ \quad (\mathit{IsPageExecutable} \land \mathit{MakePageNotExecutable}) \end{array}$$

An extremely useful, but generic, operation is the following. It allocates n pages at the same time:

$$
\begin{array}{l}
AllocateNPages \mathrel{\widehat=} \\
\quad (\forall p : 1 \mathinner{\ldotp\ldotp} numpages? \bullet \\
\qquad (p = 1 \land AllocatePageReturningStartVAddress[startaddr?/vaddr!]) \\
\qquad \lor (p > 1 \land \\
\qquad\quad (\exists vaddr : VADDR \bullet \\
\qquad\qquad AllocatePageReturningStartVAddress[vaddr/vaddr!])))
\end{array}
$$

As it stands, this operation is not of much use in this model. The reason for this is that it does not set page attributes, in particular the read-only, execute-only and the read-write attributes. For this reason the following operations are defined. The collection starts with the operation for allocating n executable pages.

The following allocation operation is used when code is marked as executable and read-only. This is how Unix and a number of other systems treat code.

$$
\begin{array}{l}
AllocateNExecutablePages \mathrel{\widehat=} \\
\quad (\forall p : 1 \mathinner{\ldotp\ldotp} numpages? \bullet \\
\qquad (p = 1 \land AllocatePageReturningStartVAddress[startaddr?/vaddr!] \\
\qquad\quad \land MarkPageAsCode) \\
\qquad \lor (p > 1 \land \\
\qquad\quad (\exists vaddr : VADDR \bullet \\
\qquad\qquad AllocatePageReturningStartVAddress[vaddr/vaddr!] \land \\
\qquad\qquad MarkPageAsCode)))
\end{array}
$$

Similarly, the following operations allocate n pages for the requesting process. It should be remembered that the pages might be allocated on the paging disk and not in main store.

$$
\begin{array}{l}
AllocateNReadWritePages \mathrel{\widehat=} \\
\quad (\forall p : 1 \mathinner{\ldotp\ldotp} numpages? \bullet \\
\qquad (p = 1 \land AllocatePageReturningStartVAddress[startaddr?/vaddr!] \\
\qquad\quad \land MarkPageAsReadWrite) \\
\qquad \lor (p > 1 \land \\
\qquad\quad (\exists vaddr : VADDR \bullet \\
\qquad\qquad AllocatePageReturningStartVAddress[vaddr/vaddr!] \land \\
\qquad\qquad MarkPageAsReadWrite)))
\end{array}
$$

$$
\begin{array}{l}
AllocateNReadOnlyPages \mathrel{\widehat=} \\
\quad (\forall p : 1 \mathinner{\ldotp\ldotp} numpages? \bullet \\
\qquad (p = 1 \land AllocatePageReturningStartVAddress[startaddr?/vaddr!] \\
\qquad\quad \land MarkPageAsReadOnly) \\
\qquad \lor (p > 1 \land \\
\qquad\quad (\exists vaddr : VADDR \bullet \\
\qquad\qquad AllocatePageReturningStartVAddress[vaddr/vaddr!] \land \\
\qquad\qquad MarkPageAsReadOnly)))
\end{array}
$$

To support the illusion of virtual storage, virtual addresses can be thought of as just natural numbers (including 0):

$VADDR == \mathbb{N}$

and the virtual store for each process can be considered a potentially infinite sequence of equal-sized locations:

$$
\begin{array}{|l}
\hline
\textit{VirtualStore} \\
\hline
vlocs : \mathrm{seq}\, PSU \\
maxaddr : \mathbb{N} \\
\hline
\#locs = maxaddr \\
\hline
\end{array}
$$

It must be emphasised that each virtual address space has *its own copy* of the *VirtualStore* schema. Page and segment sharing, of course, make regions of store belonging to one virtual address space appear (by magic?) as part of another.

The usual operations (read and write) will be supported. However, when the relevant address is not present in real store, a *page fault* occurs and the *OnPageFault* driver is invoked with the address to bring the required page into store.

For much of the remainder, a block-copy operation is required. This is used to copy data into pages based on addresses. For the time being, it can be assumed that every address is valid (the hardware should trap this, in any case).

$$
\begin{array}{|l}
\hline
\textit{CopyVStoreBlock} \\
\hline
\Delta\, VirtualStore \\
data? : \mathrm{seq}\, PSU \\
numelts? : \mathbb{N} \\
destaddr? : VADDR \\
\hline
vlocs' = (\lambda\, i : 1 \,..\, (destaddr? - 1) \bullet vlocs(i)) \\
\qquad\qquad {}^\frown\langle data?\rangle {}^\frown (vlocs \ \mathit{after}\ (destaddr? + numelts?)) \\
\hline
\end{array}
$$

If any of the addresses used in this schema are not in main store, the page-faulting mechanism will ensure that it is loaded.

Operations defining the user's view of virtual store are collected into the following class. It is defined just to collect the operations in one place. (In the next section, the operations are not so collected—they are just assumed to be part of a library and are, therefore, defined in Z.)

The class is defined as follows. The definition is somewhat sparse and contains only two operations, *CopyVStoreBlock* and *CopyVStoreFromVStore*. In a full implementation, this class could be extended considerably. The point, here, though, is merely to indicate that operations similar to those often implemented for real store can be implemented in virtual storage systems at a level above that at which virtual addresses are manipulated as complex entities.

UsersVStore

$$\upharpoonright (INIT, CopyVStoreBlock, CopyVStoreFromVStore)$$

$$vlocs : \mathrm{seq}\, PSU$$
$$maxaddr : \mathbb{N}$$

$$\#locs = maxaddr$$

INIT

$$maxaddress? : \mathbb{N}$$

$$maxaddr' = maxaddress?$$

$$CopyVStoreBlock \mathrel{\widehat{=}} \ldots$$
$$CopyVStoreFromVStore \mathrel{\widehat{=}} \ldots$$

Data can be copied into a virtual-store page (by a user process) by the following operation:

CopyVStoreBlock

$$\Delta\, VirtualStore$$
$$data? : \mathrm{seq}\, PSU$$
$$numelts? : \mathbb{N}$$
$$destaddr? : VADDR$$

$$vlocs' = (\lambda\, i : 1 \mathinner{..} (destaddr? - 1) \bullet vlocs(i)) \frown \langle data? \rangle \frown (vlocs\ after\ (destaddr? + numelts?))$$

A similar operation is the following. It copies one piece of virtual store to another. It is useful when using pages as inter-process messages: the data comprising the message's payload can be copied into the destination (which might be a shared page in the case of a message) from the page in which it was assembled by this operation:

CopyVStoreFromVStore

$$\Delta\, VirtualStore$$
$$fromaddr? : VADDR$$
$$toaddr? : VADDR$$
$$numunits? : \mathbb{N}_1$$

$$(\exists\, endaddr : VADDR \mid endaddr = fromaddr? + numunits? - 1 \bullet$$
$$vlocs' = (\lambda\, i : 1 \mathinner{..} (toaddr? - 1) \bullet vlocs(i))$$
$$\frown (\lambda\, j : fromaddr? \mathinner{..} endaddr \bullet vlocs(j))$$
$$\frown (vlocs\ after\ endaddr + 1))$$

6.4.3 Mapping Pages to Disk (and Vice Versa)

Linux contains an operation called `memmap` in its library. This maps virtual store to disk store and is rather useful (it could be used to implement persistent store as well as other things, heaps for instance).

A class is defined to collect the operations together. Again, this class is intended only as an indication of what is possible. In a real system, it could be extended considerably; for example, permitting the controlled mapping of pages between processes, archiving of pages, and so on.

$$
\begin{array}{l}
\textit{PageMapping} \\
\hline
\upharpoonright(\mathit{INIT}, \mathit{MapPageToDisk}, \mathit{MapPageFromDiskExtendingStore}) \\
\quad \mathit{usrvm} : \mathit{UserVStore} \\
\quad \mathit{pfr} : \mathit{PageFrames} \\
\quad \mathit{INIT} \\
\quad \mathit{uvm?} : \mathit{UserVStore} \\
\quad \mathit{pgfrm?} : \mathit{PageFrames} \\
\quad \hline
\quad \mathit{usrvm}' = \mathit{uvm?} \\
\quad \mathit{pfr}' = \mathit{pgfrm?} \\
\mathit{MapPageToDisk} \mathrel{\widehat=} \ldots \\
\mathit{MapPageFromDiskExtendingStore} \mathrel{\widehat=} \ldots \\
\mathit{writePageToDisk} \mathrel{\widehat=} \ldots \\
\mathit{readPageFromDisk} \mathrel{\widehat=} \ldots
\end{array}
$$

The operations in this class are all fairly obvious, as is their operation. Commentary is, therefore, omitted.

$$
\begin{array}{l}
\textit{writePageToDisk} \\
\hline
\ldots \\
\mathit{diskparams?} : \ldots \\
\mathit{pg?} : \mathit{PAGE} \\
\hline
\ldots
\end{array}
$$

$$
\begin{array}{l}
\mathit{MapPageToDisk} \mathrel{\widehat=} \\
\quad (\mathit{pgt}.\mathit{PhysicalPageNo}[\mathit{ppno}/\mathit{ppgno!}] \land \\
\qquad \mathit{pfr}.\mathit{GetPage}[\mathit{ppno}/\mathit{pageno?}, \mathit{pg}/\mathit{fr!}] \land \\
\qquad \mathit{writePageToDisk}[\mathit{pg}/\mathit{pg?}]) \setminus \{\mathit{ppno}, \mathit{pg}\}
\end{array}
$$

$$\begin{array}{|l}
\hline
\textit{readPageFromDisk} \\
\ldots \\
\textit{diskparams?} : \ldots \\
\textit{pg!} : \textit{PAGE} \\
\hline
\ldots \\
\hline
\end{array}$$

This operation extends the store of the requesting process. It is used when reading pages from disk. The disk page is added to the process' virtual storage image.

$$\begin{array}{l}
\textit{MapPageFromDiskExtendingStore} \mathrel{\widehat=} \\
\quad \textit{usrvm.AllocatePageReturningStartVAddress}[\textit{pageaddr!}/\textit{vaddr!}] \mathbin{⨾} \\
\qquad (\textit{readPageFromDisk}[\textit{pg}/\textit{pg!}] \land \\
\qquad (\exists \textit{sz} : \mathbb{N} \mid \textit{sz} = \textit{framesize} \bullet \\
\qquad\quad \textit{usrvm.CopyVStoreBlock} \\
\qquad\qquad\qquad [\textit{sz}/\textit{numelts?}, \textit{pg}/\textit{data?}, \textit{pageaddr!}/\textit{destaddr?}])) \\
\qquad\qquad\qquad\qquad \setminus \{\textit{pg}\}
\end{array}$$

There ought to be an operation to delete the page from the image. However, only the most careful programmers will ever use it, so the operation is omitted. It is, in any case, fairly easy to define.

Note that there is no operation to map a disk page onto an existing virtual-store page. This is because it will probably be used extremely rarely.

The operations in this class could be extended so that the specified disk as well as the paging disk get updated when the frame's counter is incremented. This would automatically extend the disk image. A justification for this is that it implements a way of protecting executing processes from hardware and software failure. It can be used as a form of journalling.

This scheme can also be used on disk files. More generally, it can also work on arbitrary devices. This could be an interesting mechanism to explore when considering virtual machines of greater scope (it is an idea suggested by VME/B). Since this is just speculation, no more will be said on it.

6.4.4 New (User) Process Allocation and Deallocation

This section deals *only* with user-process allocation and deallocation. The general principles are the same for system processes but the details might differ slighty (in particular, the default marking of pages as read-only, etc.).

When a new process is created, the following schema is used. In addition, the virtual-store-management pages must be set up for the process. This will be added to the following schema in a compound definition.

UserStoreMgr

$\upharpoonright(INIT, MarkPageAsReadOnly, MarkPageAsReadWrite, MarkPageAsCode,$
$AllocateNPages, AllocateNExecutablePages, AllocateNReadWritePages,$
$AllocateNReadOnlyPages, CopyVStoreBlock, CopyVStoreFromVStore,$
$AllocateNewProcessStorage, ReleaseSharedPages,$
$FinalizeProcessPages, AllocateCloneProcessStorage)$

$usrvm : UserVStore$
$pgt : PageTables$

INIT

$uvm? : UserVStore$
$ptbl? : PageTables$

$usrvm' = uvm?$
$pgt' = ptbl?$

$MarkPageAsReadOnly \mathrel{\widehat=} \ldots$

$MarkPageAsReadWrite \mathrel{\widehat=} \ldots$

$MarkPageAsCode \mathrel{\widehat=} \ldots$

$AllocateNPages \mathrel{\widehat=} \ldots$

$AllocateNExecutablePages \mathrel{\widehat=} \ldots$

$AllocateNReadWritePages \mathrel{\widehat=} \ldots$

$AllocateNReadOnlyPages \mathrel{\widehat=} \ldots$

$CopyVStoreBlock \mathrel{\widehat=} \ldots$

$CopyVStoreFromVStore \mathrel{\widehat=} \ldots$

$AllocateNewProcessStorage \mathrel{\widehat=} \ldots$

$ReleaseSharedPages \mathrel{\widehat=} \ldots$

$FinalizeProcessPages \mathrel{\widehat=} \ldots$

$AllocateCloneProcessStorage \mathrel{\widehat=} \ldots$

AllocateNewProcessStorage

$p? : APREF$
$codepages? : \mathrm{seq}\, PSU$
$codesz?, stacksz?, datasz?, heapsz? : \mathbb{N}$

$(\exists sg : SEGMENT;\ codeszunits : \mathbb{N} \mid$
$\qquad sg = \mathsf{code} \land codeszunits = \#codepages? \bullet$

$$
\begin{array}{l}
\quad AllocateNExecutablePages \\
\qquad\qquad [codesz?/numpages?, addr/startaddr!] \mathbin{;} \\
\qquad usrvm.CopyVStoreBlock \\
\qquad\qquad [codepages?/data?, codeszunits/numelts?, \\
\qquad\qquad\quad addr/destaddr?]) \\
(\exists sg : SEGMENT \mid sg = \mathsf{data} \bullet \\
\quad AllocateNReadOnlyPages[datasz?/numpages?]) \\
(\exists sg : SEGMENT \mid sg = \mathsf{stack} \bullet \\
\quad AllocateNReadWritePages[stacksz?/numpages?]) \\
(\exists sg : SEGMENT \mid sg = \mathsf{heap} \bullet \\
\quad AllocateNReadWritePages[heapsz?/numpages?])
\end{array}
$$

ReleaseSharedPages

$$
\begin{array}{l}
\Delta(smap) \\
p? : APREF \\
\hline
(\forall ps : PAGESPEC \mid ps \in \mathrm{dom}\, smap \land pgspecpref(ps) = p? \bullet \\
\quad (\exists s : PAGESPEC \mid (ps, s) \in smap \bullet \\
\qquad smap' = smap \setminus \{(ps, s)\})) \\
\land (\forall ps : PAGESPEC \mid ps \in \mathrm{ran}\, smap \land psgpecpref(ps) = p? \bullet \\
\quad (\exists s : PAGESPEC \mid (s, ps) \in smap \bullet \\
\qquad smap' = smap \setminus \{(s, ps)\}))
\end{array}
$$

Once this schema has been executed, the process can release all of its pages:

$$
\begin{array}{l}
FinalizeProcessPages \mathrel{\widehat=} \\
\quad pgt.RemovePageProperties \land pgt.RemoveProcessFromPageTable
\end{array}
$$

AllocateCloneProcessStorage

$$
\begin{array}{l}
p? : APREF \\
clonedfrom? : APREF \\
stacksz?, datasz?, heapsz? : \mathbb{N} \\
\hline
(\exists sg : SEGMENT \mid sg = \mathsf{code} \bullet \\
\quad ShareLogicalSegment[clonedfrom?/ownerp?, p?/sharerp?, \\
\qquad\qquad sg/ownerseg?, sg/sharerseg?]) \\
(\exists sg : SEGMENT \mid sg = \mathsf{data} \bullet \\
\quad AllocateNReadOnlyPages[datasz?/numpages?]) \\
(\exists sg : SEGMENT \mid sg = \mathsf{stack} \bullet \\
\quad AllocateNReadWritePages[stacksz?/numpages?]) \\
(\exists sg : SEGMENT \mid sg = \mathsf{heap} \bullet \\
\quad AllocateNReadWritePages[heapsz?/numpages?])
\end{array}
$$

This works because of the following argument. The first of the two schemata (*ReleaseSharedPages*) above first removes the process from all of the pages that it shares but does not own. Then it removes itself from all of those shared pages that it does own. This leaves it with only those pages that belong to it and are not shared with any other process.

If a child process performs the first operation, it will remove itself from all of the pages it shares with its parents; it will also delete all of the pages it owns. The parent is still in possession of the formerly shared pages, which might be shared with other processes. As long as the parent is blocked until all of its children have terminated, it cannot delete a page that at least one of its children uses. Thus, when all of a process' children have terminated, the parent can terminate, too. Termination involves execution of the operations defined by *ReleaseSharedPages* and *FinalizeProcessPages*.

The only problem comes with clones. If the clone terminates before the original, all is well. Should the original terminate, it will delete pages still in use by the clone. Therefore, the original must also wait for the clone to terminate.

An alternative—one that is possible—is for the owner to "give" its shared pages to the clone. Typically, the clone will only require the code segment and have an empty code segment of its own. If the code segment can be handed over to the clone in one operation (or an *atomic* operation), the original can terminate without waiting for the clone or clones. Either is possible.

The allocation of child processes is exactly the same as cloning. The difference is in the treatment of the process: is it marked as a child or as a completely independent process? Depending upon the details of the process model, a child process might share code with its parent (as it does in Unix systems), whereas an independent process will tend to have its own code (or maybe a copy of its creator's code). In all cases, the data segment of the new process, as well as its stack, will be allocated in a newly allocated set of pages. In this chapter's model, data and stack will be allocated in newly allocated segments. The mechanisms for sharing segments of all kinds have been modelled in this chapter, as have those for the allocation of new segments (and pages). The storage model presented in this chapter can, therefore, support many different process models.

6.5 Real and Virtual Devices

There is often confusion between real and virtual devices. It is sometimes thought that the use of virtual store implies the use of virtual devices. This is not so. In most operating systems with virtual store, the devices remain real, while in some real-store operating systems, devices are virtual.

Virtual devices are really interfaces to actual, real ones. Virtual devices can be allocated on the basis of one virtual device to each process. The virtual device sends messages to and receives them from the device process. Messages are used to implement requests and replies in the obvious fashion. Messages to the real device from the virtual devices are just enqueued by the device process and serviced in some order (say, FIFO).

The interface to the virtual device can also abstract further from the real device. This is because virtual devices are just pieces of software. For example, a virtual disk could just define read and write operations, together with return codes denoting the success of the operation. Underneath this simple interface, the virtual device can implement more complex interfaces, thus absolving the higher levels of software from the need to deal with them. This comes at the cost of inflexibility.

This model can be implemented quite easily using the operations already defined in this book. Using message passing, it can be quite nicely structured.

There is another sense in which devices can be virtualised. Each device interface consists of one or more addresses. Physical device interfaces also include interrupts. Operations performed on these addresses control the device and exchange data between device and software. The addresses at which the device interface is located are invariably fixed in the address map. However, in a virtual system, there is the opportunity to map the pages containing device interfaces are mapped into the address space of each process. (This can be done, of course, using the sharing mechanism defined in this chapter.) This allows processes directly to address devices. However, some form of synchronisation must be included so that the devices are fairly shared between processes (or virtual address spaces). Such synchronisation would have to be included within the software interface to each device and this software can be at as low a level as desired.

A higher-level approach is to map standard addresses (by sharing pages) into each address space but to include a more easily programmed interface. Again, the mechanisms defined in this book can be used as the basis for this scheme.

6.6 Message Passing in Virtual Store

At a number of points in this chapter, the idea of using shared pages (or sets of shared pages) to pass messages between processes has been raised. The basic mechanisms for implementing message passing have also been defined.

When one process needs to send a message to another, it will allocate a page and mark it as shared with the other process. Data will typically be placed in the page before sharing has been performed. The data copy operation can be performed by one of the block-copy operations, *CopyVStoreBlock* or *CopyVStoreFromVStore* (Section 6.5.1).

The receiving process must be notified of the existence of the new page in its address space. This can be achieved as either a synchronous or an asynchronous event—the storage model is completely neutral with respect to this. In a system with virtual storage, message passing will be implemented as system calls, so notification can be handled by kernel operations. For example, the synchronous message-passing primitives defined in Chapter 5 can easily be modified to do this. What is required is that the message call point to a page and not to a small block of storage. Equally, the asynchronous mechanism outlined in Chapter 3 can be modified in a similar fashion.

Message passing based on shared pages will be somewhat slower at runtime than a scheme based upon passing pointers to shared storage blocks (buffers), even when copying buffers between processes is required. The reason for this is clear from an inspection of the virtual storage mechanisms. For this reason, it would probably be best to implement two message-passing schemes: one for kernel and one for user messages. The kernel message scheme would be based on shared buffers within kernel space; user messages would use the shared-page mechanism outlined above.

In some cases, additional system processes are required in addition to those executing inside the kernel address space and they will be allocated their own virtual store. In order to optimise message passing between these processes and the kernel, a set of pages can be declared as shared but not *incore* (i.e., not locked into main store). The set of pages can be pre-allocated by the kernel at initialisation time, so no new pages need to be allocated. All that remains is for the pages to be given to the processes. This can be achieved using the primitives defined in this chapter.

6.7 Process Creation and Termination; Swapping

Process creation, activation and termination are unaffected by the virtual storage mechanisms. The virtual storage subsystem must be booted before any processes are created, so all processes, even those inside the kernel, are created in virtual address space. The primitives to allocate and deallocate storage have been defined above (Sections 6.5.1 and 6.5.3). The operations to create and delete processes can be implemented in a way analogous to those defined in Chapter 4 (and assumed in Chapter 5), with the virtual-store primitives replacing those handling real store. The most significant difference between the two schemes is that the virtual-store allocation operations are not as limited in the amount of store they can allocate. The virtual storage operations are only limited by the number of pages permitted in a segment and not be the size of main store.

Virtual store also has advantages where swapping is concerned. It is possible to include a swapping system in a virtual-store-based system. As with the scheme defined in detail in Chapter 4, the swapper will transfer entire process images between main store and the swap disk (or swap file). Under virtual storage, the swapper treats the page as the basic unit for transfer. The swapper reads the page table and swaps physical pages to disk. Not all segments need be swapped to disk; code segments might be retained in main store while there are active child processes. The process is, however, complicated by the fact that a process image is likely to be shared between the paging disk and main store.

7

Final Remarks

Sic transit Gloria Swanson
– *Anon.*

7.1 Introduction

Rather than just end the book with the virtual storage model, it seems appropriate to add a few concluding remarks.

The chapter is organised as follows. In the next section, the models of this book are reviewed and some omissions mentioned. In Section 7.3, the whole idea of engaging in the formal modelling activity is reviewed. Finally, Section 7.4 contains some thoughts about what to do in the future.

7.2 Review

The formal models of three operating systems have been presented. All three kernels are intended for use on uni-processor systems. They are also examples of how the classical kernel model described in Chapter 1 can be interpreted; it should be clear that the invariants stated in Chapter 1 are maintained by the three kernels.

The first model (Chapter 3) is of a simple kernel of the kind often encountered in real-time and embedded systems. The system has no kernel interface and does not include such things as ISRs and device drivers. The user of this kernel is expected to provide these components on a per-application basis. This is common for such systems because the devices to which they are connected are not specified and are expected to vary among applications.

The first kernel can be viewed as a kind of existence proof. It shows that it *is* possible to produce a formal model of an operating system kernel. However, the kernel of Chapter 3 should not be considered a toy, for it can be refined to real working code.

The second kernel is for a general-purpose system. The model includes a number of device drivers, in particular a clock process that is central to the process-swapping mechanism. The kernel uses semaphores for synchronisation and as the basic inter-process communication mechanism (here, shared memory). The kernel uses a time-based mechanism for multiplexing main store between processes; the kernel supports more processes than can be simultaneously maintained in main store. A storage-management subsystem is also provided to manage main store. It does so in a fairly rudimentary fashion, based upon the allocation of relatively large chunks of store for each process (the actual division of process store is left undefined because it is often determined by the compiler—GNU C's approach was at the back of the author's mind while producing this model). The chapter contains the proofs of many kernel properties, and includes a proof of the correctness of the model for semaphores.

The second kernel is of approximately the complexity of kernels such as those built by Digital Equipment for the excellent operating systems running its PDP-11 series of minicomputers in the 1970s. It is of approximately the complexity of the kernel of Tannenbaum's MINIX [30] system (minus signals, file system and terminal interface). Indeed, MINIX was a significant influence on the models in Chapters 4 and 5.

The third kernel is not presented in its entirety. It is a variation on the second one. The two differ in that the third uses message passing for IPC. The message-passing primitives are modelled, as is a generic ISR based on the use of messages for the unblocking of drivers. All communication and synchronisation in this kernel is based upon synchronous message exchange. The various device drivers and the process-swapping subsystem are outlined as message-passing processes. A kernel interface is also outlined. The interface implements system calls as messages and a library of system calls is presented. The chapter contains a number of proofs of properties of the message-passing mechanisms and also contains a proof that only one process can be in the kernel at any one time.

The final exercise is in the modelling of virtual storage. This was included because many systems today use virtual store for system and user processes. There are issues in the construction of virtual storage systems that are not covered in detail in standard textbooks (they must be confronted without much support from the literature). In a sense, it is necessary to have virtual store in order to construct it. Virtual storage affords a number of benefits including automatic storage management at the page level, management of large address spaces and support for more processes than will simultaneously fit into main store without having to resort to the all-or-nothing techniques exemplified by the swapping mechanisms in the previous kernels. Message passing is also assisted by virtual storage, as is device-independent I/O (although it is not considered in detail in Chapter 6)—more will be said on these matters in the last section of this chapter (Section 7.4).

It has been pointed out (in Chapter 1) that file systems are not considered part of the kernel. File systems are certainly part of the operating system but not part of the kernel. They are considered *privileged* code that can directly access kernel services such as device drivers, but they are not considered by the author to be kernel components—they rely upon the abstractions and services provided by the kernel. File systems do provide an abstraction: the abstractions of the file and the directory. However, it is not necessary for a system to have a file system, even in general-purpose systems—consider diskless nodes in distributed systems and, of course, real-time systems, and there have been a number of attempts to replace file systems with databases; Mach, famously, relegates the file system to a trusted process outside the kernel. In keeping with the designers of Mach, the author believes that the inclusion of file systems in kernels should be resisted as an example of "kernel bloat" (the tendency to include all OS modules inside the protected walls of the kernel, as is witnessed by many familiar kernels).

It can be argued that this approach to file systems restricts the task of the kernel. This cannot be denied. It also restricts the services expected of the kernel. This, again, cannot be denied. Indeed, the author considers both points to be positive: the kernel should be kept as small as possible so that its performance can be maximised. Furthermore, by restricting the kernel in this way, it is easier to produce formal kernel models and to perform the kind of modelling activity that has been the subject of this book. This has the side-effect that, should the kernel be implemented, it can be supported by correctness arguments of the kind included above and its implementation can be justified by formal refinement.

As far as the author is concerned, the most significant omissions are:

- initialisation;
- asynchronous signals.

The initialisation operations for each kernel can be inferred from remarks in the models as well as the formal structure of the classes (modules) that comprise them. The modelling of the initialisation routines for each kernel should be a matter of reading through the models; the idle process and the basic processes of the kernels must be created and started at the appropriate time. Initialisation, even of virtual store, poses no new problems as far as formal models are concerned.

Asynchronous signals should be taken as including such things as the actions taken by the system when the user types control-c, control-d, etc., at a Unix (POSIX) console. From experience with Unix, it is clear that there is not much of an in-principle difficulty, just a practical one of including it in the models[1]. Asynchronous signals need to be integrated with ISRs and with the interrupt scheme for the system (it can be done in a device-independent

[1] For this book, there were time and length constraints that mitigated against the inclusion of such a component.

fashion, as the Unix and Linux kernels show). It is just a matter of producing the models and showing that they are correct (the latter is a little more of a challenge, of course).

A more detailed model of a complex interrupt structure would also be of considerable interest. This should be taken as an initial step in the formal modelling of the lowest level of the kernel. Such a model would have to be hardware specific and would have to be undertaken during the refinement to code of a model of a kernel at the level of this book.

7.3 Future Prospects

In this section, a number of possible projects are suggested. The author is already refining two formal models to implementation, so the issue of refinement is being attempted.

The first area is to employ formal models in the definition and exploration of non-classical kernel models. For example, some embedded systems are event driven. This has implications for IPC, device handling and process structuring. As a first step, the author is working on a tiny event-based kernel but more work is required. It is clear that the benefits of formal models can be demonstrated most graphically in the embedded and real-time areas, areas in which the highest reliability and integrity are required.

In a similar fashion, the formal approach can assist in the production of more secure systems. After all, if hackers can gain unauthorised access to a kernel, they can control the entire system (as tools such as Back Orifice demonstrate). There are many areas in the kernels modelled in this book that need attention as far as security is concerned. Many of these areas were identified during the modelling process but nothing has been done to plug the holes.

The extension of formal techniques to multi-processor systems is clearly of importance, particularly with the advent of multicore processor chips. It is natural, then, also to consider distributed operating systems from a formal viewpoint, at kernel and higher levels. Within the area of distributed systems, there are systems that support code and component mobility. The classical position is that the kernel must support a basic set of features and that the rest can be relegated to servers; this needs to be questioned, particularly because the basic set of features can look like a specification of a rather large classical kernel. There is a need for IPC and networking, as well as storage management but what else should be supported, particularly when components are mobile?

As Pike [24] has pointed out, there is very little new in operating systems research these days. Most "new" systems look like Unix or Windows. Pike makes the point that the domination of these systems serves to reduce the level of innovation in operating systems in particular and systems research in general. The two major systems have their own ways of doing things and there is a tendency to believe that they will be there for all time. This leads to a reluctance to think of genuinely new ways of doing things. In addition, the existence of such giants and their established user communities implies that the cost and risk of developing new system concepts are just not worthwhile.

There is, though, a need to look for new approaches to operating system design. New concepts are appearing in other areas that will impact upon operating systems (mobility is a case in point, as is ubiquitous computing) and it is unlikely that systems designed in the 1960s, 1980s or 1990s will be able to form an adequate basis for their full exploitation. The whole area of computing is changing: networks are established as a structure and are always becoming cheaper. Networks suggest distributed applications, mobility and ubiquity.

There are also hardware developments (multicore processors have already been mentioned—they will offer genuine parallel processing within a single box). Prompted by the appearance of 64-bit processors (and why not have 128- or even 1024-bit address spaces), there has been some work on systems with very large address spaces. In these systems, persistence can become a reality, not an add-on. The idea is that, with a sufficiently large address space, there is never any need to delete or destroy anything. The use of storage networks is another development in support of this, as is the idea that storage devices autonomously handle all storage and retrieval. The interfaces to such devices deal mostly with naming. If objects are never deleted, there is not only a naming problem, but the problem of determining which object to retrieve—it will hardly be possible to remember the names of *all* the objects stored in a space that can potentially hold 2^{128} objects. It will be necessary to introduce new ways to access these objects and in reasonable time, too.

The classical models have served us well, but it is not necessarily the case that they will do so in the future, given the demands of huge address spaces, large networks and mobility. Formal techniques can help in these research areas for reasons stated earlier in this chapter: they constitute a method by which systems can be designed and experimented with without implementation. Promising ideas can be explored in a real scientific and engineering manner, and with less ambiguity.

References

1. Baseten, J. C. M., *Applications of Process Algebra*, Tracts in Theoretical Computer Science, No. 17, Cambridge University Press, Cambridge, England, 1990.
2. Bevier, W., *A Verified Operating System Kernel*, Ph. D. Dissertation, University of Texas, Austin, 1987. (Ftp: `ftp.cs.utexas.edu/pub/boyer/diss/bevier.pdf`.)
3. Birrell, A. D., Guttag, J. V., Horning, J. J. and Levin, R., Synchronistaion Primitives for a Multiprocessor: A Formal Specification, *ACM Operating Systems Review*, 1987.
4. Boret, Daniel P. and Cesati, Marco, *Understanding the Linux Kernel*, O'Reilly and Associates, Sebastopol, CA, 2001.
5. Brinch Hansen, Per, *Operating Systems Principles*, Prentice-Hall, Englewood Cliff, NJ, 1973.
6. Brinch Hansen, Per, *The Architecture of Concurrent Programs*, Prentice-Hall, Englewood Cliffs, NJ, 1977.
7. Cavalcanti, Ana, Sampaio, Augusto and Woodcock, Jim, A Refinement Strategy for CIRCUS, *Formal Aspects of Computing*, Vol. 15, Nos. 2 and 3, pp. 146–181, 2003.
8. Cleveland, Rance, Li, Tan and Sims, Steve, *The Concurrency Workbench of the New Century*, North Carolina State University and SUNY, 2000. (Available from `http://www.cs.sunysb.edu/~cwb`)
9. Comer, Douglas, *Operating Systems Design, The Xinu Approach*, Prentice-Hall, Upper Saddle River, NJ, 1984.
10. Craig, I. D., *Formal Models of Advanced AI Architectures*, Ellis Horwood, Chichester, England, 1991.
11. Deitel, H. M., *Operating Systems*, 2nd ed., Addison-Wesley, Reading, MA, 1990.
12. Duke, Roger and Rose, Gordon, *Formal Object-Oriented Specifications using Object-Z*, Macmillan, Basingstoke, England, 2000.
13. Elphinstone, Kevin, Future Directions in the Evolution of the L4 Microkernel, in [23], 2004.
14. Fowler, S., *Formal Analysis of a Real-Time Kernel Specification*, Real-Time Systems Research Group, Dept. of Computer Science, University of York, York, UK, February, 1996.
15. Hayes, I., ed., *Specification Case Studies*, Prentice-Hall, Hemel Hempstead, England, 1987.

16. Hoare, C.A.R., *Communicating Sequential Processes*, Prentice-Hall, Hemel Hempstead, England, 1985.
17. Iliffe, J. K., *Basic Machine Principles*, 2nd ed., MacDonald/American Elsevier Computer Monographs, London, 1972.
18. Labrosse, Jean J., *MicroC/OS-II, The Real-Time Kernel*, Miller Freeman Inc., Lawrence, KS, 1999.
19. McKeag, R. M., T. H. E. Multiprogramming System, in [20], pp. 145–184.
20. McKeag, R. M. and Wilson, R., *Studies in Operating Systems*, Academic Press, New York, 1976.
21. Milner, R., *Communication and Concurrency*, Prentice-Hall, Hemel Hempstead, England, 1989.
22. Milner, R., *Communicating and Mobile Systems: The π-calculus*, Cambridge University Press, Cambridge, England, 1999.
23. NICTA OS Verification Workshop, 2004, NICTA, Canberra, Australia, 2004.
24. Pike, Rob, *Systems Software Research Is Irrelevant*, 2000. (`http://herpolhode.com/rob/utah2000.pdf`.)
25. Rubini, A., *Linux Device Drivers*, O'Reilly and Associates, Sebastopol, CA, 1998.
26. Silberschatz, A., Galvin, P. and Gagne, G., *Applied Operating System Concepts*, John Wiley, New York, 2000.
27. Smith, Graeme, *The Object-Z Specification Language*, Kluwer Academic Publications, Boston, MA, 2000.
28. Spivey, J. M., *The Z Notation: A Reference Manual*, 2nd ed., Prentice-Hall, Hemel Hempstead, England, 1992.
29. Tannenbaum, A., *Modern Operating Systems*, Prentice-Hall, Englewood Cliffs, NJ, 1992.
30. Tannenbaum, A., *Operating Systems: Design and Implementation*, Prentice-Hall, Englewood Cliffs, NJ, 1987.
31. Tuch, Harvey and Klein, Gerwin, Verifying the L4 Virtual Memory System, in [23], 2004.
32. Walker, B. J., Kemmerer R. A. and Popek, L., Specification and Verification of the UCLA Unix Security Kernel, *Communications of the ACM*, Vol. 23, No. 2, pp. 118–131, 1980.
33. Walker, D. and Sangiorgi, D., *The pi-calculus*, Cambridge University Press, Cambridge, England, 2001.
34. Wilson, R., The TITAN Supervisor, in [20], pp. 185–263.
35. Wirth N. and Gutknecht, J., *Project Oberon*, Addison-Wesley, Reading, MA, 1989.
36. Wirth N. and Gutknecht, J., The Oberon System, *Software Practice and Experience*, Vol. 19, No. 9, 1989.
37. Zhou, D. and Black, Paul E., Formal Specification of Operating Systems Operation, *Proc. IEEE TC-ECBS Working Group WG10.1*, pp. 69–73, IEEE, Washington, DC, 2001.

List of Definitions

Index